Animal Migration

Animal Migration

A Synthesis

EDITED BY

E.J. Milner-Gulland,
John M. Fryxell,
and
Anthony R.E. Sinclair

OXFORD

UNIVERSITY PRESS

Great Clarendon Street, Oxford OX2 6DP

Oxford University Press is a department of the University of Oxford.
It furthers the University's objective of excellence in research, scholarship,
and education by publishing worldwide in

Oxford New York

Auckland Cape Town Dar es Salaam Hong Kong Karachi
Kuala Lumpur Madrid Melbourne Mexico City Nairobi
New Delhi Shanghai Taipei Toronto

With offices in

Argentina Austria Brazil Chile Czech Republic France Greece
Guatemala Hungary Italy Japan Poland Portugal Singapore
South Korea Switzerland Thailand Turkey Ukraine Vietnam

Oxford is a registered trade mark of Oxford University Press
in the UK and in certain other countries

Published in the United States
by Oxford University Press Inc., New York

British Library Cataloguing in Publication Data

Data available

Library of Congress Cataloging in Publication Data

Data available

Typeset by SPI Publisher Services, Pondicherry, India
Printed in Great Britain
on acid-free paper by
CPI Antony Rowe, Chippenham, Wiltshire

ISBN 978–0–19–956899–4 (Hbk.)
      978–0–19–956900–7 (Pbk.)

1 3 5 7 9 10 8 6 4 2

Contents

Contributors

David J. Agnew, Department of Life Sciences, Imperial College London, UK

Susanne Åkesson, Department of Biology, Ecology Building, Lund University, SE-223 62 Lund, Sweden

Silke Bauer, Netherlands Institute of Ecology (NIOO-KNAW), PO Box 1299, 3600 BG Maarssen, The Netherlands *and* Swiss Ornithological Institute, Luzernstrasse, 6204 Sempach, Switzerland

Roy H. Behnke, The Odessa Centre, Great Wolford, UK

Luca Börger, Department of Integrative Biology, University of Guelph, Guelph, Ontario, Canada, N1G 2W1

Melissa S. Bowlin, Department of Biology, Ecology Building, Lund University, SE-223 62 Lund, Sweden

Annette C. Broderick, Centre for Ecology and Conservation, University of Exeter, Cornwall Campus, Penryn, TR10 9EZ, UK

Patrick J. Butler, Centre for Ornithology, School of Biosciences, The University of Birmingham, UK

Jason W. Chapman, Plant and Invertebrate Ecology, Rothamsted Research, Harpenden, Herts AL5 2JQ, UK

Iain Couzin, Princeton University, Department of Ecology & Evolutionary Biology, Princeton, NJ 08544, USA

Katherine A. Cresswell, Center for Stock Assessment Research, Mail Stop E2, Jack Baskin School of Engineering, University of California, Santa Cruz, California, 95064, USA

Maria E. Fernandez-Gimenez, Department of Forest, Rangeland and Watershed Stewardship, Colorado State University, Fort Collins, CO, USA

John M. Fryxell, Department of Integrative Biology, University of Guelph, Guelph, Ontario, Canada, N1G 2W1

Jarl Giske, University of Bergen, Department of Biology, Postboks 7803, 5020 Bergen, Norway

Brendan J. Godley, Centre for Ecology and Conservation, University of Exeter, Cornwall Campus, Penryn, TR10 9EZ, UK

Anders Hedenström, Department of Biology, Ecology Building, Lund University, SE-223 62 Lund, Sweden

Ricardo M. Holdo, University of Missouri, Columbia, MO 65211, USA

Robert D. Holt, Department of Biology, University of Florida, Gainesville, FL 32611, USA

Niclas Jonzén, Department of Biology, Ecology Building, Lund University, SE-223 62 Lund, Sweden

Endre Knudsen, Centre for Ecological and Evolutionary Synthesis (CEES), Department of Biology, University of Oslo, PO Box 1066, Blindern, 0316 Oslo, Norway

Jason Matthiopoulos, Scottish Oceans Institute, School of Biology, University of St Andrews, St Andrews, Fife, KY16 8LB, Scotland, UK

Edward McCauley, National Center for Ecological Analysis and Synthesis, Santa Barbara, CA 93101, USA

E.J. Milner-Gulland, Imperial College London, Silwood Park Campus, Buckhurst Road, Ascot, SL5 7PY, UK

Juan Manuel Morales, Laboratorio Ecotono, INIBIOMA-CONICET, Universidad Nacional del Comahue, Quintral 1250, 8400 Bariloche, Argentina

Ran Nathan, Department of Evolution, Systematics and Ecology, Alexander Silberman Institute of Life Sciences, The Hebrew University of Jerusalem, Edmund J. Safra Campus at Givat Ram, 91904 Jerusalem, Israel

Bart A. Nolet, Netherlands Institute of Ecology (NIOO-KNAW), PO Box 1299, 3600 BG Maarssen, The Netherlands

Nir Sapir, Department of Evolution, Systematics and Ecology, Alexander Silberman Institute of Life Sciences, The Hebrew University of Jerusalem, Edmund J. Safra Campus at Givat Ram, 91904 Jerusalem, Israel

Bernt-Erik Sæther, Centre for Conservation Biology, Department of Biology, Norwegian University of Science and Technology, NO-7491 Trondheim, Norway

William H. Satterthwaite, Center for Stock Assessment Research, Mail Stop E2, Jack Baskin School of Engineering, University of California, Santa Cruz, California, 95064, USA

Jennifer L. Shuter, Department of Integrative Biology, University of Guelph, Guelph, Ontario, Canada, N1G 2W1

Anthony R.E. Sinclair, Department of Zoology, University of British Columbia, Vancouver, BC, Canada, V6T 1Z4

Florian Stammler, Arctic Centre, University of Lapland, PL 122, FIN 96101 Rovaniemi, Finland *and* Scott Polar Research Institute, University of Cambridge, Lensfield Road, Cambridge, CB2 1ER, UK

Gregory A. Sword, School of Biological Sciences, The University of Sydney, Sydney, NSW, 2006 Australia

Simon Thirgood, Macaulay Land Use Research Institute, Craigiebuckler, Aberdeen, UK

Matthew D. Turner, Department of Geography, 160 Science Hall, 550 Park Street, University of Wisconsin, Madison, WI 53706, USA

Martin Wikelski, Department of Migration & Immuno-ecology, Max Planck Institute for Ornithology, Germany *and* Chair of Ornithology, Konstanz University, 78457 Konstanz, Germany

Acknowledgements

We thank the UK Natural Environment Research Council's Centre for Population Biology for funding and hosting the workshop upon which this book is based.

N.S. was supported by the US–Israel Binational Science Foundation (BSF grant numbers 229/2002 and 124/2004), the Robert Szold Fund for Applied Science, the Ring Foundation, and Rieger-JNF fellowships. P.J.B. was supported by BBSRC (grant number BB/F015615/1 awarded to himself and Dr C.M. Bishop). RDH and JF thank NSF and NSERC for supporting their research in this general area for a number of years. RDH thanks the University of Florida Foundation for support. EJMG acknowledges the support of a Royal Society Wolfson Research Merit award.

We thank Penny Hancock for valuable discussions and input during the preparation of the book, and Tal Avgar and Cortland Griswold for thoughtful comments on Chapter 3. EJMG thanks Franck Courchamp for hosting her in his lab during the final stages of the book.

We thank Ian Sherman and Helen Eaton at OUP for all their support and patience throughout.

CHAPTER 1

Introduction

John M. Fryxell, E.J. Milner-Gulland, and Anthony R.E. Sinclair

'...to resolve this intricate issue [migration] it is essential that many people in various places on Earth make careful observation and report them to the learned world'

Linnaeus (1757)

Migration, like fine art, is something that everyone knows about, a subject on which everyone has a strong opinion, yet nobody can precisely define. Some scientists have chosen to focus on individual behavioural attributes (Kennedy 1985; Baker 1978), defining migration as persistent, un-distracted movements with consistent directional bias (Taylor 1986; Gatehouse 1987). Others suggest that a key attribute is that migratory movements are on a larger spatial scale, involving moves of longer duration and greater displacement than those seen in more trivial daily patterns of movement (Baker 1978; Dingle and Drake 2007). Other authors take a more formal ecological view, defining migration as a round-trip journey between discrete home ranges (Sinclair 1983; Fryxell and Sinclair 1988b). A myriad of potential consequences of migration can be readily imagined: to take advantage of spatially distributed resources; to evade predators; to gain access to rich, but temporary resources; to find suitable habitats for different life stages with markedly different physical needs and ecological constraints. All agree that migration is a fascinating phenomenon that might well contribute in a major way to the fundamental structuring of ecosystems, while at the same time raising a host of ecological and evolutionary questions and challenging conservation issues.

From the earliest writings of Aristotle (350 BC) and Cajus Plinius Secundus (AD 77), natural historians have pondered the mysterious disappearance of birds in the waning summer and dramatic re-appearance with the first stirring of spring (T. Alerstam, pers. comm.). Indeed, the Middle Ages witnessed remarkable speculation that swallows, martins, and other birds simply hibernated at the bottom of lakes, to emerge in the spring once again (Magnus 1555). Avian migration was a primary focus of Linnaeus' (1757) doctoral thesis, which probed the seasonal rhythms, myriad causal factors, and even the practical implications of migration. Although subsequent decades have witnessed dramatic changes in our understanding of natural history, such that we can describe the distribution and variation in migratory patterns for a range of organisms, there is still a dearth of underlying ecological and evolutionary theory, and a particular lack of general quantitative theory.

We are now on the brink of a major breakthrough in the understanding of the evolutionary and ecological dynamics of migration, scaling up from the individual to the landscape level. This is driven by two technological advances—the improvement in computer power and the concomitant increase in our ability to carry out detailed simulation modelling at an individual level, and the improvement in tracking devices that allow us to collect unparalleled amounts of data about animal location and behaviour, also at an individual level. There has also been a major recent increase in understanding of the mechanisms and adaptive significance of learning, and of methods for scaling between genetic, individual and population-level dynamics.

In a recent *Science* special issue (August 2006), 'movement ecology' was highlighted as an emerging discipline. However, most of the cited theoretical and empirical work in this field was focused on dispersal rather than on migration. Migration presents additional challenges and interest, because of its potential role in community structuring, the long-range nature of most migrations, the questions that it raises about the evolution of adaptations to environmental variation and its role in animal life histories, and the difficulties in conceptualizing individual-level decision-making in time and space when animals make directional long-range movements. It is a widespread phenomenon, but populations and species that are closely related taxonomically or geographic neighbours can display contrasting strategies.

With a keen sense of anticipation, we coerced The Centre for Population Biology at Silwood Park, Imperial College London into hosting a symposium. Our objective was to explore the potential for emerging insights arising from mathematical modelling and empirical research for a new, unified understanding of the causes and consequences of migration. Much published work had already been done on the topic and we saw little point in revisiting the same well-worn paths that had already been so clearly articulated. Indeed, each participant was given a scant 5 minutes to summarize a lifetime of work. Rather, we invited an eclectic mix of scientists, whose only common ground was familiarity with new ways of measuring and modelling movement. Some were botanists, others zoologists, some had only a passing familiarity with real ecosystems, others had spent decades studying the most minute details of migration behaviour. More audaciously, we also proposed an accompanying volume to synthesize the discussions that ensued.

Based on the natural staging points that emerged over 48 hours of heated discussion at Silwood, not to mention periodic adjournment for liquid refreshment, we identified four themes. The first theme involves the evolution of migration. Much has been written about the facts of migration, yet surprisingly few mathematical models have been developed to explain the key factors guiding its repeated evolution in so many different taxa, often in disparate ecological conditions. Rarer still are suitable empirical tests. In Chapter 2, Sword *et al.* seize the challenge of interpreting the broad swathe of empirical evidence for an adaptive basis for migration. Using the comparative method as a dissecting instrument, they explore the commonalities and (equally important) the differences between migratory taxa. By contrast,

in Chapter 3, Holt and Fryxell reduce migration to its basic essentials, identifying the conditions under which migration might be expected to be an evolutionarily stable strategy in a seasonal, spatially-heterogeneous world.

The second major theme reflects perhaps the most rapidly expanding area of recent work—how to migrate? New forms of instrumentation have been developed, allowing researchers for the first time precisely to measure and experimentally test the physical and energetic constraints that shape migratory movement. Hedenström *et al.* lead off in Chapter 4—outlining the major aerodynamic and hydrodynamic principles that constrain migratory movement in either air or water. Their effort is well-matched by Sapir's treatment of the physiological adaptations of migrants, developed in Chapter 5. Through elegant experimental trials, many of these basic principles have been tested in recent years under suitably-controlled laboratory conditions. Even more remarkably, field testing has begun in earnest, using micro-electronics in ways that were scarcely imagined a decade ago. Physiological constraints have limited meaning without deep understanding of the cues and decision rules that guide both the proximate and ultimate goals of migration, probed in some depth in a comparative analysis by Bauer *et al.* in Chapter 6.

In Theme 3, we consider the issues of spatial and temporal scale. Nomadism and migration are obviously first cousins, yet few scientists have formally considered their common roots in spatial and temporal heterogeneity. This is the formidable task taken up by Jonzén *et al.* in Chapter 7. Börger *et al.* bridge the gap between empirical and theoretical studies in discussing how models, combined with modern data sources and statistical methods, can be used to test different hypotheses about the causes of migration. In Chapter 8, they provide a useful road map to a rapidly developing armada of new methodologies for data analysis and movement model evaluation.

In Theme 4, we look at migration in a broader context. In Chapter 9, Holdo *et al.* use a spatially-explicit model of one of the most intensively studied ecosystems on Earth, the Serengeti, to consider the importance of migration for key attributes of Serengeti structure and ecosystem function. In Chapter 10,

Behnke *et al.* evaluate the intriguing parallels and crucial differences between wild migratory species and transhumant systems of migratory livestock husbandry. We conclude this theme by examining the difficult question of how best to manage migratory species, in Chapter 11. By definition, such species cannot be conserved effectively in static reserves; instead their needs must be incorporated into dynamic conservation planning at the landscape level. Shuter *et al.* demonstrate the breadth of current threats to wild migratory species, often through the multiplicity of habitats that they require.

These latter two chapters strike a chord with many other issues in contemporary conservation including climate change and the need to integrate human and wildlife needs in multi-use landscapes. They also run slightly contrary to current paradigms of community-based conservation, suggesting that a new paradigm for the link between people and wildlife is needed for conservation of migratory species. Finally, we revisit these four disparate themes in Chapter 12, to evaluate how this new migratory synthesis enhances our understanding of (i) new theoretical and technological developments, (ii) the common features of causes, mechanisms and consequences of migration across taxa, (iii) the impacts of migrations on ecosystems, (iv) the interactions of humans and migrating species and requirements for the conservation of these taxa, and finally (v) future developments and directions.

PART 1

The evolution of migration

Steelhead trout.
Picture by Claudia Makeyev

CHAPTER 2

Understanding the evolution of migration through empirical examples

Katherine A. Cresswell, William H. Satterthwaite, and Gregory A. Sword

Table of contents

2.1 Introduction

Migration is one of the most widely observed phenomena in biology, probably in no small part because movement itself is a defining characteristic of animals. Migratory movements occur across a variety of invertebrate and vertebrate species (Dingle 1996). Yet, despite its seeming ubiquity, there has been considerable debate among scientists about the evolution of migration, its maintenance in populations, the physiological and genetic mechanisms involved, and even the basic definition of the term. Understanding migration and how it evolves in response to the environment is of critical importance due to its obvious implications for human endeavours such as agriculture, resource management and conservation. Though much remains to be learned,

some key generalizations have emerged about the evolution of migration that provide an important foundation for future empirical studies of both the ecology and evolution of this remarkable process. For the purposes of this chapter (and in keeping with the rest of the book), we take a broad view of migration. We adopt the framework developed by Kennedy (1961, 1985), which defines migration as individual movement that (i) is straightened out or persistent, (ii) is actively undertaken either through locomotion or through use of a medium such as water or air currents for transport, and (iii) is undertaken while temporarily suppressing some otherwise routine 'station keeping' behaviours such as foraging. We acknowledge that this definition is not iron clad and that there will invariably be exceptions. However, by taking an individual-based as opposed to pattern-based perspective, we intentionally do not limit our treatment of migration to systems that only meet specific criteria, such as coordinated movements, large spatial scales, return trips, or predictable timing; although in many cases migratory movements exhibit these characteristics.

In this chapter we will briefly define what we mean by the evolution of migration, and then provide an overview of the study of migration as an adaptation. We look at some examples of the circumstances under which migration may evolve, and outline the potential costs and benefits to an animal with a migratory lifestyle. We examine documented examples of variations in migration as a means of gaining a better understanding of the ways in which it may have evolved, and also look at the relative contribution of genetic versus environmental factors. Finally, we use all of the above to discuss if and how it may be possible to predict changes in species' migratory patterns in response to future changes in climate. There have been a number of synthetic treatments of migration and its evolution, including some excellent recent reviews (e.g. Dingle 1996; Åkesson and Hedenström 2007; Dingle and Drake 2007; Pulido 2007; Ramenofsky and Wingfield 2007; Roff and Fairbairn 2007; Salewski and Bruderer 2007; Nathan *et al.* 2008). Our intention here is to set the stage for the chapters that follow by outlining some of the key conceptual advances that have shaped our understanding of migration as an adaptation, and focus on some of the major questions and themes of empirical investigation that have provided insight into its evolution.

2.2 Migration as an adaptation—an overview

Central to understanding migration and its evolution is the recognition that migration is a complex adaptation arising as a result of interactions between individuals, their genes, and the environment. With respect to the environment, it is widely accepted that migration can evolve in response to spatiotemporal variation in resources (Southwood 1962, 1977; Dingle and Drake 2007) or threats (Mangel and Clark 1988; Bollens and Frost 1989; Hebblewhite and Merrill 2009; McKinnon *et al.* 2010). Individuals using resources whose availability or quality varies

in time and space within their lifespan may benefit by exploiting different habitat patches for survival and reproduction. Migration may also arise as a result of changing resource needs as animals move from one life-history stage to another, for example from a sedentary young stage to a more mobile and reproductive adult stage. The fact that individual fitness can vary as a result of migratory behaviour has helped focus experimental approaches to the evolution of migration squarely upon analyses of individuals and variation between them (Kennedy 1985; Dingle 1996). Considering migration as an individual property in this manner provides a useful benchmark for empirical studies to compare migrants with non-migrants and to test hypotheses about the mechanisms underlying individual behaviour and movement. At the same time, considering the behaviour of individuals also enables the process of migration to be distinguished from its population-level consequences in terms of demography, population dynamics and spatial distribution patterns (Dingle 1996).

Another key aspect is the recognition that migration is an attribute of an individual and is not itself a specific trait (Zink 2002). The act of migrating by an individual is the manifestation of any number of underlying behavioural, physiological, or morphological traits, each with their own respective genotypes and regulatory mechanisms (Åkesson and Hedenström 2007; Pulido 2007; Ramenofsky and Wingfield 2007; Roff and Fairbairn 2007). Thus, migration can, at least in principle, be broken down into its component traits, the evolution, expression and resultant effects of which are amenable to empirical study. However, it is important to acknowledge that neither the evolution nor proximate expression of any one trait occurs in isolation from other traits in the genome (Pigluicci 2003). For example, trade-offs within an individual affect the relative proportions of energy or nutrients that can be devoted to any given trait (Sih *et al.* 2004; Dingle 2006), and genetic correlations between traits involved in migration may simultaneously affect multiple phenotypic traits in response to selection on any one in isolation (Pulido 2007). The notion of a migratory phenotype being a combination of many integrated traits that function together within an individual has been encapsulated by the term, *migratory syndrome* (Dingle 1996). Migratory syndromes typically include characteristic behavioural, physiological and morphological traits involved in the conduct and regulation of migratory movements. The variety of traits functioning as part of a migratory syndrome may include the suppression of station keeping or maintenance behaviours, orientation ability, intentional use of wind or water currents, fat deposition, hormonal regulation, along with changes in morphology and life-history traits such as fecundity (Dingle 2006).

Migration appears to have independently evolved a myriad times across an array of unrelated taxa (e.g. Alerstam *et al.* 2003). This suggests that sufficient variation for the expression of traits involved in migration has rarely if ever been a phylogenetic constraint for the initial origin of migration (Zink 2002; Piersma *et al.* 2005; Bruderer and Salewski 2008). Not only has migration as a process repeatedly evolved, but the underlying traits comprising migratory syndromes are also quite similar across organisms. Dingle (2006) documented considerable similarities in behavioural, physiological, morphological and life history traits implicated as part of migratory syndromes across insects, fish and birds. Thus natural selection has clearly favoured convergent solutions to common problems facing migrants such as timing, fuelling and orientation, albeit with different underlying mechanisms in unrelated taxa.

Another important consideration arising out of the observation that migration has repeatedly evolved across lineages is the distinction between the analysis of the origin of migration and its subsequent modification and maintenance. The widespread distribution of migration across taxa suggests that it was present very early on in the diversification of the lineages we currently observe. Thus, in attempting to account for variation in migration across extant taxa, we are primarily concerned with the modification and maintenance of traits associated with migration rather than their initial phylogenetic origin (Zink 2002; Salewski and Bruderer 2007; Bruderer and Salewski 2008). Phylogenetic hypotheses are critical tools that can form the basis of comparative studies to examine variation in the traits and mechanisms associated with migration and biogeographical patterns (e.g. Zink 2002; Outlaw and Voelker 2006; Boyle and Conway 2007).

It is in the comparison of ancestral and derived character states that we may find crucial unexpected differences between taxa, revealing the importance of developmental systems and genetic constraints (Harvey and Pagel 1991). Similarly, a phylogenetic context is critical in comparative studies that seek to examine the causal relationship between changes in a trait relative to changes in other traits or the environment (Harvey and Pagel 1991).

2.3 A brief history of thought about the evolution of migration

The study of the evolution of migration was dominated early on by considerations of avian migration. Thomson (1926) and Lincoln (1939) summarized the early thoughts on the evolutionary origin of bird migration into two main groups: the northern ancestral home hypothesis, suggesting that present migratory forms were initially permanent residents in pre-glacial northern areas that were forced south by glaciers; and the southern ancestral home hypothesis, suggesting that present migrants were originally permanent residents in southern areas that moved north for breeding following glacial retreat. These positions have been both supported (Berlioz 1950; Dorst 1961) and criticized (Mayr and Meise 1930; Cox 1968; Salewski and Bruderer 2007), with the main criticism being that extreme crowding of birds in tropical and subtropical areas during glacial maxima would have been unlikely. Recent modelling work instead proposes a 'shifting home' model, where the ancestral breeding sites of long-distance migrants were not 'northern' or 'southern' but shifted across high and middle latitudes while migrations emerged through winter range shifts (Louchart 2008). Mayr and Meise (1930) suggested intra-specific competition and intensification of migratory drive were responsible for the 'leapfrog' patterns, where populations breeding to the furthest north also spend the winter furthest south. This developed into the idea of intra- and interspecific competition acting as selective agents in the evolution of migration, where selection should favour genetic mechanisms leading to the migration of resident species into seasonally favourable adjacent areas if the associated gain in survival or reproduction exceeds the migration cost (Cox

1968). Migratory distance and periodicity should grow, from partial into long-distance migrations, as populations spread into increasingly seasonal environments (Gauthreaux 1982; Cox 1985; Milá *et al.* 2006). To date, results from studies of the role of competition in shaping migrant–resident communities remain largely unconvincing (Salewski and Jones 2006).

Bell (2000) suggests that selection favours migratory movements when there is predictable seasonal variation in the direction and/or severity of environmental gradients across the range of a sedentary population. The tendency for the centre of activity to shift directionally outside of the breeding season, added to the tendency of individuals to return to a successful breeding site, produces a behavioural pattern that bears all the distinctive characteristics of migration. This pattern is supported by comparisons of distribution and phylogeny in some shorebirds (Joseph *et al.*, 1999), as well as recent modelling work by Barta *et al.* (2008) and in Chapter 3 of this book, which show that even small levels of seasonality in food availability can lead to migration.

An important expansion in the study of migration across taxa came through the study of insect movements in relation to spatial and temporal variation in habitat quality, thereby providing a more general ecological framework within which to consider the evolution of migration (e.g. Southwood 1962, 1977, 1981; Dingle 1996). In insects, the need to colonize new habitats is considered to be a major selective force because areas suitable for growth and reproduction, such as the host plants of herbivorous insects, are often ephemeral. This hypothesis is widely supported by both theoretical and empirical studies, where we see that when resources are temporary in relation to generation time, migration and/or dormancy are required strategies (Southwood 1962; Dingle 1989; Roff 1990; Dingle 1996). Thus, migration can reasonably be summarized as an adaptive response to spatial and temporal heterogeneity in resources (Chapter 3). This intuitive and simple explanation for the evolution of migration may seem relatively straightforward. However, as we shall see in this chapter and others in this book, understanding the precise nature of the trade-offs underlying the evolution of migration, not to mention the mechanisms underlying traits involved in migratory syndromes, is no simple task (Alerstam *et al.* 2003; Holland *et al.* 2006b; Roff and Fairbairn 2007).

2.4 The costs and benefits of migration

Lack (1954) was the first to apply a theoretical cost–benefit approach to the evolution of migration; if the benefits of moving exceed the costs of staying, then migration will be favoured by natural selection. The main benefits of migration are predicted to be escape from inter- and intra-specific competition for resources in saturated habitats, but they can also be the avoidance of predators and parasites (Gauthreaux 1982; Cox 1985; Alerstam *et al.* 2003; McKinnon *et al.* 2010). In addition, there are long-term fitness benefits to migration in the colonization of new favourable habitats (Roff and Fairbairn 2007). One of the most obvious predicted costs of migration is increased metabolic costs. Some forms of migratory locomotion are more efficient than others (Schmidt-Nielsen, 1975; Chapter 4) and the media in which these movements take place are probably heavily constrained by phylogeny. Some animals may face the additional cost of constructing the migratory apparatus, such as the development of wings for flight in insects (Rankin and Burchsted 1992). In addition, before setting off on their journeys most animals must store large amounts of fat, which is a large cost in energy and in time taken to build stores, with many animals showing a positive relationship between the amount of fat stored and the distance travelled (Dingle 1996; Chapter 5). Other potential costs that have been identified include the increased risk of predation en route, delayed reproduction, shortened lifespan, decrease in overall fecundity, settlement costs in the new habitat, and the risk of not finding a suitable habitat at all (Rankin and Burchsted 1992; Dingle, 1996).

There are a variety of examples across taxa that illustrate the many different types of fitness trade-offs involved in migration. In aquatic systems, studies on migration have mainly been focused on commercially important fish species, such as the migration of salmonids from streams to the sea (Thorpe 1988; Økland *et al.* 1993; Dodson 1997). Benefits that arise in the shift from freshwater to marine environments (anadromy) include improved feeding opportunities, and higher growth rate, asymptotic body length and expected reproductive success (Gross 1987). Costs involve the predicted energetic cost of physiologically preparing for the change from fresh water to salt water, and in many cases a greatly increased risk of mortality at sea (Mangel and Satterthwaite 2008). There are also potential benefits in moving in the opposite direction, from marine to freshwater environments (catadromy) and other combinations of fresh and marine environments (Myers, 1949). In freshwater fish, seasonal mass migrations have been observed, as in cyprinids that migrate into streams and wetlands in the fall and return to the lake in spring (Brönmark *et al.*, 2008). A study on 1800 individually marked cyprinids showed that their migration patterns followed the response to seasonal changes predicted if fish were balancing the cost-to-benefit ratio between predation mortality and growth rate (Brönmark *et al.*, 2008).

On land, a different type of trade-off between mortality risk and habitat use has recently been found to drive the mass migration of millions of Mormon crickets (*Anabrus simplex*) in western North America—with a sinister underlying mechanism. Mormon crickets can form huge migratory bands containing millions of flightless individuals marching in cohesive groups across the landscape

(Fig. 2.1). Migratory band members are protected from predation relative to individuals separated from the group (Sword *et al.* 2005). Individual band members are subject to increased intra-specific competition for nutritional resources, namely protein and salt (Simpson *et al.* 2006). They are also notoriously cannibalistic and their propensity to cannibalize is a function of the extent to which they are nutritionally deprived. As a result, individuals within the band that fail to move risk being attacked and cannibalized by other nutritionally deprived crickets approaching from the rear (Simpson *et al.* 2006). Group movement, in this case, arises as a result of the trade-off between the anti-predator benefits of being in a group versus the need to move to find fresh resources and to avoid being cannibalized (Simpson *et al.* 2006). The same trade-off with cannibalism as the underlying mechanism has been shown to account for mass movement in migratory bands of locusts as well (Bazazi *et al.* 2008; Bazazi *et al.* 2010).

Other massive terrestrial migrations are most notably undertaken by various species of ungulates (Dingle 1980; Sinclair 1983; Fryxell and Sinclair 1988b). Although the animals are widely regarded to be responding to seasonality in resource availability, the actual fitness trade-offs that have led to the evolution of migration in these species are still poorly understood (Bolger *et al.* 2008). Both direct

(a)

(b)

Figure 2.1   (a) A migratory band of Mormon crickets (*Anabrus simplex*) crossing a dirt road in NE Utah, USA. This band contained millions of insects and it marched across the road at a similar density for several days. (b) Cannibalism in response to limited nutritional resources is rife in these migratory bands (Photos: G. Sword).

and indirect energetic costs of travel can be substantial for these large terrestrial animals, with the additional risk of predation or accidental injury during the journey. However, the potential benefits of migration to these animals in terms of positive effects on fitness are less clear (Bolger *et al.* 2008).

2.5 Variation in migration strategies

Variation in migration strategies occurs on a number of scales and can provide clues as to how migration probably evolved under different scenarios. Variation can occur at the level of the individual in several ways (Dingle and Drake 2007). Migrants can be classified as either obligate or facultative, depending on whether they always migrate or do so only in an environmentally-determined manner, typically mediated by cues associated with the deterioration of local habitat conditions. In partial migration, as its name suggests, only a fraction of the population migrates, while the remainder stay either in the breeding or non-breeding area. Variation between individuals can also result from differences in the migration patterns of older versus younger individuals, or differences between the sexes. Variation also occurs over time, where seasonal versus irruptive migration may occur.

In high latitudes, migrations are primarily synchronized to the predictable warm–cold seasonal variations of the annual cycle (Dingle and Drake 2007). Changes to the timing of climatic events therefore predictably have the greatest overall effect on temperate species. For example, various temperate bird migrants show a decrease in population size if the climate changes and affects the timing of resource availability (Newton and Dale 1996a,b; Both *et al.* 2006). The chipping sparrow *Spizella passerina* exhibits both long-distance migratory behaviour in temperate North America and sedentary behaviour in Mexico and Central America, suggesting that migration evolved as a result of northern population expansion into temperate North America since the last glacial maximum 18 000 years ago (Milá *et al.* 2006). This is the strongest evidence to date that historical climate patterns have played a role in the rapid evolution of avian migration in natural populations. In these cases, the known effect of changes in climate at a specific latitude can lead to a better understanding of how

migration evolved and the environmental cues that trigger the migratory syndrome.

Variation in migration can also occur via a range expansion, a decrease in migration, or even the establishment of a new migratory pattern. In a changing environment, directional selection favours the expression or suppression of migratory activity such that an initially sedentary population can become migratory or vice versa (Salewski and Bruderer 2007). Examples of a discontinuation of migration are known, for example, in the lesser black-backed gull *Larus fuscus*, great crested grebes *Podiceps cristatus* and the blackbird in Europe (Lack 1968; Adriaensen *et al.* 1993; Berthold 2001; Fiedler 2003; Bearhop *et al.* 2005). Gull populations that migrated prior to 1940 became sedentary within a few decades, now feeding on refuse dumps outside towns during the winter. In this instance, humans provided a new abundant source of food, available consistently around the year, and the gulls changed their behaviour to utilize it (Lack 1968). Contrastingly, the eastern house finch is one of very few documented cases of the establishment of a new large-scale pattern of migratory behaviour under natural conditions (Able and Belthoff 1998). The house finch (*Carpodacus mexicanus*) was introduced on to Long Island, New York around 1940, probably from southern California where around 80% of individuals are sedentary (Elliot and Arbib 1953; Able and Belthoff 1998). By the early 1960s, 36% of the eastern population of house finches was performing migratory movements, with the proportion fluctuating between 28 and 54% ever since. In this example, we observe several different types of variation in the migratory pattern: (i) they are partial migrants at the population level, (ii) individuals that migrate one year may not migrate in the following year, and (iii) younger birds are more likely to migrate than older birds (Able and Belthoff 1998). The overall movement of these birds between their winter and breeding seasons reflected a seasonal round-trip migration typical of the flow of songbird migration in the north-eastern United States, suggesting that the machinery for migratory behaviour pre-existed in the parent population (Able and Belthoff 1998).

Other examples of changes in migratory behaviour include its loss in several land-locked populations of formerly migratory salmon and trout, although only some such populations truly lose their migratory behaviour while others migrate between freshwater habitats (Quinn 2005). In contrast, a migratory strategy has reappeared in introduced rainbow trout coming from what may have been purely non-migratory ancestral stock (Pascual *et al.* 2001; Thrower *et al.* 2004; Ciancio *et al.* 2008).

The modification of a migration pattern may be relatively fast or slow, in terms of the number of generations involved, with different factors potentially affecting the rate of change. Slow modifications have been observed in some cases where current migration routes do not seem to be optimal. For example, individuals may not migrate to the closest possible wintering area, or others such as emperor penguins may take an unexpectedly long route (Crick 2004; Pulido 2007). In contrast, there is evidence of fast modifications from captive breeding studies and artificial selection in songbirds, where the suite of traits accompanying migratory behaviour can be established in, or eliminated from, a population in as few as three to six generations (Berthold 1996). In the field, the European blackbird has been observed over the last 50 years to extended its wintering to areas much further north than previously recorded, with evidence that birds wintering further north also produce larger clutches and fledge more young (Bearhop *et al.* 2005). Although there was no evidence for morphometric or genetic differences between the populations that use more northern versus more southern wintering sites in the Bearhop *et al.* (2005) study, this example shows how underlying differences in fitness can potentially drive a rapid change in migratory strategy. It also serves as an example of how higher levels of connectivity may influence adaptation to the breeding areas and lead to speciation (Bearhop *et al.* 2005).

2.6 The role of genetics versus the environment

Phylogenetic studies fail to reveal a deeply embedded ancestral pattern for migration (Piersma *et al.* 2005). Rather, the evidence implies that migratory adaptations are simple extensions of traits that already exist to function in other contexts, such as

flight (Grinnel 1931; Alerstam 2006; Dingle and Drake 2007). The phylogenetic distribution of swarming in *Schistocerca* locusts provides a good example of the potential roles that both genetic and environmental effects can play in the expression of migratory behaviour, as well as the need for a phylogenetic framework to differentiate between the contributions of genetics versus the environment in explaining extant instances of migration across taxa. Phylogenetic evidence based on mtDNA indicate that all New World *Schistocerca* locusts probably share a mass migrating common ancestor whose swarms once flew across the Atlantic ocean from Africa, but mapping swarming onto phylogeny suggests that it has been lost and then independently evolved at least twice in the New World (Lovejoy *et al.* 2006). However, mapping the propensity to migrate in swarms in this manner onto a phylogenetic tree belies the fact that swarming itself is not a specific phenotypic trait, but rather an outcome of the expression of underlying traits that can lead to swarm formation and mass migration (Simpson and Sword 2009). In other words, swarming is an attribute, not a trait (*sensu* Zink 2002). Swarming locusts are defined by the expression of a form of phenotypic plasticity referred to as 'phase polyphenism'. Phenotypic changes in locusts are mediated by changes in local population density, with crowded rearing conditions inducing the production of swarming gregarious phase individuals that can be behaviourally, morphologically and physiologically distinct from the relatively sedentary phenotypes produced under low population density conditions (Simpson and Sword 2008, 2009; Pener and Simpson 2009). Many of these density-dependent phenotypic changes can be interpreted as components of a migratory syndrome, and as in other such cases (Dingle 2006), different component traits of the syndrome can be independently regulated and are genetically uncoupled (Simpson and Sword 2009).

To explain the existence of both swarming and non-swarming *Schistocera* species, we must therefore consider the evolution of the underlying plastic traits that comprise the migratory syndrome and their relationship to swarming across taxa. Is the lack of swarming observed in many of the New World taxa due to the loss of genetic variation for the expression of one or more density-dependent traits involved in the migratory syndrome? Or alternatively, do non-swarming species retain latent genetic variation for the expression of traits involved in the migratory syndrome, but simply exist in habitats and ecological conditions that are not conducive to the increases in population density that lead to swarm formation (Sword 2003; Simpson and Sword 2009)? Answering questions such as these about the role of genetics versus the environment in migration will require phylogenetically-controlled comparative studies that assess the relationship between genetic variation in the expression of traits comprising the migratory syndrome and relevant environmental factors (e.g. habitat, climatic or demographic variables) across migratory and non-migratory taxa.

The investigation of genetic variation in migratory traits has often been split into two different approaches: quantitative genetics (the heritability of migratory traits); and reaction norms (or environmental effects on trait expression). In other words, we often investigate whether phenotypic differences among individuals living in different environments are due to genetic variation in constitutively-expressed traits or whether they are plastic and environmentally-determined (Pulido and Berthold 2003; van Noordwijk *et al.* 2006). A number of studies have shown that differences in migratory behaviour are genetically-determined and are largely considered to be constitutively-expressed as obligate strategies (Gwinner 1969; Berthold and Querner 1981; Widmer 1999; Déregnacourt *et al.* 2005; Helm *et al.* 2005). Alternatively, migratory phenotypes may differ between individuals due to the effects of the environment on their expression. Importantly, these plastic or facultative traits themselves have an underlying genetic basis with specific responses to the environment described by a reaction norm (a range of phenotypes expressed by a single genotype as a function of a specified environmental variable; van Noordwijk *et al.* 2006). In both of these cases, evolutionary changes in the frequency of alleles for migratory phenotypes in a population can occur by selection, drift, gene flow, or mutation (to a much lesser extent), with the caveat that phenotypic plasticity can serve as a buffer against selection by effectively hiding genetic

variation in the face of environmental variation (Wright 1931; Schlichting and Pigliucci 1998).

Changes in migratory behaviour may appear very quickly in facultative migrants where the decision of individuals to migrate results from a plastic genotype × environment interaction. For example, in the steelhead trout, *Oncorhynchus mykiss*, non-migratory residents may have progeny that are anadromous, and anadromous fish may have offspring that spend their entire lives as freshwater residents (Pascual *et al.* 2001; Thrower *et al.* 2004; Olsen *et al.* 2006; Ciancio *et al.* 2008). Importantly, constitutively expressed phenotypes can arise from an initially plastic ancestral state via reaction norm evolution and genetic assimilation (Pigliucci and Murren 2003). In fact, there is mounting evidence to suggest that genetic assimilation of once plastic phenotypes is not only common, but can proceed so rapidly within populations that it has simply been overlooked as a factor in phenotypic evolution (Pigluicci and Murren 2003). Thus, genetic variation for facultative migration strategies not only provides individuals with the ability to express alternative phenotypes in response to short-term environmental variation, but also provides a means for the rapid evolution of migration as an obligate strategy in response to changes in the environment.

The genetic basis underlying traits in a migration syndrome has been described as a complex that incorporates both genes and genetic architecture (van Noordwijk *et al.* 2006; Pulido 2007; Roff and Fairbairn 2007). The genetic basis of these traits can be examined in breeding and artificial selection experiments, with a majority of such studies using insects, but recently involving birds (Pulido and Berthold 2003). Despite clear evidence for a genetic basis to the traits involved in migration, there are no data to indicate that genetic change is required either to produce the first migratory individual from a sedentary population, or the reverse. In fact, Rappole (2003) suggest that environmental change comes before the genetic change. It is most likely that environmental and genetic impacts on behaviour are not mutually exclusive. Genes play a central role in defining the potential range of responses to environmental cues, and phylogenetic studies in conjunction with physiological and ecological investigations will be central in determining the relative importance of genetic versus environmental effects in specific taxa (Rappole and Jones 2002; Ricklefs 2002; Helbig 2003; Rappole 2003).

2.7 How might migratory species respond to climate change?

'It has been hypothesized that seasonality and environmental gradients are key drivers in the evolution and maintenance of migratory behaviour. If so, climate change is likely to result in changes in the frequency and magnitude of migratory behaviours. The currently observed and predicted future effects of climate change on the migratory patterns of birds have been particularly well studied. We can only touch on some highlights of this work, referring the reader to Gordo (2007) and the entire issue of *Climate Research* within which it is published for a more complete treatment of the subject.

In many, but not all, species with breeding grounds in Europe and North America, birds have arrived on breeding grounds earlier as temperatures have warmed. Several proximate and ultimate reasons have been hypothesized for these changes, and it is likely that the dominant explanation may vary with species. Among migratory birds, changes in photoperiod are of clear importance in determining departure date (Berthold 1996), but in most species photoperiod interacts with environmental conditions in the feeding grounds, either directly or through indirect effects on food supply and physiological status (Studds and Marra 2007). Thus, changes in the feeding grounds may initiate earlier departure and corresponding earlier arrival at the breeding grounds, although it is unclear to what extent the changes observed to date result from genetic changes versus the impacts of changing cues on fixed genotypes.

Additionally, but not exclusively, climate change may change conditions along the migration route, resulting in earlier arrival at the breeding grounds even with an invariant departure date. Climate change may directly affect the cost (and therefore speed) of travel through alterations in the prevailing winds or precipitation patterns (which can force delays in flight). In addition, earlier warming may increase the availability of food en route to the breeding grounds, allowing faster travel with fewer and/or shorter stops to replenish food supplies.

The mechanisms described above would allow for changes in migration timing through plastic responses alone, with birds responding only to conditions and information available to them from their immediate surroundings. Over the long term, one might expect additional factors to select for genetic changes in migratory behaviour as well, for instance changing a bird's response to photoperiod such that, all else being equal, it is prepared to migrate earlier in the year. This outcome could reasonably be expected because of alterations in the benefits and costs associated with early arrival at the breeding grounds. Early arrival offers many benefits, including access to the best territories, increased time to find a mate, and increased time for multiple clutches (Jonzén *et al.* 2007 and references therein). However, these benefits must be traded off against the costs of arriving too early, when low temperatures and low food availability entail substantial mortality risk. As the breeding grounds' warm and suitable conditions exist earlier in the year, this trade-off would shift to favour earlier arrival, thereby favouring genotypes that are predisposed toward earlier migration.

The effects of climate change on migration in other vertebrate taxa have received less attention, particularly in terms of empirical documentation of actual changes. Mangel (1994) offered a modelling framework to predict the effects of climate change on migratory behaviour in anadromous fish, but did not explicitly parameterize the model to represent a specific population or species. Satterthwaite *et al.* (2009, 2010) applied a similar model to specific populations, laying a framework for predicting how changes in growth rate (as might be driven by climate change) would be predicted to change migratory behaviour.

As with vertebrates, a variety of changes in insect species distributions and abundances in response to climate change have been documented (Parmesan 2003). Although changes in insect migration per se have so far received little direct attention, changes in either the seasonal cues or resource distribution and abundance patterns resulting from climate changes would be predicted to influence the timing, pattern and frequency of insect migration. For example, both long- and short-term patterns of climate variation have been associated with outbreaks of the swarming migratory locust, *Locusta migratoria*, over the last 1000 years in China. These outbreaks are driven primarily by the frequency-dependent effects of temperature, flood and drought on locust population dynamics (Stige *et al.* 2007). There is also evidence from insects, and vertebrates as well, indicating that they have undergone climate-related genetic changes, most notably in traits relating to the timing of seasonal events or season length (Bradshaw and Holzapfel 2006). These findings highlight the fact that climate change has already led to evolutionary changes in traits that could quite feasibly be part of the migratory syndromes of many different species. Some migratory taxa, such as insects or other small animals with short generation times and large populations, will probably be able to adapt and persist in the face of rapidly changing environments. However, larger animals with longer life cycles and smaller population sizes may not fare as well (Bradshaw and Holzapfel 2006).

2.8 Future directions

We have provided an overview and some examples of the major themes and conclusions in the study of the evolution of migration. Considerable progress has been made in understanding the variety of ecological factors, selective pressures and fitness consequences that can be involved in the evolution and maintenance of migration, particularly within specific systems. Yet, some very general and fundamental questions about migration as a movement strategy remain unanswered. Does selection for migration differ or not from selection for movement in general? And if there are differences, what are they? As evidenced by the enormous amount of variation in migratory strategies both within and across taxa just in the limited examples we have provided, it seems unlikely that a consensus can be reached on a unifying explanation of the factors underlying the evolution of migration that would be generally applicable across taxa.

What is clear is that the study of migration is a vast multidisciplinary pursuit. Specific questions relating to migration and its evolution are relevant across the basic and applied sciences, and conducted by investigators ranging from physiologists and ecologists to biophysicists and engineers. More cross-disciplinary studies, aided in part by some key ongoing technological advances, are likely to play major roles in the further development of our understanding of the evolution of migration. The rapidly expanding catalogue of whole genome sequences and the explosion of other '-omics' approaches, many of which are no longer limited to model organisms (see Mitchell-Olds *et al.* (2008) and the accompanying special issue of *Heredity*), will provide unprecedented insight into the genetic basis and regulatory pathways underlying the expression of traits comprising animal migratory syndromes. Similarly, advances in tracking technologies will enable the movements of a wider array of animals to be tracked with greater precision over greater distances and longer time periods (e.g. Wikelski *et al.* 2007). Technologies are also improving our ability to monitor the physiological state of migrants along their journey (Cooke *et al.*, 2004; Chapter 5). Importantly, the improvement and taxonomic extension of tracking technologies can facilitate the conduct of manipulative experiments in which the relative performance and fitness of individuals can be quantified and compared in order empirically to test specific hypotheses about mechanisms underlying the evolution of migration (e.g. Sword *et al.* 2005)—all while providing novel insights into the movement patterns themselves.

When incorporated into a phylogenetic framework, an improved understanding of the genetic basis of migration and its phenotypic consequences in terms of individual movement patterns will ultimately enable us to determine if differences in migration among species are due to genetic effects (from the genes themselves or other regulatory loci), environmental effects, or particular combinations of the two. We will also gain insight into the widespread convergent evolution of migration and potentially determine whether the same or different genetic mechanisms produce similar migratory phenotypes (Arendt and Reznick 2008).

CHAPTER 3

Theoretical reflections on the evolution of migration

Robert D. Holt and John M. Fryxell

Table of contents

3.1   Introduction

Migration in nature occurs against a complex, shifting backdrop of kaleidoscopic changes through time due to many distinct forces of variation acting over different time scales, from stochastic daily fluctuations in the weather, to multi-annual oscillations in the abundance of resources and natural enemy populations, to the stately march of climate change over geological time. Migrants move across landscapes that are complex over many spatial scales (Chapter 7). Organisms themselves are of course comparably complex, with plastic, multifaceted adaptive strategies for contending with variation in the environment across space and time (Chapter 2).

The theoretical fitness consequences of migration have been considered many times and in many ways. Most of these models, however, have been built around specific taxa or ecosystems. To gain a conceptual handle on the evolutionary forces that have generated and today maintain migration, we suggest it might be equally useful to start with a simple idealized organism living in an equally idealized world. Here we consider the conditions under which non-migratory home range behaviour is vulnerable to invasion by a migratory phenotype; we also consider the converse conditions under which a migratory population can be invaded by non-migratory individuals. Our larger goals are to clarify the logic underlying migratory habitat selection in a seasonal environment, help to define conditions for what we might call the 'evolutionary statics' of migration, and set the stage for more complex evolutionarily dynamic models. At the end of this chapter, we will consider how our simple model compares with other models of the evolution of migration in the published literature.

3.2   An idealized organism in an idealized world

The organism we consider is one that has haploid or clonal inheritance with no age or stage structure. It lives in a world with two distinct habitats, and two distinct seasons in each habitat, but no inter-annual variability. The model is deterministic, and so we are ignoring the impact of both environmental and demographic stochasticity. Within a given habitat type, all individuals are assumed to have the same fitness. Genetic variation, if present, influences the propensity to move between the two habitats, but not fitness within each habitat. We follow the usual protocol of analyses of evolutionary stability and adaptive dynamics, which is to assume that initially the species is fixed for one strategy (i.e., one clone), and then ask if this strategy can be invaded by another, rare clone. The habitats are assumed to be far enough apart that an imposed perturbation in population density in one habitat does not at the same time alter the fitnesses of the individuals in the other habitat. In other words, density dependence, if it occurs, is entirely within-habitat.

Our measure of 'fitness' is very simple. If, in the absence of movement, at the start of an annual cycle in generation t there are $N(t)$ individuals of a given clone in habitat i ($i = $ A,B), and at the end of season 1, of length τ, there are $N(t + \tau)$ present, then average 'seasonal fitness' is measured as the per capita contribution of each individual (including itself) to the population present at $t + \tau$, or

$$R_{i1} = \frac{N(t+\tau)}{N(t)}.$$

Likewise, for each individual present at the beginning of season 2, of length $1 - \tau$, the expected number it will leave to start the next generation, $t + 1$, (its seasonal fitness for this second season) is

$$R_{i2} = \frac{N(t+1)}{N(t+\tau)}.$$

The fitness over a complete annual cycle in habitat i is the multiple of these:

$$F_{Tannual} = R_{i2}R_{i1} = \frac{N(t+1)}{N(t)},$$

the usual measure of fitness for an organism with discrete generations synchronized to an annual cycle. What we have done is express overall annual fitness as a product of what we are calling 'seasonal fitnesses'. The proposition that we will explore is that the patterning of seasonal fitnesses can be examined to characterize when non-migration and complete migration are respectively evolutionarily stable strategies, or ESSs for short (Maynard-Smith 1982).

We can generalize this approach to consider two habitats, where R_{ij} denotes the seasonal fitness of an individual in habitat i during season j. Any of these seasonal fitnesses can be functions of density. Over any given generation, the system is thus defined by four seasonal fitnesses (two seasons in each of habitats A and B). In this snapshot, there are four combinations of seasonal fitnesses possible over an annual cycle ($R_{A1}R_{A2}$, $R_{B1}R_{B2}$, $R_{A1}R_{B2}$, and $R_{B1}R_{A2}$). The first pair of annual fitnesses describes individuals who stay in either habitat over the annual cycle, whereas the second pair pertains to individuals who migrate, in each of the two possible directions (habitat A to B versus habitat B to A), over a given year.

We are here interested in populations that persist and are naturally regulated by density dependence occurring somewhere in the life cycle, in at least one habitat at one season. We assume that our initial population is genetically homogeneous and in demographic equilibrium. Our general models make no specific assumptions about density dependence, except making the implicit assumption that it occurs, and that the population is initially at its demographic equilibrium. When we use a specific model to provide some numerical examples to accompany the analytical results, for simplicity we assume that density dependence is in a single season. Future work should extend such specific models to incorporate density dependence across both seasons.

3.3   When is a non-migratory species vulnerable to invasion by migratory strategies?

We first start with a population that is non-migratory, and ask if a migratory strategy can colonize. For non-migration to be in equilibrium, without loss of generality we can assume that the initial habitat occupied is habitat A. In other words, for this case, the initial condition is for the species to occupy a restricted geographical range, within which seasonal variation occurs. When is this restricted distribution an ESS, relative to a rare invasive migratory strategy? We assume for now that there are no costs to movement (this is, of course, a huge assumption).

For the non-migratory specialist in habitat A to be in demographic equilibrium requires $R_{A1}R_{A2} = 1$, hence the 'rate' of geometric growth must be unity. We will denote fitness at population equilibrium by asterisks ($R_{A1}^{*}R_{A2}^{*}$). If an individual moves to habitat B, and stays there (i.e., a rare dispersal event, rather than migration), for the initial condition to persist (and thus for the restricted range to be ecologically stable) it requires that

$$R_{A1}^{*}R_{A2}^{*} > R_{B1}R_{B2} \quad \text{or} \quad 1 > R_{B1}R_{B2}. \tag{1}$$

This simply restates the fact that, in a closed population, there is a geometric growth rate criterion for population persistence.

If now a mutant individual arises that regularly shuttles back and forth between the two habitats, it can experience two possible net growth rates, depending upon the order of seasons it experiences. For non-migration to be an ESS (i.e., a rare migratory clone which attempts to invade, and migrates completely between the two habitats within each year, then declines towards extinction), we must have both

$$R_{A1}^{*}R_{A2}^{*} > R_{A1}^{*}R_{B2} \quad \text{and} \quad R_{A1}^{*}R_{A2}^{*} > R_{B1}R_{A2}^{*}.$$

The fact that there are two conditions reflects the fact that there are logically two distinct patterns of complete migration between two habitats. Based on our assumption of demographic equilibrium for the resident population, this pair of inequalities simplifies to

$$1 > R_{A1}^{*}R_{B2} \quad \text{and} \quad 1 > R_{B1}R_{A2}^{*}.$$

Without loss of generality, assume season 2 in habitat A is the bad season. Because $R_{A1}^{*}R_{A2}^{*}=1$, we can eliminate R_{A1}^{*} from our equation by letting $R_{A1}^{*} = 1/R_{A2}^{*}$. So, the conditions for non-migration to be an ESS are:

$$1 > (1/R_{A2}^{*})R_{B2} \quad \text{or} \quad R_{A2}^{*} > R_{B2}, \quad \text{and}$$
$$1 > R_{B1}R_{A2}^{*} \quad \text{or} \quad 1/R_{B1} > R_{A2}^{*}.$$

We can put these two inequalities together in a joint inequality, defining the necessary and sufficient demographic conditions for non-migration to be an ESS, as follows:

$$1/R_{B1} > R_{A2}^{*} > R_{B2}. \tag{2}$$

Note that Equation 2 implies Equation 1, which was earlier deduced to be the condition for ecological stability of habitat specialization (to habitat A) in a seasonal world. So a necessary condition for non-migration to be an ESS, with specialization to habitat A, is that colonization of habitat B (without back-migration) fails.

But this is not sufficient. In other words, there can be habitats that cannot sustain a population, on their own, but which could foster the evolution of migration. In particular, if $R_{A2}^{*} < R_{B2}$, then the non-migratory condition is not an ESS, and complete migration can invade. Put simply, if the seasonal fitness during the worst season in the habitat initially occupied (habitat A) is less than seasonal fitness in the other unoccupied habitat (habitat B), an individual that begins a migratory shuttle between habitats, leaving habitat B after season 2 then back to habitat A for season 1, enjoys the best of both, and can invade. A comparable condition arises in analyses of the utilization of stable sink habitats when source habitats fluctuate in fitness, as part of spatial bet-hedging strategies (Holt 1997).

If the necessary condition in Equation 1 does not hold, then habitat 2 should be colonized. In the continued absence of migration, each habitat should then equilibrate over time such that its geometric mean fitness is unity, or

$$1 = R_{A1}^{*}R_{A2}^{*} \quad \text{and} \quad 1 = R_{B1}^{*}R_{B2}^{*}. \tag{3}$$

We have now added asterisks to the seasonal fitnesses in habitat B to denote the requirement for population equilibration. The reason is that after invasion of habitat B, numbers must grow there until density dependence occurs, such that fitness over the annual cycle is unity. This requires that one or both seasonal fitnesses must be functions of density, so realized seasonal fitness for at least one season at equilibrium is depressed over the initial seasonal fitness at the time of invasion.

Is this distribution, where a species occupies two habitats but does not migrate between them, stable against invasion by a completely migratory genotype? As a limiting case, we assume that there is no cost to such migration. The migratory genotype has two possible fitnesses, depending upon its order of movement during the annual cycle.

$$R_{A1}^* R_{B2}^* \quad \text{or} \quad R_{B1}^* R_{A2}^*.$$

For both of these to be less fit than the resident non-migratory type requires that both

$$R_{A1}^* R_{B2}^* < 1 \quad \text{and} \quad R_{B1}^* R_{A2}^* < 1. \tag{4}$$

Again, we can use the assumption of demographic equilibrium to eliminate some terms. At demographic equilibrium in the non-migratory population,

$$R_{A2}^* = 1 / R_{A1}^* \quad \text{and} \quad R_{B2}^* = 1 / R_{B1}^*. \tag{5}$$

Substituting the left-hand equality of (5) into the right inequality of (4) leads to

$$R_{B1}^* < R_{A1}^*. \tag{6}$$

Substituting the right-hand equality of (5) into the left inequality of (4) R_{B2}^* gives

$$R_{B1}^* > R_{A1}^*. \tag{7}$$

Inequalities (6) and (7) are contradictory.

Ergo, the above line of reasoning thus leads to the simple conclusion that:

> If a species can persist across two habitats without movement between them, and those two habitats fluctuate through time, in a manner that is reflected in 'seasonal fitness', then the species is always vulnerable to invasion by a completely migratory genotype.

The migratory strategy in effect exploits the fact that in each habitat, there is a season where seasonal fitness is greater than one (a necessary consequence of temporal variation in fitness, combined with the assumption of demographic equilibrium in each habitat). Shuttling regularly between habitats, when rare, then clearly implies an initial geometric mean fitness greater than one. Interestingly, this may hold even if it implies that the migratory type (which we recall is rare) moves into a habitat during the season when seasonal fitness there is lower than in the alternate season, and the two habitats experience synchronous and in-phase variation in seasonal fitnesses.

A numerical example may suffice to illustrate this point. For simplicity, we assume that density dependence occurs in the same season in each habitat, which is the time of year when a single bout of reproduction occurs, with no density dependence in the other season. Fitness in the birth season in both habitats is given by a Ricker formulation (1954), with density dependence arising from the summed density of all clones that are found in a given habitat. Fitness in the non-breeding season is determined by the magnitude of a density-independent rate of survival. We can represent morph dynamics with the following system of equations:

$$N1(t+1) = N1(t) \cdot \exp(r_A \cdot [1 - N1(t) - N3(t)] + s_A), \tag{8a}$$

$$N2(t+1) = N2(t) \cdot \exp(r_B \cdot [1 - N2(t) - N4(t)] + s_B), \tag{8b}$$

$$N3(t+1) = N3(t) \cdot \exp(r_A \cdot [1 - N1(t) - N3(t)] + s_B), \tag{8c}$$

$$N4(t+1) = N4(t) \cdot \exp(r_B \cdot [1 - N2(t) - N4(t)] + s_A), \tag{8d}$$

where $\exp(r_j)$ is the maximal birth rate of individuals in habitat j during the breeding season, $\exp(s_j)$ is the exponential survival rate of individuals in habitat j in the non-breeding season, and $Ni(t)$ is the population density of individuals of behavioural morph i at time t. In this formulation, morph 1 represents individuals that are selective for the best year-round habitat A; morph 2 represents individuals that are selective for the worst year-round habitat B; morph 3 represents individuals that reproduce in the best habitat, but migrate to the other habitat in the non-growing season (dubbed 'logical migrants'); and, morph 4 represents individuals that reproduce in the poorer habitat, but migrate to the other habitat in the non-growing season (labelled 'perverse migrants'). We have scaled density so that if survival is guaranteed through the non-breeding season, carrying capacity (viz., equilibrial density) is set at unity. Morphs that co-occur are competitively equivalent, as measured by density dependence in births.

This is perhaps the simplest mathematical representation possible for depicting seasonal migration, demarcating a period of density-dependent growth from a season of density-independent mortality, and contrasting the fates of clones that interact equivalently with each other but display different migratory propensities.

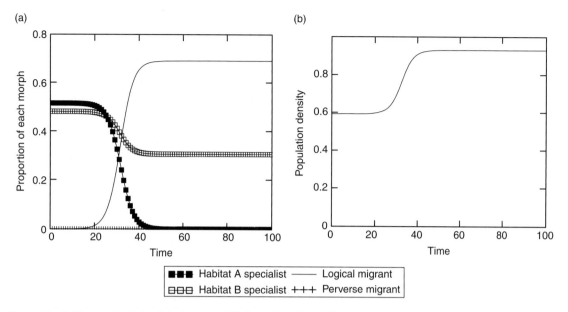

Figure 3.1 Variation over time in the relative frequency of (a) behavioural morphs and (b) total population abundance in a system with specialists in two distinct habitats that is invaded by rare morphs that migrate between habitats on a seasonal basis ($r_A = 1.0$, $r_B = 0.5$, $s_A = -0.69$, $s_B = -0.36$). Habitat A is best during the growing season, whereas habitat B is best during the non-growing season.

Fitness maxima (i.e. the per capita annual rate of change as $N \to 0$, which is equal to the product of the seasonal fitnesses) can be readily calculated as $R_{A1} R_{A2} = \exp(r_A + s_A)$ for sedentary (i.e., non-migratory) specialists in habitat A, $R_{B1} R_{B2} = \exp(r_B + s_B)$ for sedentary specialists in habitat B, $R_{A1} R_{B2} = \exp(r_A + s_B)$ for those migrants that use the best habitat to breed and $R_{B1} R_{A2} = \exp(r_B + s_A)$ for the perverse migrants that use the worst habitat to breed. For any of these four morphs, its maximal fitness must exceed unity, or it will surely disappear. In isolation, the equilibrium abundance of specialists in the best habitat = $1 + s_A/r_A$, whereas the equilibrium abundance of specialists in the worst habitat = $1 + s_B/r_B$. For these quantities to make sense (i.e., have a non-zero population) requires that $|s_i| < r_i$. We will assume that these general conditions for potential viability of each strategy hold.

Assume as an example that a non-migratory species occupies habitat A, where $r_A = 1$ and $s_A = -0.69$. These demographic parameters imply an equilibrium density of 0.31 (where $N_{eq} = 1 + s_A/r_A$). Seasonal fitness in habitat A fluctuates between $R_{A1} = \exp(r_A[1 - N_{eq}]) = 2$ in the growing season and $R_{A2} = \exp(s_A) = 0.5$ in the non-growing season. Fit-

ness over the year is the product of these two numbers, 1, so the population in habitat A is in demographic equilibrium. A non-migratory population in habitat B, where $r_B = 0.5$ and $s_B = -0.36$, is in demographic equilibrium at a population density of 0.28 (where $N_{eq} = 1 + s_B/r_B$). While breeding success is lower at equilibrium in habitat B, such that $R_{B1} = \exp(r_B[1 - N_{eq}]) = 1.43$, animals residing there enjoy a more benign environment in the non-breeding season, such that $R_{B2} = \exp(s_B) = 0.7$. Good and bad seasons are synchronized across space. If a migratory genotype now arose, which resided in habitat A in the good season, and habitat B in the bad season, its annual growth rate when rare would be (2)(0.7) = 1.4, so it would be selected and increase when rare (Fig. 3.1).

In this example, one would expect to see the evolution of at least some migration into and out of the poorer habitat. Paradoxically, however, some individuals should leave just as conditions begin to improve locally in habitat A. In turn, the evolution of logical migration in this example makes it inevitable that non-migratory morphs would also persist in the poorer habitat, because their increase in the growing season more than compensates for

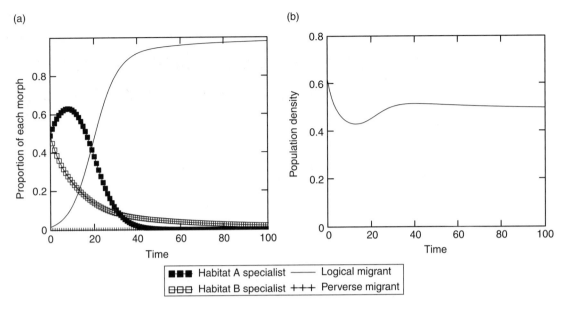

Figure 3.2 Variation over time in the relative frequency of (a) behavioural morphs and (b) population abundance in a system with specialists in two distinct habitats that is invaded by rare morphs that migrate between habitats on a seasonal basis ($r_A = 1.0$, $r_B = 0.5$, $s_A = -0.69$, $s_B = -0.51$). Habitat A is best during the growing season, whereas habitat B is best during the non-growing season. Unlike the system in Fig. 3.1, habitat B is a sink if occupied year-round. Initial abundance of both specialists is 0.31.

their density-independent losses during the non-growing season. This of course hinges on the assumption that the non-migrant can escape from any density-dependent effects experienced from the much larger resident population, when they co-occur during the non-growing season. If habitat B is a sink (were it to be occupied on a year-round basis) then only migrants can persist in the system (Fig. 3.2), echoing a pattern seen in earlier models of habitat-mediated dispersal (McPeek and Holt 1992; Holt 1997). Provided that the fitness in alternating habitats is high enough, it is readily possible to construct a system in which both habitats are sinks, yet permitting persistence of migrants, although such migrants could only arrive via colonization from elsewhere.

So far we have assumed that habitats alternate seasonally in terms of fitness advantage. This need not be the case, of course; habitat A might well yield both the higher rate of growth and the higher survival in the non-growing season. In this case, non-migratory individuals obviously would have an advantage over all other morphs and would predominate (Fig. 3.3). However, this situation opens

an opportunity for a perverse migrant to evolve, one that chooses to breed in the poorer habitat, but then moves into the better habitat during the non-growing season. As before, this outcome may depend on our assumption of no competition during the non-growing season.

3.4 When is a migratory species vulnerable to invasion by non-migratory strategies?

We now start, in effect, at the other end of the spectrum of migratory behaviours, and assume that the species is initially completely migratory, thus abandoning each habitat in turn over the course of the annual cycle. When can this species be invaded by a completely non-migratory strategy? Another question that should be considered is whether this migratory strategy can be invaded by a countermigratory strategy (i.e. individuals that go in the opposite direction to the general spatial flow of the population)?

To address these questions, we can assume that the initial condition is such that individuals use the

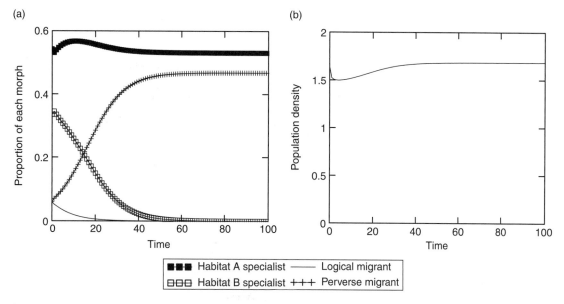

Figure 3.3 Variation over time in the relative frequency of (a) behavioural morphs and (b) population abundance in a system with specialists in two distinct habitats that is invaded by rare morphs that migrate between habitats on a seasonal basis ($r_A = 1.0$, $r_B = 0.5$, $s_A = -0.11$, $s_B = -0.22$). Habitat A is best in both seasons.

best habitat (A) in the breeding season, but migrate to the other habitat (B), during the non-breeding season. The realized annual fitness over this cycle (again ignoring the potential cost of movement itself) is

$$F_{migration} = R_{A1}^* R_{B2}^* = 1, \qquad (9)$$

assuming demographic equilibrium (hence the use of asterisks). Because breeding occurs in A we can assume that

$$R_{A1}^* > 1 > R_{B2}^*. \qquad (10)$$

If a few individuals stay behind in habitat A, or habitat B, respectively, they will experience little or no density dependence, and so have expected fitnesses of

$$F_1 = R_{A1}^* R_{A2} \quad \text{and} \quad F_2 = R_{B1} R_{B2}^*. \qquad (11)$$

The asterisks indicate that in the respective seasons, there may be density dependence in that habitat. The absence of asterisks indicates that, in that season, there should be no density dependence, when

a novel non-migratory clone is initially rare (because the migratory population has completely left that habitat, for the other one).

For complete migration to be an ESS, each of these fitnesses for non-migrants must be less than unity:

$$R_{A1}^* R_{A2} < 1 \quad \text{and} \quad R_{B1} R_{B2}^* < 1. \qquad (12)$$

Multiplying these two inequalities together leads to

$$R_{A1}^* R_{A2} R_{B2}^* R_{B1} < 1. \qquad (13)$$

Substituting (9) into (13), then

$$R_{B1} R_{A2} < 1. \qquad (14)$$

The inequality in (14) represents the annual growth rate of a perverse migratory genotype, going in the opposite direction to the resident type, when it is initially rare and hence experiences no density dependence. So if non-migrants cannot invade (i.e., the conditions in (12) hold), then counter-migration is simply unviable.

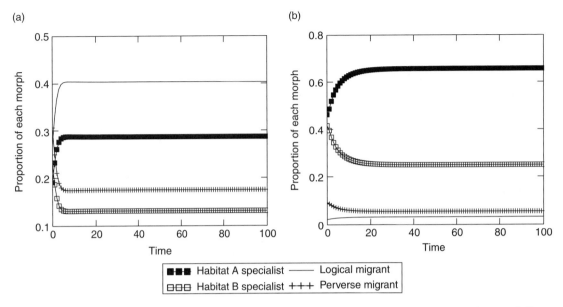

Figure 3.4 Variation over time in the relative frequency of behavioural morphs in a system with specialists in two distinct habitats that is invaded by rare morphs that migrate between habitats on a seasonal basis ($r_A = 1.0$, $r_B = 0.5$, $s_A = -0.36$, $s_B = -0.36$). Both habitats yield equal fitness in the non-growing season, although habitat 1 provides higher fitness in the growing season. The two sub-plots show simulations with differences in initial conditions: (a) population densities for both habitat specialists set initially at equilibria, (b) habitat specialist A was initiated at equilibrium, whereas habitat specialist B was initiated ~ 10% below its equilibrium. Migrants started lower in (b).

What about non-migration? We can re-write (12) as

$$R_{A2} < \frac{1}{R_{A1}^*} \quad \text{and} \quad R_{B1} < \frac{1}{R_{B2}^*}. \quad (15)$$

It is clear after substitution from (10) into the left side of (15) that a non-migratory strategy in the better habitat A is excluded, provided that seasonal fitness in the 'off'-season (when the migrants are elsewhere) is sufficiently less than unity.

Intriguingly, exclusion does not require low fitness in the poorer habitat B. Because $R_{B2}^* < 1$, it is possible for $R_{B1} > 1$, yet for a resident strategy remaining in habitat B nonetheless to be excluded. The reason is that even though seasonal fitness in the absence of migrants exceeds one in season 1, fitness averaged over the entire annual cycle has to take into account fitness in the other season. If the intrinsic fitness of habitat B in season 2 is sufficiently low, then non-migration may be excluded even in the absence of density dependence (an example of exclusion consistent with this effect is shown in Fig. 3.2). Alternatively, strong negative density dependence from the migrant may sufficiently lower fitness that a

non-migrant is excluded—even though a non-migratory population could persist just fine in isolation For example, Fig. 3.1 shows a system where each habitat is initially occupied by a non-migratory species. When migratory morphs are introduced, they rapidly replace one of the specialists.

If survival rates in the non-growing season are identical across habitats ($R_{A2} = R_{B2}$), then there is no fitness advantage between pairs of competing morphs, even if there is pronounced habitat-mediated variation in fitness during the growing season. Habitat A specialists and migrants have equivalent annual fitness, since $R_{A1}^* R_{B2} = R_{A1}^* R_{A2}$. The same is true of habitat B specialists and perverse migrants, since $R_{B1}^* R_{B2} = R_{B1}^* R_{A2}$. Constant mortality across habitats leads to all four morphs coexisting in the two habitats, at frequencies dictated by initial conditions (Fig. 3.4), provided that $R_{A1}R_{A2} > 1$ and $R_{B1}R_{B2} > 1$. A very similar process of neutral selection was demonstrated previously in models of habitat-mediated dispersal (McPeek and Holt 1992), where many combinations of dispersal strategies are neutral, relative to each other.

More typically, however, one might expect to see differences in demographic parameters across different habitats. Under these more general conditions, in systems that settle down to demographic equilibrium, at most two out of the four discrete behavioural morphs that we are contrasting can coexist in a system with two habitats, due to direct density-dependent selection between pairs of competing morphs (habitat A specialists vs. logical migrants or habitat B specialists vs. perverse migrants). Only migrants by themselves, or a pair of migratory morphs (logical and perverse migrants), can persist when both habitats are sinks (as assessed by the fate of non-migratory morphs, when alone).

Our general conclusion is that if a species is widely distributed, and the world fluctuates through time (an ecological truism if there ever were one), then if one does NOT observe migration, it must be because migration itself is costly. Such costs are not explicitly built into the logic of fitness in the seasonal environments that we have presented above. However, one can use the general approach sketched above to 'titrate' such costs, so as to determine threshold conditions for when migration would not be favoured because of its intrinsic costs. In particular, assume that there is a multiplicative decrement in fitness of c (for cost) for each bout of migratory movement, expressed as realized seasonal fitness in each habitat. Assume that the highest such fitness is $F_{potential} \equiv R_{A1}^* R_{B2}^* > 1$, so migration should be favoured (the asterisks indicate that there should be density dependence, given that both habitats are occupied, and the population persists without moving between them). If migration is not in fact favoured, that must be because it is costly to move, as measured in a fitness decrement. Incorporating a fitness decrement of c equal in magnitude for each year (habitat A to B, and habitat B back to A), the actual annual fitness is

$$F_{actual} = R_{A1}^* R_{B2}^* - c. \qquad (16)$$

The threshold between migration being favoured and not corresponds to a cost of

$$c = R_{A1}^* R_{B2}^* - 1. \qquad (17)$$

Extending the numerical example discussed earlier, if rare migrants invade a system in which fitness during the growing season is $R_{A1} = 2$ and fitness during the non-growing season in the alternative habitat B is $R_{B2} = 0.7$, then the threshold cost each way required to impede the evolution of migration $c = (2 \times 0.7) - 1 = 0.4$. To illustrate this effect, we apply a travel cost slightly exceeding the threshold for the migratory system modelled in Fig. 3.1. The imposition of a minor energetic cost makes the system vulnerable to re-invasion by the non-migratory specialist in habitat A (Fig. 3.5). Such a cost that (just) prevents migration from being favoured in a temporally variable environment within an occupied geographical range (in our case, two discrete habitats) is determined by the geometric mean of the better seasonal fitnesses across both habitats. The threshold cost per move cannot be assessed just by looking at one habitat, but involves an assessment of fitness benefits across the entire migratory cycle.

So far, we have considered only outcomes in which equilibria are locally stable. It is well known that discrete time models with this structure exhibit population cycles or even deterministic chaos at elevated rates of growth (May 1976). Variation in fitness over time has the potential to influence evolutionary dynamics in any system, including those with spatial structure (McPeek and Holt 1992; Holt and McPeek 1996), so unstable systems are important to consider. Increase in the rate of population growth for our two habitat system typically induces fluctuations in both total population abundance and relative frequencies of each morph (Fig. 3.6). Peaks and troughs in population density are positively correlated with variation in the relative frequency of migrants.

A variation on this expected theme arises when both habitats offer similar growth rates ($r_A = r_B$), but differ in survival rates ($s_B > s_A$). When maximum per capita growth rates are large, migrants and habitat B specialists often show a complex pattern of out-of-phase fluctuation as the migrant behavioural morph invades the ecosystem, but these fluctuations settle down to constant proportions over time, despite the fact that overall population abundance remains highly unstable (Fig. 3.7).

Interestingly, this outcome is sensitive to initial conditions. Slight modifications in the initial

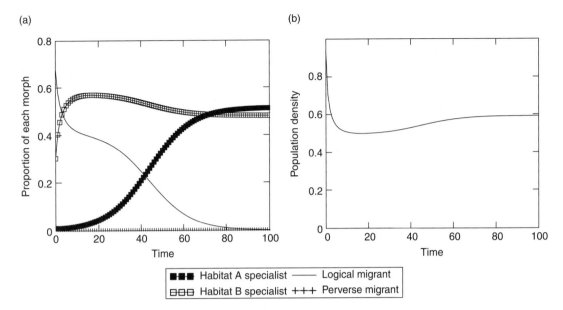

Figure 3.5 Variation over time in the relative frequency of (a) behavioural morphs and (b) population abundance in a system with specialists in two distinct habitats that is invaded by rare morphs that migrate between habitats on a seasonal basis ($r_A = 1.0$, $r_B = 0.5$, $s_A = -0.69$, $s_B = -0.36$). Habitat A is best during the growing season, whereas habitat B is best during the non-growing season. A demographic cost of 0.44 was applied at each habitat transition for migrants. Initial population densities of habitat B specialists and migrants were based on equilibria from the simulation depicted in Fig. 3.1.

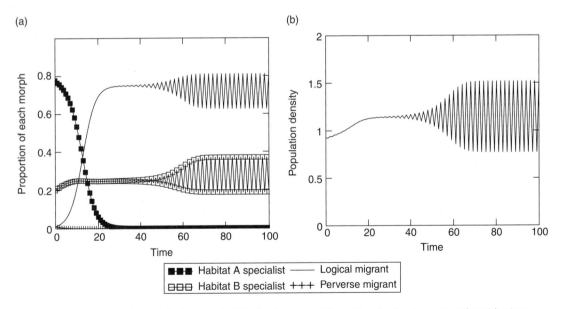

Figure 3.6 Variation over time in the relative frequency of (a) behavioural morphs and (b) population abundance in a system with specialists in two distinct habitats that is invaded by rare morphs that migrate between habitats on a seasonal basis ($r_A = 2.5$, $r_B = 0.5$, $s_A = -0.69$, $s_B = -0.36$). Habitat A is best in the growing season whereas habitat B is best during the non-growing season.

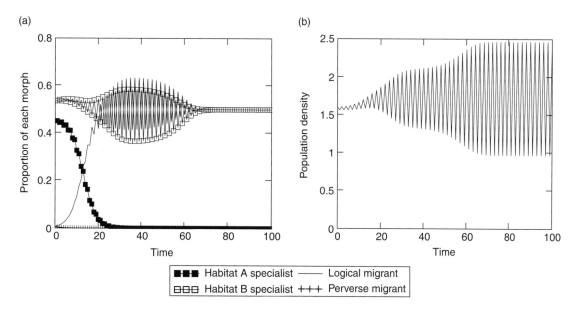

Figure 3.7 Variation over time in the relative frequency of (a) behavioural morphs and (b) population abundance in a system with specialists in two distinct habitats that is invaded by rare morphs that migrate between habitats on a seasonal basis ($r_A = 2.5$, $r_B = 2.5$, $s_A = -0.69$, $s_B = -0.36$). Both habitats are equal during the breeding season whereas habitat B is best during the non-breeding season.

abundance of specialists before the mutant migrant morph invades can produce a mirror-image: population abundance that stabilizes over time, yet with morph frequencies that fluctuate violently over time (Fig. 3.8). In other words, dynamic instability in migratory systems can be expressed through variation in either population abundance or behaviour.

This example suggests that a ripe area for future work will be to examine the interplay of seasonality and intrinsic population instabilities, and how this can both promote migration and potentially lead to counter-intuitive results. Only more detailed analyses, across a range of models, will be able to determine if these intriguing patterns are merely curiosities, or instead arise as surprising outcomes in a broad array of circumstances. The interplay of temporal variation and spatial processes often leads to surprising results, and migration may at times amplify the dynamic complexities inherent in many ecological systems.

3.5 Discussion

Our simple two habitat model captures some of the key biological characteristics that recur in virtually

all migratory ecosystems. First and foremost, the evolution of migration requires some interplay between seasonal and spatial variation in fitness (Lack 1968; Fryxell and Sinclair 1988a; Lundberg 1988). Such variation is, of course, nearly ubiquitous in the natural world. One would be hard-pressed not to find spatial variability in critical ecological characteristics across the habitable range of most organisms, particularly those of larger body size. Seasonality, as well, is a hard fact of life, even in the tropics. Alternation between periods of breeding and non-breeding is the norm, rather than the exception, in nature, even if the absence of seasonality is the norm in most ecological theory. Even environments that at first glance seem to be devoid of seasonality (e.g., caves, the deep sea) can be influenced by seasonal variation, if they are coupled to external, variable environments—which they almost always are.

Our model is admittedly—and unashamedly—rather crude. It is useful to summarize these limitations in the model, each of which we suggest represents a potential avenue for further theoretical exploration and refinement of the evolutionary theory of migration.

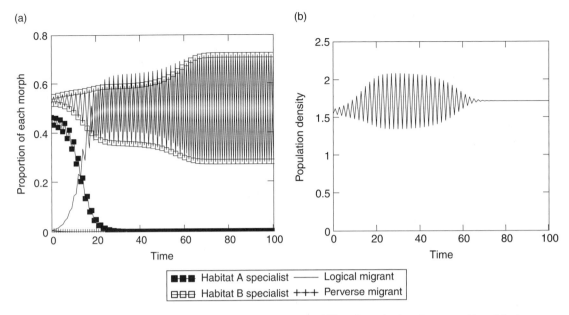

Figure 3.8 Variation over time in the relative frequency of (a) behavioural morphs and (b) population abundance in a system with specialists in two distinct habitats that is invaded by rare morphs that migrate between habitats on a seasonal basis ($r_A = 2.5$, $r_B = 2.5$, $s_A = -0.69$, $s_B = -0.36$). Both habitats are equal during the growing season whereas habitat B is best during the non-growing season. This figure differs from Fig. 3.7 only in a slight difference in initial conditions: In Fig. 3.7, (a) initial population densities set at their equilibria for both habitat specialists; in this figure, habitat specialist A was initiated at equilibrium, whereas habitat specialist B was initiated ~10% below its equilibrium.

We did not account for the ecological and genetic complexities that no doubt occur in real organisms, because we assumed clonal inheritance. Yet most species that migrate are sexual, and migratory behaviour is likely to be under rather complex genetic control (Chapter 2). Even for clonal inheritance, we have not paid attention to the potential for mixed strategies, such as partial migration, say with different propensities to migrate in each habitat. Broadening the range of migratory strategies that are competing with each other, and how they are generated by alternative rules of inheritance, is clearly an important step that would go beyond the models we have presented here. Griswold *et al.* (2010) have developed an excellent template for exploring the intricacies of genetics and demography in migration models of the sort that we have described here.

We have also ignored the crucial role sensory cues can play in the actual mechanisms generating migratory behaviour (Chapter 6). Our simple model presumed in effect that organisms have no ability to anticipate fitness in either their current or alternative habitats and make appropriate decisions. This seems inconsistent with well-documented instances of migratory organisms reversing their migratory circuit under unusual environmental conditions, such as wildebeest returning to the Serengeti plains in years with exceptional dry season rainfall (Maddock 1979).

Space is implicit and admittedly rudimentary in our models. Real migrants must use constrained movement modes to traverse rugged landscapes, with dynamic interplay between navigational capacity, social pressures, and complex motivational goals (Mueller *et al.* 2008; Nathan *et al.* 2008; Schick *et al.* 2008). This in turn makes it likely that some habitats could remain part of a migratory repertoire even if, in terms of the strict calculus of natural selection, such habitats should be ignored.

Our models also, and crucially, do not explicitly include ecological interactions other than direct density-dependent competition between behavioural morphs. Ecological interactions may be implicitly contained, however, depending upon the details. For instance, if one ecological dominant species migrates, this automatically sets up a seasonal driver in the lives of other, more subordinate species.

Some ecological interactions may need to be specifically modelled, because there are direct feedbacks between the migratory species and these other players in the ecological system (Chapter 9). In our model, there is no explicit resource dependence, predation or disease risk, nor social or agonistic interactions within or among morphs. Incorporating such interactions could permit a more refined assessment about which environments might foster migration, and for which taxa. Moreover, many of these interactions operate with a time lag. Depending on the timing of these lags, they could either magnify, or dampen, the impact of external seasonal drivers on fitness, and thus alter the relationship between seasonality and migration.

Moreover, our model organisms have a simple life history, whereas real organisms can employ state- and age-dependent decisions to maximize lifetime reproductive success, which we know can considerably alter optimal evolutionary strategies (Clark and Mangel 2000, Stephens *et al.* 2007). Even simple models such as ours can employ carryover effects from one season to the next, which can have important population dynamic properties (Ratikainen *et al.* 2007).

Despite all these caveats, a simple model such as the one we have presented is valuable, because it is tractable enough to allow analytical approaches as well as straightforward simulation, a mixture that is often powerful in understanding the full range of possible outcomes of ecological models. In our experience, many insights from simple models such as the ones we have presented provide crucial yardsticks for assessing the importance of the various factors we have listed as potential caveats and complicating factors in the above paragraphs.

There is a substantial published literature on the evolutionary dynamics of dispersal in systems with two or more habitat patches. Limitations on space preclude a detailed review here. Suffice it to say that formal models of invasion dynamics often suggest the coexistence of two or more dispersal morphs in a metapopulation setting (McPeek and Holt 1992; Doebeli 1995; Doebeli and Ruxton 1997; Parvinen 1999). Population dynamics play a key role in main-

taining behavioural polymorphism in meta-populations (McPeek and Holt 1992; Doebeli 1995; Doebeli and Ruxton 1997; Parvinen 1999), as does stochastic variation in extinction risk across the ensemble of patches (Heino and Hanski 2001). We have found in our numerical simulations of the seasonal Ricker model that likewise coexistence of alternative migratory morphs can occur. Note that our clonal model can also be interpreted as a model for interactions between two competing species, which are equal with respect to density dependence, when they co-occur. What our model results suggest is that the combination of seasonality and migration can permit species coexistence, despite their competitive equivalence within patches.

It has long been appreciated that geographic variation in survival rates outside the breeding season can contribute importantly to the selective advantage of migration (Lundberg 1988). Lack (1968) pointed out that temporal variation in survival rates could lead to balanced long-term fitness of migrants vs. residents. Similar arguments underlie von Haartman's (1968) state-dependent evolutionary arguments for the evolution of avian migration; resident birds obtain compensatory reproductive advantage balancing the higher over-wintering costs relative to migrants. Both these treatments were well ahead of the development of ESS theory, so it is not surprising that they did not consider conditions for successful invasion by other phenotypes. In a prescient theoretical study, Cohen (1967) developed a formal model of invasion dynamics in the special case of a population with geometric growth with inter-annual variation in λ. Population dynamic effects alone could not create opportunities for invasion by alternate phenotypes, such as those we have shown. As a simplifying assumption, Cohen presumed identical reproductive rates of migrants vs. residents, focusing purely on variation in over-winter survival. Theoretical treatments of migration evolution rarely consider the possibility that residents could exist as an ancestral condition in both habitats.

More recent models on the evolution of migration have typically used more highly structured models and less generic contrasts between habitats (Lundberg 1987; Kaitala *et al.* 1993; Kokko and Lundberg 2001; Griswold *et al.* 2010). For example,

Kaitala *et al.* (1993) evaluate conditions under which over-wintering migration is an ESS in an age-structured population of birds that must breed in one habitat, developing a theme first formally developed by Lundberg (1987) in a graphical model. Over-wintering survival is assumed to be density-independent in migrants, but linearly density-dependent in non-migrant individuals. These conditions are reasonable for many passerine birds, and demographic parameters were chosen accordingly. Kaitala *et al.* (1993) found that mixed strategies (i.e. partial migration) were selected for under conditions of density-dependent over-winter survival. Griswold *et al.* (2010) linked genetic effects with habitat-mediated variation in seasonal fitness, using paired habitats in a similar manner to our model, with an important difference: either reproduction or survival was assumed impossible in one habitat. Griswold *et al.* (2010) found that the evolution of partial migration depended on the genetic basis for behaviour and in which season habitats were shared, demonstrating a clear need for proper linkage of genetic and demographic dynamics in future modelling efforts. Our more generic formulation, with milder seasonality, predicts partial migration as a common outcome except under exceptional conditions of source–sink dynamics. Echoing Kaitala *et al.*'s findings, partial migration requires some form of density dependence, but this can be expressed either for the breeding or non-breeding season.

Where spatial and seasonal variation in fitness does occur, our simple model predicts that migration should often prove selectively advantageous. This assertion may explain why migration has repeatedly evolved in a wide variety of biomes and taxa (Chapter 2). The conditions favouring migration are common, so it is not surprising that this lifestyle has evolved countless times in evolutionary history. Indeed, the conditions favouring migration are so general, it is perhaps more relevant to ask why migration isn't ubiquitous?

One answer to this paradox suggested by our model is that the demographic cost of migration itself may exceed the benefits in some systems. This cost could be expressed in myriad ways. It could be a simple energetic debt that simply cannot be repaid at the end of each arduous journey. Many migrants respond to this challenge through stopover sites en route, used to restore depleted energy reserves (Alerstam *et al.* 2003). Nonetheless, energetic costs may be of sufficient magnitude to compromise individual fitness, particularly for females that must choose wisely between investment of energy reserves in movement vs. production of offspring. In probing this question, it might well prove instructive to compare energetic costs of migration across different taxa and modes of locomotion, particularly in relation to body size (Chapter 4).

The cost of migration could be social, particularly in species where social dominance and therefore access to potential mates depend on acquisition and control of scarce resources, via territoriality or semi-exclusive home range use (Lundberg 1987; Kaitala *et al.* 1993; Kokko and Lundberg 2001). Individuals that reposition each season may lose priority access to favoured patches of real estate, which is untenable in the long-run. For example, territorial lions are seemingly unable or unwilling to track the seasonal migration of large herbivores from the Serengeti plains to the northern woodlands, possibly because of the risk of losing fiercely-defended group territories to neighbouring prides (Mosser *et al.* 2009). It is intriguing to speculate that social costs of migration may be more prevalent in organisms placed at the top of food-webs, where territoriality and complex social dominance are the norm, rather than the exception.

The demographic cost of migration could be expressed through increased exposure to predators, particularly if migrants are easier to find, readily visible, more vulnerable to attack, or less capable of predator avoidance than non-migrants. Grizzly bears congregating along salmon spawning runs or crocodiles lurking at traditional river crossings used by wildebeest instantly leap to mind. There are certainly well-documented examples of humans lying in wait along migratory routes to ambush migratory ungulates, fish or waterfowl (Chapter 11).

Whatever these costs may be, our simple model suggests that they need to be compared with the demographic benefits of migration to realize a full Malthusian accounting. Indeed, our model suggests that it is not enough to think about the ecological characteristics of used habitats—these are only meaningful in comparison with habitats passed up by migrants. Demographic assessment could be quite challenging, admittedly, if migratory individuals choose not to live in some habitats. One can even imagine experimental titration of the demographic costs vs. benefits of migration allowing rigorous new testing of alternative constraints on the evolution of migration.

Our model suggests that alternate behavioural morphs should be sought as coexisting strategies, as a mixed ESS, in systems with habitat and seasonal structure. The favoured mix of coexisting strategies depends, however, on the magnitude of spatial and temporal variation in vital rates. Systems with seasonal alternation in the habitats with optimal fitness select for migrants that shift seasonally between the best habitats available and non-migrants that specialize in the poorest breeding sites. Systems in which one habitat is always optimal select for habitat specialists in the ideal habitat, mixed with perverse migrants that move away to breed in poor sites before returning to share the non-breeding season with their specialist brethren. Where neither habitat is best during the non-breeding season, anything goes; all combinations of migrants and non-migrants can coexist. We should caution that the particular model we used to illustrate these points assumes that density dependence occurs entirely in one habitat, in one season. Future studies should examine what happens when there is density dependence in both habitats and in both seasons. We surmise that our results will prove to be robust to at least weak density dependence in the second habitat.

It is intriguing that even in our simple model, coexisting migratory strategies are found to robustly persist. Where close study has been conducted, a mixture of migratory strategies often occurs (Lundberg 1988). For example, some elk in the Rocky Mountains migrate seasonally between high and low elevations, yet others in close proximity remain rooted within a small home range

year-round (Hebblewhite and Merrill 2007; Hebblewhite et al. 2008). Similarly, although most Canada geese migrate between northern latitudes and the southern United States, resident populations also thrive alongside more typical migrants in many northern sites. Such observations are what we might expect, based on the general outcomes of our models. Rarely do we know, however, which pairs of strategies are represented; is a migrant behavioural type exploiting the best of all possible habitat combinations, or is it eking out a second-rate coexistence on poorer breeding habitat? Are non-migrants evolutionary winners or losers, inevitably on their way out, but caught at present in a transient snapshot of temporary coexistence? These alternate possibilities, predicted by our simple model, call for a fresh look at the fitness of migrants vs. residents. We caution again that our specific model assumes that individuals have fixed migratory strategies, and that there are few operative constraints or migratory costs. In some of these empirical examples, individuals may be making the best of a bad situation, for instance, and the evolutionarily stable strategy is in fact a fixed, conditional strategy.

The range of mixed ESS outcomes suggested by our models suggests that it may be more relevant to ask not how migration behaviour has evolved, but rather why we don't always see a mixture of strongly contrasting movement strategies. In our model, a migratory strategy is the only pure ESS provided at least one habitat is a sink, unable to sustain a non-migratory population at all, and the other habitat has sufficiently large temporal variation in fitness or fitness components. Relating this prediction to empirical data will require information both on spatial variation in fitness and temporal variability. This is daunting, but a nettle that must be grasped, if we are ever to understand at a deep level the evolutionary basis of migration.

PART 2

How to migrate

White storks (*Ciconia ciconia*) above Israel during their migration.
Photographer: Lior Kislev (http://www.tatzpit.com)

CHAPTER 4

Mechanistic principles of locomotion performance in migrating animals

Anders Hedenström, Melissa S. Bowlin, Ran Nathan, Bart A. Nolet, and Martin Wikelski

Table of contents

4.1 Introduction

How and why do animals migrate? These questions, which can be answered at many levels of analysis, are deceptively simple. Aside from the obvious answer 'because the benefits outweigh the costs', research has yet to provide concrete answers. In addition, the answers may vary from species to species. Thanks to the advent of new technologies that allow us to examine the year-round behaviour and physiology of migrants, however, we are now closer than previously to being able to answer these questions—and ultimately to understanding how a life-history strategy like migration evolves in the first place. In this chapter, we discuss what theory has to tell us about 'how' animals should migrate and whether or not they behave the way that theory predicts; we leave the 'why' questions for elsewhere in the book (e.g. the Theme 1 chapters).

The success or failure of various migratory strategies depends on two major mechanistic components of the migration process: the mode of locomotion and the navigational capacity of migrants. Both components have physical constraints. They interact with both the environment and the internal state of the individual (Nathan *et al.* 2008). Among migratory vertebrates, locomotion modes include running/walking, swimming, and flying. The physical principles of these modes are quite different, but some features are surprisingly similar and allow us to make general predictions about migratory behaviour. In this chapter we introduce the basic equations governing the locomotion performance of migratory organisms and outline some key predictions that stem from these models. We refer the readers to Chapter 6 for a discussion of the navigational aspects of migration. This chapter focuses on birds because the majority of these models were created with birds in mind and most attempts to test them have been made in avian systems. It is not yet clear how well these models will predict the behaviour of other types of organisms.

The theoretical models we present here are not entirely new, but they represent the state of the art in the field. In some cases the models agree with empirical data and need no improvement, but more often they have few if any empirical data available to test them. As technology advances and we begin to collect more empirical data, we will need to refine the models; in a few cases we may need to redefine them completely. We anticipate that many previously-overlooked variables, such as atmospheric stability for flying migrants (Bowlin and Wikelski 2008) or the temperature–depth relationship for endothermic swimming migrants (Teo *et al.* 2007), will need to be added to these models to increase their predictive power. Ultimately, we hope that these models will help us understand how animals migrate the way they do.

First, we examine the range equation, which predicts how far an animal can travel on a given amount of fuel. We then consider the implications of this equation for optimal migration strategies in animals, examine the role of body size, and finish by examining the effects of moving in fluids and in three dimensions.

4.2 The basics

Before we can define 'optimal' migration strategies, we need to consider two variables. The first is how far a migrant can move assuming a certain type of locomotion, fuel load, speed, and other relevant variables. For example, at first glance, it may appear that the more fuel a migrant has, the better off it is—but for many animals, carrying additional fuel can be expensive. The second issue concerns what we mean by 'optimal'. From an evolutionary standpoint 'optimal' means 'conferring the highest lifetime reproductive success', but animals do not reproduce during migration, and measuring survival during migration is often difficult or even impossible. We address these two issues in this section.

4.2.1 The range equation

Given a certain amount of fuel at departure (usually fat in migratory animals) the animal will be able to migrate a certain distance without refuelling. A flying bird, for example, must support its weight and propel itself through a medium (air). Because of the pull of gravity it costs more to lift a heavy fuel load than a small load; the bird will spend more energy flying when it is heavy than when it is light. In addition, adding fuel to the body increases its volume; the frontal area that is exposed to the oncoming airflow thus increases in proportion to the fuel load. An increase in the projected frontal area means increased drag; additional power is required to overcome this 'parasite' drag (drag on body parts that are not generating hydrodynamic or aerodynamic lift), since power is proportional to drag times speed. The combined effect of these two factors yields a range equation as follows (see also Fig. 4.1; Alerstam and Lindström 1990):

$$Y = c(1 - (1 + f)^{-1/2}), \qquad (1a)$$

where Y is the range and f is the relative fat load of the individual expressed as the fuel load mass divided by lean body mass (Table 4.1). The constant c has the dimension distance and includes several factors, such as the energy density of the fuel, conversion efficiency of fuel into mechanical work, acceleration due to gravity, wing and body morphology, and the aerodynamic drag coefficient (Alerstam and Hedenström 1998). For certain predictions based on the range equation, such as the optimal departure fuel load, c is cancelled out from the derivation because the above factors are constants.

Contrary to initial assumptions, in birds it appears that the frontal area does not increase in direct proportion to the fuel load when they accumulate fat (as if the body were a cylinder with fat added uniformly to the cylinder's width; Hedenström 1992). Instead, passerine birds tend to accumulate fat deposits in certain areas, particularly near the front (tracheal pit) or the rear (abdomen), which changes the shape of the body but does not increase the frontal area as much as it would have if the fat had been

deposited uniformly (Wirestam *et al.* 2008). If we assume that the only additional flight cost is due to carrying the extra fuel load (i.e. without an increased drag due to increased frontal area), then the range equation becomes:

$$Y = \frac{c}{2}\ln(1 + f). \qquad (1b)$$

The true range probably lies somewhere in between these two equations (Fig. 4.1).

A quite different, more mechanistic, approach to deriving a flight range equation is to assume that the bird consumes a certain proportion, p, of its body mass per unit of flying time, which means that the instantaneous amount of fuel being consumed decreases as the bird gets lighter during the course of a flight. Empirical support for this assumption comes from birds flying in a wind tunnel (Kvist *et al.* 1998). Regardless, this approach yields the total flight time,

$$T = \frac{1}{p}\ln(1 + f), \qquad (1c)$$

which, when multiplied with the flight speed, U, gives the flight range as:

$$Y = \frac{1}{p}U\ln(1 + f). \qquad (1d)$$

These range equations have been derived for birds, but animals using other modes of locomotion at similar Reynolds numbers (dimensionless numbers that describe the relative effects of friction and inertia on an organism moving through a fluid) will have similar range equations. For example, the migration range using terrestrial locomotion (running or walking), although based on different physical principles, also follows a natural logarithm function (Hedenström 2003a) because the main cost of movement in running and walking is weight support (Kram and Taylor 1990). Swimmers, which are nearly neutrally buoyant in water, do not pay a cost to support weight, but do pay the cost of parasite drag. If one assumes that the body length is unaffected by fuel accumulation (which may not be valid in all swimmers), the range is then:

$$Y \propto \ln(1 + f). \qquad (1e)$$

Table 4.1 List of symbols

Symbol	Definition
c	Proportionality coefficient in range equations
D	Detour distance
f, f_1, f_2, f_b	Relative fuel load to lean mass
f_0	Search/Settling fuel cost
g	Acceleration due to gravity
k, k_1, k_2	Daily fuel accumulation rate as proportion of lean mass
M, m	Body mass
p	Fuel consumption as hourly proportion of body mass
P	Power required to fly
P_{dep}	Fuel deposition rate
S	Instantaneous speed of migration
U	Air speed
U_{loc}	Speed of locomotion
U_{migr}	Overall speed of migration
U_{mp}	Minimum power speed
U_{mr}	Maximum range speed
t	Stopover duration
t_0	Search/Settling time cost
Y	Migration range
Y_{max}	Maximum migration range

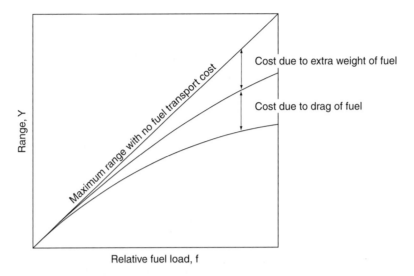

Figure 4.1 Migration range (Y) in relation to relative fuel load (f). Depending on the mode of locomotion, an animal pays a cost of carrying additional fuel that decreases its potential range compared with that of the linear utility function (which represents no additional costs). This cost arises due to increased drag because of increased projected body frontal area in relation to fuel load (swimmers, flyers), and/or due to the added gravitational pull on the extra weight (flyers, runners). Animals that only need to overcome the gravitational forces (swimmers and runners) follow the intermediate curve, while flyers pay both these costs and follow the lower curve. Based on Alerstam (2001).

In an immediate way the range equation determines how far an animal can go without refuelling, but its uses are more far reaching than that, because it is a fundamental component of the optimality models of migration that we will discuss in the next section.

4.2.2 Currencies of migration

Because migration is a non-reproductive activity, it is hard to relate it in a straightforward way to direct measures of fitness such as lifetime reproductive success. Yet lifetime reproductive success is the product of (a) the probability of survival to reproduction, (b) the probability that surviving adults reproduce, and (c) the reproductive output. In several avian migratory species, the most critical factor in this equation is survival during migration (Sillett and Holmes 2002; Jones *et al.* 2004). Furthermore, in many avian migrants arrival timing on the breeding grounds, which can be a direct result of migration performance, is a strong predictor of variables such as territory quality and annual reproductive success (Smith and Moore 2005). Therefore, migratory behaviour ultimately helps determine fitness. By

migrating, animals can escape periods of food shortage and low survival, which is thought to be one of the major factors driving the evolution of migration (Alerstam *et al* 2003, but see Holland *et al.* 2006b).

Migration behaviour can be performed in a way that maximizes survival en route, so safety could be a useful currency that animals strive to optimize (Alerstam and Lindström 1990). Alternative surrogate currencies for fitness often used in studies of migration are energy and time. Energy minimization should be used if food availability is limited along the migratory route. Conversely, in a situation where animals compete for some resource at the destination, for example high quality breeding or wintering territories, nesting sites, etc., it may be advantageous to migrate as fast as possible, which is referred to as time minimization (Alerstam and Lindström 1990). Time minimization here refers to the combined behaviour of fuelling and transport that result in a maximum overall travel rate, i.e. including time for fuel acquisition and transport. Safety, energy use and rate of migration are all variables that researchers can use as proxies for lifetime reproductive success when no direct fitness measurements are available.

Real animals may not be optimizing just one of these single currencies, but rather some combination of them (Houston 1998). For example, feeding in a way that maximizes the rate of fuel accumulation may compromise safety by exposing an animal to elevated predation risk, but avoiding all risk of predation may provide such a low rate of fuel accumulation as to prevent migration altogether. Thus, migrants may (and almost certainly do) opt for a strategy somewhere in between these two extremes. More generally migrants may rely on simple decision rules that approximate optimal decisions under most circumstances but do not always lead to optimal behaviour (Chapter 6; Cochran and Wikelski 2005).

4.3 Stages of migration

A migration may include many cycles of fuelling followed by transport, but in some cases only one such cycle is needed to complete the migration. An extreme example of the latter scenario is the autumn migration of Alaskan bar-tailed godwits *Limosa lapponica breuri* between Alaska and New Zealand; some individuals may fly non-stop for about 8 days in order to cover the 11 000 kilometres between the two locations (Gill *et al.* 2009). Generally, however, migrants stop along the way. This lends itself to various optimality arguments about the consequences of behavioural decisions during migration.

4.3.1 The optimal departure fuel load

First we address the question of when it is adaptive to leave a fuelling site and commence the next migration step, neglecting (for a while, see Section 4.5.2) considerations of external factors such as favourable meteorological conditions for subsequent cross-country flights. Here we focus on birds because they have been studied most regarding this issue, but in principle the arguments may be applied to other animals for which we have a range equation. Let us assume that there is a search/settling time cost (t_0) when a bird arrives at a new stopover site. This can be thought of as a waiting time until feeding territories become available (Rappole and Warner 1976), or the time it takes to locate food. A

typical observed pattern is that newly arrived birds actually lose body mass for the first few hours, so there may also be a search/settling energy cost (f_0) before fuel (and mass) can be gained (Alerstam and Hedenström 1998). We assume the currency is the overall time of migration, S. The fastest migration is achieved when distance divided by time is maximized, which is when:

$$S = \frac{Y(f) - Y(f_0)}{t + t_0},\qquad(2)$$

where Y is the potential movement distance according to Equation 1. The optimal departure fuel load depends on the trade-off between the flight step length and the associated search and settling costs (longer steps means fewer stops and lower total search/settling cost) versus the added locomotion cost of lifting large fuel loads when flying long distances. If we let $f = kt$ and $df/dt = k$, where k is the daily fuelling deposition rate, substitute for t in Equation 2 and differentiate, we obtain the following criterion for optimal departure fuel load:

$$\frac{dY}{df} = \frac{Y(f) - Y(f_0)}{f + kt_0}.\qquad(3)$$

The solution to this criterion can be graphically illustrated, as shown in Fig. 4.2(a), where the optimal departure fuel load is found by drawing a tangent from a point (kt_0) in the fourth quadrant to the range curve, where k is the daily rate of fuel accumulation. The effect of k on departure fuel load can be illustrated as shown in Fig. 4.2(b), where k_1 and k_2 represent a relatively low and high daily fuelling rate, respectively. The effect of an increased fuelling rate k_2 is to shift the origin of the tangent to the left in the fourth quadrant, with the effect of increasing the departure fuel load from f_1 to f_2 (Fig. 4.2(b)). Notice that this prediction assumes a scenario where birds have some knowledge about the quality of subsequent stopover sites along the migration route, and that the experience at the current site represents (such as is the case in Fig. 4.2(b)) a local deviation from the mean (Weber *et al.* 1999). However, birds may update their expectations based on the current site, which might be a more realistic assumption for migrants, such as juveniles, that have no previous experience with available stopover sites, those that do not use the same routes

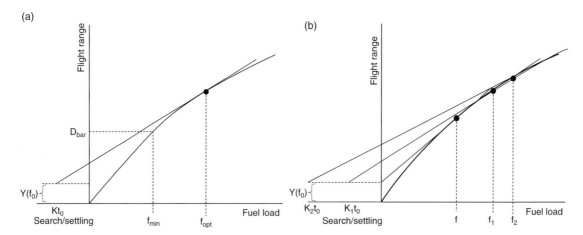

Figure 4.2   (a) A range equation for a flying animal, showing how the optimal departure fuel load (f_{opt}) is obtained according to Equation 3. In the case of a barrier at distance D_{bar} ahead of the animal, the minimum requirement is a fuel load f_{min}. t_0 and f_0 are search/settling time and energy costs, respectively, and k is the daily fuel accumulation rate. (b) Illustration of the effect of different fuel accumulation rates k_1 and k_2 ($k_2 < k_1$) on the optimal departure fuel load f_1 and f_2, and the optimal departure fuel load associated with minimization of cost of transport (f). Based on Alerstam and Hedenström (1998).

during each migration, and those that stopover in places with high environmental unpredictability. In these cases the model predicts that the bird will leave the current site before the optimum according to the local variation rule, as illustrated in Fig. 4.2(b), has been reached. However, the departure fuel load will still depend on the fuel accumulation rate, with a predicted positive relationship between departure fuel load and fuel accumulation rate. This prediction has been corroborated qualitatively on a number of occasions using an experimental approach where food was provided and the body mass of individually colour-banded birds was recorded with an electronic balance, (Hedenström 2008).

If birds are minimizing transport costs only, they should depart with the same fuel load irrespective of the fuel deposition rate experienced (Fig. 4.2(b)), but the available empirical evidence does not support this strategy. Alternatively, they could attempt to minimize the total energy cost of migration (Hedenström and Alerstam 1997), which includes an element of time because the longer the migration is, the more energy is spent while in transit (notice that here we minimize all energy, i.e. maintenance and transport). When minimizing this currency there is also a positive relationship between departure fuel load and fuel deposition rate. For a small

passerine bird species, the energy spent during migration is divided into actual flight cost and energy spent at stopovers in an approximate ratio of 1:2 (Hedenström and Alerstam 1997), a prediction that has received support from empirical measurements (Wikelski *et al* 2003).

The above models assume that the 'terrain' the animal moves over is constant; this is rarely the case. Some areas an animal moves through or over may prevent refuelling or may increase travel costs in other ways. In the real world, birds may have a set of discrete, suitable stopover locations available to them along the flyway, or they may have to pass a wide stretch of unsuitable habitat, such as seas, deserts or mountains, that effectively form an ecological barrier because they cannot stop and/or refuel. If birds are minimizing the energy cost of transport and have perfect information about their environment (or have evolved in a perfectly predictable environment), they should fly between successive stopover sites and accumulate exactly the amount of fuel required to reach the next site. For time minimization, alternate strategies may be optimal depending on the relative food abundance among the potential sites along the route. The key to analysing this problem is to rearrange the flight range equation (Equation 1) to express the fuel load as a function of flight distance, $f(Y)$, and differentiate the range equation with respect to stopover time. This yields an expression for the instantaneous migration speed, S, as a function of flight range (see Alerstam and Hedenström 1998 for details):

$$S = \frac{c}{2}k\left(1 - \frac{Y}{c}\right)^3.$$ (4)

This relationship can be illustrated for a hypothetical set of stopovers along the flight route to demonstrate different possible optimal policies for a situation representing spring migration (Fig. 4.3). The optimal policy is to leave a staging site as soon as another site can be reached that provides a higher instantaneous speed S. Note how S is a diminishing marginal rate utility function of added distance (or fuel). While adding fuel, the bird is accumulating potential flight distance, and the problem is when and how much potential distance to convert into real distance. The fuel deposition rate may be such that the current site provides a higher rate than the next site, and the best policy may therefore be to exploit the current site and depart with an overload that allows the skipping of the next site (Site 2, Fig. 4.3(a); Gudmundsson *et al.* 1991); this phenomenon has been demonstrated in Bewick's Swans (*Cygnus bewickii*) and Barnacle Geese (*Branta leucopsis*; Beekman *et al.* 2002, Eichorn *et al.* 2009). If the fuel gain is sufficiently different between potential sites a situation may occur where it is advantageous to leave the current site (Site 1) with an overload (i.e. more fuel than is required to reach Site 2; Fig 4.3b), and resume refuelling at Site 2 for the next migration stage to the breeding site (in this case Site 3 is skipped, Fig 4.3b; Gudmundsson *et al.* 1991). For such a migration scenario the possible overloading strategy depends on the exact form of the range equation (e.g. 1a or 1b; Weber and Houston 1997). It is only with Equation 1a, in which frontal area, and therefore parasite drag, increases with fuel load, that overloads will theoretically occur; overloads will never be optimal for an animal where Equation 1b is valid and increased fuel loads do not increase parasite drag.

Most of the models discussed above assume either full knowledge of the conditions at future stopover sites, or a highly predictable environment, which will be discussed in the detour section below. Prior knowledge of future conditions may be a realistic

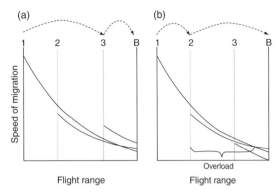

Figure 4.3 Schematic figure showing a set of discrete stopover sites (1, 2, 3) and the breeding site (B) representing a spring migration. The curves show the instantaneous migration speed (*S*) at different sites and show how the optimal policy (maximizing migration speed) leads to (a) skipping (of Site 2) and (b) overloading (at Site 1). The different migration strategies depend on the relative quality (fuelling rate) among the sites; in the cases illustrated Site 3 has a lower quality in case (b) than in case (a). Based on Gudmundsson *et al.* (1991).

assumption for certain long-lived migrants such as sandpipers and geese, where entire populations can concentrate at a few key stopover sites, but many migrants may never have encountered particular stopover locations. Individuals that have previously migrated may have considerably more information than first-time migrants, although it may be low-quality information (e.g. 'at this latitude and time of the year conditions are good' or 'there is a lot of good habitat in this general area'), which may lead to age-dependent fuelling strategies. Thus, we might expect first-time migrants to update their expectations based on current site conditions, while adults may use mean conditions from previous migrations to estimate stopover site quality. However, a theoretical investigation of this question (Erni *et al.* 2002) suggested that juvenile birds could approximate optimal behaviour by using a simple rule of thumb, in which case we would not expect refuelling strategy to vary much with age. More empirical data are required before we can address this question.

4.3.2 Detours

Migrants often face changing habitats along their migration route, and sometimes they have to cross ecological barriers. In such cases the animal has two options: either cross the barrier directly by accumulating sufficient energy reserves, or move around the barrier in order to cross it where it is shorter or to avoid it altogether. Any such detour must by necessity comprise a longer distance than the distance it takes to cross the barrier directly. In the case of a direct flight across a barrier, the optimal fuel load as calculated according to the criteria above may be invalid, because the minimum fuel load is of course that required to cover the barrier (Fig. 4.2(a)). As the marginal utility in terms of potential range decreases with added fuel in bird flight (Equation 1), an animal saves energy by migrating with short steps associated with small fuel loads rather than long steps with large fuel loads (Alerstam 2001). Therefore, an animal could avoid a barrier by migrating along a longer path within an ecologically suitable habitat allowing multiple stopovers, and still use less energy for the whole migration compared with the direct flight (Fig. 4.1). Alerstam (2001) derived predictions on break-even detour distances for migratory birds and optimal routes including some barrier crossing. He contrasted the range calculated from Equation 1 with the range attained by cumulative short movements associated with an infinitesimal fuel load. The maximum range with no fuel transport costs is $Y_{max} = cf$. The break-even detour is $D = Y_{max} - Y(f_b)$, where f_b is the fuel load required to pass directly across the barrier. Notice that this prediction is based only on the trade-off between the transport cost of having heavy fuel reserves and the additional distance that can be covered if these costs are reduced. Hence, this is an energy based criterion, independent of time minimization. In any case, Alerstam (2001) presented some empirical support that some migration routes in birds could be governed by this detour criterion.

Large ecological barriers are predictable habitat features, and it is therefore likely that individuals have evolved optimal strategies to deal with them. Indeed, there is a great deal of evidence that suggests that even first-time avian migrants detour around the Sahara Desert, for example, despite the fact that they have never encountered it before (e.g. Gwinner and Wiltschko 1978). This contrasts with the case of optimal fuel loads, where individuals who are familiar with particular stopover sites and/or general areas may have an advantage over those who have not encountered them before. Small ecological barriers, such as the Great Lakes in the United States, may fall somewhere in between these two extremes, and the response to these barriers depends on how predictable individuals' and species' routes are.

4.3.3 Migration speed

It was noted above that the overall migration speed is a currency of special interest when deriving optimal strategies associated with migration according to the time minimization hypothesis. Migration speed depends on only three variables and may be written as:

$$U_{migr} = \frac{U_{loc} \cdot P_{dep}}{P_{dep} + P_{loc}}. \tag{5}$$

where U_{loc} is the travel speed when the animal is actually moving, P_{dep} is the rate of energy deposition and P_{loc} is the power expenditure when moving. This is a general formula that is applicable to any animal using any type of locomotion. Notice that it is important to express P_{loc} and P_{dep} in the same units, which could be as multiples of the basal metabolic rate or directly in watts. The ratio $P_{dep}/(P_{dep}+P_{loc})$ expresses the fraction of time spent in actual transport taken over the whole migration. In small passerine birds this ratio is approximately 1:7 (Hedenström and Alerstam 1997), for which there is some empirical support (Hedenström 2008).

Equation 5 can conveniently be displayed in graphical form (Fig. 4.4). Here the point in the first quadrant represents a location on a curve relating power expenditure to locomotion speed. In the case of vertebrate flight there is a U-shaped function between power expenditure and airspeed (dotted curve in Fig. 4.4), so the maximum migration speed (U_{migr}) is found where a tangent from a point on the downwards extended y-axis, representing the fuel deposition rate, to the 'power curve' intersects with the (speed-) x-axis. Notice that the optimal flight speed is given by the tangent, which is a bit higher than the maximum range (U_{mr}) speed obtained by drawing a tangent from the origin to the curve. A time-minimizing flying animal should therefore fly faster than an animal minimizing the energy cost of transport. Often these alternative characteristic

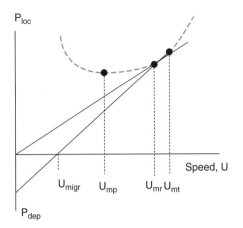

Figure 4.4 Graphical illustration of the overall migration speed (U_{migr}) as a function of power required for locomotion (P_{loc}), rate of energy accumulation and locomotion speed (U). The dashed curve shows a hypothetical relationship between the power required to fly and airspeed for a flyer (bird or bat). The characteristic speeds of minimum power (U_{mp}), maximum range (U_{mr}), and that associated with time minimization migration (U_{mt}) have been labelled.

speeds are so close to each other that it is hard to distinguish between them by measuring flight speeds of wild birds. The third characteristic speed, the minimum power speed (U_{mp}), which is given by $dP/dU = 0$, is expected in situations when the animal aims at maximizing the time spent airborne on a given amount of fuel. This speed would be expected in situations such as display flight, or possibly if a migratory bird is disoriented during a migratory flight and flies until some directional cue is obtained. Empirical support that birds actually do select flight speed appropriately according to predictions comes from the skylark (*Alauda arvensis*), which flies near 5 m/s during song flight and above 10 m/s during migration (Hedenström and Alerstam 1996). However, very large birds using flapping flight actually appear to migrate at U_{mp}, probably because they cannot sustain flying faster (Pennycuick *et al.* 1996).

4.4 Scaling of migration across taxa

Migration costs may change as an animal increases in size. For example, in aerial locomotion the power required to fly is expected to scale as $P \propto M^{7/6}$ in geometrically similar animals (Pennycuick 1975). This means that the mass specific flight cost increases

as animal size increases. In terrestrial locomotion and swimming the cost does not increase with size in such a dramatic way.

If the effect of scaling on locomotion speed is taken into account, a dimensionless index of 'cost of transport' can be defined as:

$$C = P_{loc}/mgU, \qquad (6)$$

where P_{loc} is the power expenditure of locomotion, mg is the weight and U is the locomotion speed. This particular equation is valid for any animal, regardless of Reynolds number, and will therefore apply equally to zooplankton, birds, bats, mammals, insects and whales. In a classic study, Schmidt-Nielsen (1972) plotted C against body size for animals that swim, fly and run for organisms ranging in body mass from about 10^{-6} to 10^3 kg. In this size range, for a given body size, the cost of transport differs so that swimming is cheapest and running is most expensive, with flight intermediate between swimming and running. While the same pattern probably holds for organisms operating at low Reynolds numbers, we do not yet have data confirming this hypothesis. In any case, Schmidt-Nielsen's relationship explains why it is among the swimmers and flyers, such as whales, shorebirds and terns, that we find the longest migrations (Alerstam *et al.* 2003). Admittedly, there are many other factors that influence migration distance in addition to the mode of locomotion, such as topographical constraints in the form of seas and mountains that can be difficult for terrestrial animals to cross.

Evaluating Equation 5 with respect to body size shows how migration speed, i.e. the overall speed of migration including time of stopovers (fuelling), varies among swimmers, flyers and runners across body size. Migration speed scales as $\alpha M^{1/24}$ in animals that swim, as $M^{1/11}$ in animals that run, and as $M^{-1/4}$ in animals that fly by powered flapping flight and as about $M^{0.22}$ in animals using soaring flight (Hedenström 2003b). From these observations, we can infer that, assuming there is selection for high migration speed, there should be selection for large size in animals that migrate by swimming, running or *soaring* flight (whales, large mammals such as ungulates, birds of prey and storks are all relatively large). However, in animals using *flapping* flight there is a premium on small fliers according to this criterion, and we should expect long-distance migrants to be relatively small. For example, among the three species of swans in Europe, migration distance is inversely related to body size (Hedenström 2003b); this pattern may be due to time constraints of fitting the migration into the annual budget selecting for smaller-bodied individuals. However, some small to medium-sized birds such as the European Bee-eater (*Merops apiaster*) utilize both flapping and soaring-gliding flight modes during migration (Sapir 2009). In any case, Alexander (2002) showed how these relationships can be used to predict body size thresholds for certain migration distances in each taxon.

4.5 Migration in moving fluids

Swimmers and flyers move through a medium, either water or air, that itself is moving. This can create opportunities for or pose constraints on migrating animals that have no obvious parallels in the terrestrial sphere. For example, large soaring birds can use vertical updrafts ('thermals'), complex three-dimensional atmospheric structures, to gain altitude and thereby reduce energetic costs (see Section 4.6.4). On the other hand, anadromous salmonids must swim upstream against a current in order to spawn, using a great deal more energy than they would in still water (although there is some evidence that they may also take advantage of local variation in flow strength and direction [Standen *et al.* 2004]).

4.5.1 Addition of velocity vectors

When an animal moves in a fluid, the vector (T) describing its movement relative to a stationary point outside the fluid (such as the ground), is the sum of the vectors describing the animal's movement relative to that stationary point (H) and the movement of the fluid relative to that stationary point (W; Fig. 4.5). For birds and bats the speed of the air (wind) is typically of the same magnitude as the flight speed through the air; hence tailwinds can double the migration range, while a strong headwind can prevent progress (Cochran 1972: Liechti

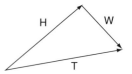

Figure 4.5 The triangle of velocities. H is heading, W is wind and T is track (the animal's movement relative to the ground). The track velocity (T) is always the sum of the wind vector (W) and the bird's movement relative to the air (H).

2006). For most insects, wind speeds are much higher than air speeds; as a consequence many migrating insects use passive wind transport or fly at low altitudes to avoid high winds (Srygley and Dudley 2008: Chapman *et al.* 2008a,b). Although ocean and river currents tend to be much slower than wind speeds, the swimming speeds of migrants (Sato *et al.* 2007; Shillinger *et al.* 2008) are proportionately lower than flying speeds of migratory bats and birds (Pennycuick 1997), so swimming migrants' ranges can also be doubled or reduced to zero, depending on the velocity of the currents they encounter.

4.5.2 Timing movement with favourable environmental conditions

The onset of migration is associated with certain meteorological and other environmental conditions experienced by birds (Richardson 1978, 1990; Cochran and Wikelski 2005), fish (Quinn *et al.* 1997), sea turtles (Sherrill-Mix *et al.* 2007), butterflies (Meitner *et al.* 2004) and moths (Chapman *et al.* 2008). In birds, the relative timing of migratory flights and stopovers is determined not only by fuel load (Section 4.3.1), but also by various environmental factors favouring long-distance flights (Alerstam and Hedenström 1998; Liechti and Bruderer 1998). Passerine birds, for example, preferentially depart for migration on days of favourable tailwind conditions (Åkesson and Hedenström 2000). A study of migratory European bee-eaters suggests that a scale-dependent response to meteorological conditions determines departure timing from staging sites (Sapir 2009). These birds respond to the ambient temperatures and barometric pressures of the previous 1–3 days, coordinated with the passage of synoptic (large-scale) weather systems.

The bee-eaters then fine-tune their departures at a resolution of 2–3 hours to coincide with the most favourable temperature, presumably so that they maximize their encounters with thermals generated by surface heating during the subsequent cross-country flight.

4.5.3 To drift or not to drift: an important question

Swimming and flying migrants generally do not encounter fluid flows that are perfectly aligned with their preferred movement direction (e.g. perfect head- or tail-winds). Instead, the fluid vector usually intersects with the animal's movement vector at an angle, such that the animal ends up not only closer to or further from its goal but also to one side or the other of the position to which it would have flown or swum in still air or water. This phenomenon is termed 'lateral drift' and is commonly referred to as 'wind drift' in birds (e.g. Alerstam 1979; Cochran and Kjos 1985; Green *et al*. 2004; Liechti 2006). Migrants do not have to drift passively with the fluid; they can alter their headings such that they compensate for lateral drift, although this affects the distance they can move along their chosen course. They can also choose to take off only when winds are blowing in the approximate direction they wish to go: the resulting correlation between birds' track directions and wind directions has been termed 'pseudodrift' (Evans 1966; Alerstam 1978).

Lateral drift may seem like a negative consequence of flying in a moving fluid, but many flying migrants actually take advantage of it. For example, birds taking off from north-eastern North America in autumn fly to the east of the direction they need to go to reach South America (Richardson 1979). However, the birds can reach their destination without changing their headings because the trade winds near the equator drift the birds to the west, toward the continent (Stoddard *et al*. 1983). Without the initial tailwind the birds obtain by taking off after a cold front passes and the lateral drift from the trade winds, the birds would probably die over the Atlantic before they reached South America.

Lateral drift can also be neutral, particularly when a migrant is still far from its goal (Alerstam

1979; Green *et al*. 2004). In fact, when there is a strong tail-flow associated with drift, the migrant may end up closer to its goal than it would have had it flown or swum in still air or water (Alerstam 1979). Although it is difficult to obtain data on the speeds of both migrants and flows, evidence is accumulating that migrants do not compensate for drift when drift is neutral but do compensate when it is unhelpful. For example, both Green *et al*. (2004) and Bäckman and Alerstam (2003) found that, as predicted, radar-tracked birds partially or completely compensated for lateral drift when close to their goal but did not compensate when further away. Sea turtles also appear to allow lateral drift early in their migratory journeys when drift is hypothesized to be neutral (Luschi *et al*. 2003), but speed up to avoid drift when it would result in them missing their destinations completely (Shillinger *et al*. 2008). Migratory butterflies partially or fully compensate for drift when crossing large bodies of water (Srygley and Dudley 2008); in this case the eventual goal may be far away but it may be important for these butterflies to reach land (and therefore food and cover) as quickly as possible. Chapman *et al*. (2008) showed that migratory moths failed to compensate for drift when the angle between the moths' preferred headings and the wind vector was small, but did compensate when it was large. They also actively selected altitudes with favourable wind speeds and directions.

Migrants may not always respond to drift optimally. They may, for example, be unable to compensate because of the strength of the fluid flow or because their navigational capacities do not allow them to do so. Many insects' flight speeds are so low that they must either fly close to vegetation where wind speeds are extremely low or passively drift with the wind (Srygley and Dudley 2008). Similarly, Cochran and Kjos (1985) found that small (~30 g) *Catharus* thrushes flying over the central United States did not compensate for lateral drift or only compensated slightly. However, the researchers did not know how close the individual thrushes were to their destinations, so it was not clear whether or not compensation was adaptive at the time. A different pattern was observed in large raptors: Juvenile ospreys (*Pandion haliaetus*) and honey-buzzards (*Pernis apivorus*) did not correct for drift in one recent study even though they appear to be physiologically capable of compensating for lateral drift, since adults can and do compensate (Thorup *et al*. 2003). These authors argued that juveniles were not aware that they were drifting due to their limited navigational capacities, whereas adults were aware.

Overall, we do not yet have enough information about lateral drift in flying migrants to predict which species and/or age classes are capable of using drift optimally. We urge those with satellite- or radio-tracking data or those with radar data on animals of known species and/or age to collect information on winds aloft to measure compensation for wind drift in their organisms.

Swimming migrants also face a choice between letting themselves drift with the current or actively moving against it. We do not yet know how these migrants, particularly those in the open ocean, recognize drift, since they have few or no visual references with which to compare their movement. However, it seems likely that their navigational abilities allow them to determine their position on the Earth and correct for any unwanted displacement (Lohmann *et al*. 2008a). The effects of currents on swimming migrants have been studied less extensively than the effects of wind on flying migrants, but swimming migrants appear to behave as optimality reasoning predicts when these hypotheses have been tested (e.g. Luschi *et al*. 2003). Atlantic bluefin tuna (*Thunnus thynnus*) might be an exception to this rule as they do not appear to use currents optimally when leaving the Gulf of Mexico, but this may be because they must avoid overheating (Teo *et al*. 2007).

In the oceans' circular surfaces, current gyres allow migrants to drift with the currents and return to their starting positions, although migrants typically do not drift passively since they might be drifted out of the gyres if they did (Lohmann *et al*. 2001). Sea turtles not only swim forward but also orient themselves such that they remain within the North Atlantic gyre, effectively compensating for non-adaptive lateral drift.

So-called 'loop migrations' in birds may be the result of a phenomenon similar to oceanic gyres. Loop migrations occur when birds use a path during spring migration to the east or west of the path they used during autumn migration; classic examples include the willow warbler (*Phylloscopus trochlius*) and the red-backed shrike (*Lanius collurio*) (Hedenström and Pettersson 1987; Alerstam 1990). These 'loops' may be the result of birds being drifted by the prevailing winds, which blow in different directions in the spring and in the autumn (Strandberg 2008).

4.5.4   Using fluids for orientation and navigation

Fluids can also facilitate the orientation and navigation of migrants. For example, hatchling sea turtles use the fact that waves are typically oriented parallel to the shore to get offshore and into deep water quickly (Lohmann *et al*. 2008a). Researchers have hypothesized that animals may also be able to use wave orientations in open water or the wakes generated in the ocean by islands to orient or navigate, although both have yet to be demonstrated (Lohmann *et al*. 2008a).

Fluids can also carry information such as olfactory cues; salmon are famous for using these cues to navigate back to their natal stream. Chemical cues become less and less concentrated further away from a source in both water and air, so animals can literally 'follow their noses' up a concentration gradient in order to reach a particular scent-emitting location. When the fluid is moving, it can be even easier for an organism to determine the direction in which the source lies and to use that direction for either orientation or navigation. Although it may be possible for some organisms to detect these chemical cues thousands of kilometres from the source (Wallraff and Andreae 2000; Wallraff 2004), olfactory navigation is generally thought to be confined to shorter distances (Gagliardo *et al*. 2008; Lohmann *et al*. 2008b).

4.6   The third dimension: altitude, depth, and elevation

Migrants move in three dimensions. Generally, this is not very different from moving in two; the triangle of velocities presented in the previous section, for example, is valid for both two- and three-dimensional movement. However, it means that migrants must choose an altitude (above the ground), depth (below sea level) or elevation (of the Earth) at which to migrate. That choice can be extremely important given the variables that can change in this third dimension. Fluid movement in three dimensions can also be more complicated than movement in two, and several unique modes of transportation, such as thermal soaring, slope soaring, dynamic soaring and tidal transport, have evolved to take advantage of predictable but complex three-dimensional fluid movements.

4.6.1 Altitude

A moving fluid interacting with a flat surface results in a characteristic velocity profile; the velocity increases as one moves further from the boundary until the velocity matches that of the portion of the fluid that is unaffected by drag caused by the flat surface (Fig. 4.6; Vogel 1994). The area between the flat surface and the laminar flow is termed the boundary layer. The air moving across the surface of the Earth also has a boundary layer, which is generally turbulent since the Earth's surface is not flat. The height of this atmospheric boundary layer varies, but over land it is generally 1–3 kilometres above the Earth's surface during the day and >1 kilometre at night (Stull 1988). Radar studies show that most nocturnal migrants fly below 2000 m (e.g. Able 1970), meaning that many migrants fly in or

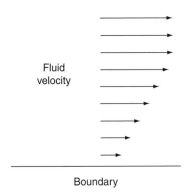

Figure 4.6 Wind or water velocities in a boundary layer. As height above the boundary surface increases, the wind or water velocity increases from zero to the average velocity of the flow immediately above it. Many migrants have evolved modes of locomotion that take advantage of the velocity gradients in the Earth's boundary layers.

near the planetary boundary layer (Cochran 1972). Nocturnal migrants that fly near the top of the boundary layer may be able to take advantage of a small amount of lift generated by the fluid shear and can avoid turbulent, slower air in the boundary layer (R. Avissar, pers. comm.), though to our knowledge researchers have yet to correlate atmospheric boundary layer height with nocturnal flight altitudes. However, the height of the boundary layer, which changes during the day, has been correlated with the flight altitudes of various soaring migrants such as pelicans (Shannon *et al.* 2002), storks (Shamoun-Baranes *et al.* 2003a) and raptors (Shamoun-Baranes *et al.* 2003b).

Both within and above the atmospheric boundary layer, wind speed usually increases with height (Kerlinger and Moore 1989). Wind direction relative to the ground also rotates, typically clockwise in the northern hemisphere and counter-clockwise in the southern hemisphere. This pattern may mean that flying migrants can predict wind directions and speeds aloft without having to sample the entire air column prior to beginning a migratory flight, although we do not know whether or not migrants are capable of doing so. It also means that flying migrants can choose between wind strata with varying amounts of tail- and cross-winds. Migrants could minimize lateral drift by choosing to fly in the atmospheric stratum with winds most closely aligned with their own heading, although *Catharus* thrushes choose higher wind speeds even when it means greater lateral drift (Cochran and Kjos 1985). Some large birds even fly in jet streams where winds are extremely high; these birds reap the benefits of strong tailwinds but could be blown off course if the streams are not blowing in the correct direction (Liechti and Schaller 1999).

Wind is not the only aspect of the atmosphere that varies with altitude. Both air density and oxygen partial pressure decrease with altitude, so the drag on a flying migrant decreases with increasing altitude, while the need to support the flight muscles with oxygen becomes an increasing problem (Pennycuick 1975). Some high-altitude migrants, such as the bar-headed goose, have haemoglobin with increased oxygen affinity in order to combat the latter problem (Weber *et al.* 1993); however, most birds do not have this adaptation. Therefore,

Pennycuick (1978) predicted that animals would begin to fly higher as a migratory flight progressed because oxygen demands would decrease with increasing mass loss. Data on flight altitudes do not appear to support this prediction, possibly because other factors have stronger effects on altitude choice (Hedenström 2003b).

One of those other factors may be water loss. Because oxygen partial pressure decreases as altitude increases, flying migrants must increase respiration rate as they gain altitude. Increased respiration rates lead to greater water loss, which could cause birds flying over desert areas to become dehydrated (Carmi *et al.* 1992). However, if energy and not water is the limiting factor in bird migration, birds should fly at higher altitudes where tailwinds are higher regardless of the possibility of dehydration (Liechti *et al.* 2000). In order to determine whether water loss or energy expenditure has a greater effect on choice of flight altitude, Liechti *et al.* (2000) analysed flight altitude distributions over desert areas using a series of models that included the effects of tailwinds, water loss with increasing altitude, and other variables. They concluded that water loss did not explain any additional variation in altitude distribution after the other factors were accounted for, suggesting that energetic concerns were more important than dehydration in determining flight altitudes.

A final consideration in the choice of flight altitude is the cost of ascent. Birds use more energy when ascending than when flying on the level (Bowlin and Wikelski 2008), so flying migrants may face a choice between having a greater tailwind during the cruising phase of migratory flight (since wind speeds increase with altitude) and using less energy in the initial stage of the flight. This also suggests that birds should not choose to sample all the available wind strata prior to selecting a flight altitude and should not take off on short flights just to sample the weather conditions as some researchers have suggested (e.g. Cochran and Kjos 1985). This may be particularly relevant for big birds that ascend slowly, and for which the effort and time to climb to high altitude would not pay off (Klaassen *et al.* 2004). Displaced Common Swifts (*Apus apus*) returning to their nests, conversely, flew much higher than they needed to fly in order to see their goal, even when they had headwinds (Gustafson *et al.* 1977).

4.6.2 Depth

In oceans, light levels decline with depth, and these levels can affect productivity, oxygen levels and temperature. Furthermore, some orientation and navigation cues may be easier to sense near the water's surface. It is, therefore, not surprising that many of the animals that have been studied so far migrate close to the surface; for example, migrating white sharks (*Carcharodon carcharias*) spend approximately 60% of their time swimming at a depth of 5 m or less (Weng *et al.* 2007). Two exceptions to this rule are tuna, which probably have difficulty thermoregulating in warm surface waters (Teo *et al.* 2007), and giant squid (*Dosidicus gigas*), which appear to tolerate the hypoxic conditions found in deeper water (Gilly *et al.* 2006). The continued use of data loggers that track depth and other variables should give us a much clearer picture of what depths are chosen by swimming migrants and why those depths are preferred.

4.6.3 Elevation

Terrestrial animals must choose an elevation at which to migrate, but comparatively few factors vary with elevation. The choice of elevation should, therefore, be driven largely by energy-minimization considerations (Alexander 2002). For example, when the slope of a hill or mountain is steep, and migrants want to get to the other side, they can minimize energy expenditure by detouring around the summit. The steeper the slope, the closer the detour should be to a contour line around the mountain. When possible, terrestrial migrants should also detour around terrain with higher costs of transport than the terrain that they are currently on (Alexander 2002). To our knowledge, these ideas have yet to be tested in migrating animals, although human paths appear to conform to some of the predictions of energy-minimization models (Minetti 1995). Furthermore, elephants appear to avoid climbing slopes altogether since climbing is so energetically costly for them (Wall *et al.* 2006).

4.6.4 Three-dimensional fluid structures and movement patterns

Fluid movement can be complicated, and many different modes of transport have evolved to take advantage of three-dimensional fluid movements. Here we discuss several forms of soaring as well as tidal transport.

Many birds (Pennycuick 1998) and even butterflies (Gibo and Pallett 1979) take advantage of atmospheric structures called thermals, which result from differential heating of the Earth's surface. Surfaces that become very hot (such as parking lots) heat the air above them, which then rises and is replaced by cold air from surrounding locations. The result is a cylindrical tower of rising warm air surrounded by an area of falling, cooler air. If migrants fly in circles within the column of rising air, they can gain altitude without flapping because the air rises faster than they sink. They can then glide to the next thermal. Thermal soaring can represent a significant energy saving for these migrants (Pennycuick 1998). A combination of thermals and tail winds allow aphids to migrate from continental Europe to England every year (Drake and Gatehouse 1995).

Turbulence does not always save organisms energy; in fact, Nisbet (1955) argued that many birds fly at night to avoid turbulence, which is lower at night than it is in the daytime. Laminar conditions also occur more frequently at night (Kerlinger and Moore 1989). Bowlin and Wikelski (2008) found that nocturnally migrating songbirds flying in conditions correlated with increased atmospheric turbulence used more energy than those flying in clear, calm conditions, although the study did not demonstrate a causal relationship.

Thermals are not the only three-dimensional air movements that benefit migrants. For example, some migrants take advantage of updrafts created by wind moving over ridges or mountains rather than those created by differential heating (Vogel 1994). These migrants do not need to move in circles, but must follow the ridge or mountain in order to stay in the updraft. This form of flight, termed 'slope soaring', can save birds a considerable amount of energy (Videler and Groenewold

1991). Both types of soaring require high turbulent kinetic energy (TKE) in the atmospheric boundary layer, which can be generated either by surface heating (thermals) or strong winds (shear-induced updrafts enable slope soaring). Turkey vultures, known as obligate soaring migrants, strongly depend on conditions of high TKE in the atmospheric boundary layer (Mandel *et al.* 2008) but appear to rely on thermals rather than shear-induced updrafts.

Flying migrants also use the velocity gradient in the Earth's boundary layer to decrease the energetic costs of flight (Vogel 1994). In 'dynamic soaring', used most frequently by pelagic seabirds, migrants rise above the ocean's surface and fly downwind, taking advantage of the high wind velocity several metres above the surface to build up a great deal of potential energy. The bird then turns in the direction it wishes to go and decreases its height, converting potential energy into kinetic energy. Because the bird is now closer to the ocean's surface, winds are much lighter; it can then fly in any direction regardless of the direction the wind is blowing. When the animal has used up all the potential energy, it gains altitude by wheeling into the wind and repeats the process.

Tidal transport in the oceans is similar to dynamic soaring. Water moving over the sea bottom has a boundary layer like wind moving over the Earth's surface, meaning that water velocities near the sea bottom are typically much lower than velocities higher up in the water column (Vogel 1994). In tidal transport, an organism moves up in the water column when the water is flowing in the direction it wants to go, and moves down nearer the sea floor when the water is flowing in the opposite direction. As a result, the organism moves only in the direction it wishes to go. Plaice (*Pleuronectes platessa*) famously use tidal transport to migrate long distances in the North Sea (e.g. Hunter *et al.* 2004).

4.7 Discussion

In this chapter, we have presented a number of theoretical models that predict how migrants 'should' behave and discussed the growing amount of evidence suggesting that migratory animals usually behave in that manner. One major caveat to this generalization is that many of these models were created to predict avian migration behaviour; likewise, most of the tests we cited were performed on birds. Thus, the predictions of these models may not be accurate for species from other clades. One major way in which birds differ from many other migrants is that very few avian species forage during migratory flight (but see Strandberg and Alerstam 1997 for an example). This allows, for example, energy-minimization models to focus on energetic output and ignore energetic input. However, many other types of migrants, including bats, sharks, whales, caribou and wildebeest, forage while making migratory movements. It is, therefore, unlikely that these animals will behave according to the simple models designed for birds; at the very least we will need to add energy inputs as well as outputs to the models in order accurately to predict these migrants' behaviour (Chapter 5). The resulting models will probably incorporate optimal foraging theory along with the directional movements we have discussed. There will be a trade-off between optimal foraging behaviour and optimal directed migratory movement. The solutions to this trade-off will vary according to the currency of migration (all else being equal, a time-minimizer faced with a choice should opt for more directed movement, whereas an energy-minimizer should opt for more foraging), which will in turn vary according to body size, foraging strategy and life history.

Although there are few tests of these models in non-avian species, most of the ones that do exist support the models' predictions. Those that do not generally fall into two categories: (i) animals, such as insects, which are not physiologically capable of behaving optimally, as in insects that fly too slowly to compensate for wind drift; and (ii) animals that respond to variables that are not incorporated into optimality models, as in tuna migrating in cooler waters, which allow them to avoid overheating. Neither issue suggests any fundamental problems with the models themselves, although we will need to collect a great deal more data on the behaviour of non-avian migrants before we can say for sure that the models are generally applicable.

Energy and time have long been accepted as important currencies during migration. However, fitness is the ultimate currency of evolution (Chapter 2). Fitness is explicitly addressed in so-called annual routine models, using dynamic optimization methods to solve for the optimal allocation of investment to different life-history stages over an entire annual cycle (McNamara and Houston 2008). At the moment we know little about how variation in rates of energy use during migration or time spent migrating causes variation in survival and/or reproductive success, although there have been some initial empirical attempts to address this question (M. Bowlin and C.W. Breuner, unpublished data). In order for migratory behaviour to evolve, it must increase either survival or reproductive success. Ultimately, data on individual survival and reproductive success can be incorporated into individual-based dynamic optimization models; currently survival or reproduction probabilities are estimated from other parameters (Hedenström et al. 2007). Coupled with research on the determinants of reproductive success, these data should allow us to begin to understand how migration evolves and what benefits organisms obtain by migrating. We encourage ecologists to link migration with direct measurements of fitness whenever possible.

Survival is not the only parameter that can be difficult to obtain in migrants; the study of migration has always been constrained by our ability to follow individual animals throughout the annual cycle. Only recently has technology, in the form of satellite transmitters and geolocators, advanced far enough to allow us to obtain locations for individual animals at any time of the year (Stutchbury et al. 2009; Egevang et al. 2010; Robinson et al. 2010). The vast majority of studies on migratory animals have, therefore, focused on a single stage in the annual cycle: breeding, wintering, stopover or migration. However, it is not possible to understand the evolution of migration without knowledge of how events in the non-migratory portion of the annual cycle affect migration and vice-versa. While we can now follow individuals year-round with satellite transmitters and geolocators, these technologies are limited. Satellite transmitters cannot be placed on the bulk of migratory species due to their size (the smallest satellite tags are approximately 10 g), and geolocators (~1 g), while much smaller, are only accurate to within ~100 km. Some researchers have focused on using stable isotopes to estimate breeding and wintering locations in small migrants, but these data also provide poor resolution. We believe geolocators and stable isotopes will become obsolete as satellite transmitters are miniaturized and new options for tracking small migrants from space become available.

It is, however, not enough simply to track individual migrants across the globe; the locations of migrants themselves do not help us to understand how and why animals migrate. To do that, we need additional information on individual behaviour and physiology throughout the annual cycle. If, for example, birds equipped with satellite transmitters are caught and weighed prior to departure from the breeding or wintering grounds, we could begin to test the accuracy of range estimates from the theoretical models presented in this chapter. One recent study measured heart rate in migratory birds and found that climate change may affect rates of energy use during flight due to altered frequency or intensity of atmospheric turbulence, something which had received very little attention in the literature (Bowlin and Wikelski 2008). An early study along the same lines in geese (Butler et al. 1998) suggested that the energetic cost of migration was lower than theoretical models predicted at the time. Another study analysed individual migration tracks in conjunction with data on winds aloft and was able to test hypotheses about the navigational capacity of adult and juvenile raptors (Thorup et al. 2003). We must begin to couple location data with data on individual behaviour and physiology throughout the annual cycle if we want to understand migration.

Another parameter we desperately need information about is mortality and when it occurs. Many populations of migratory animals are declining (Wilcove and Wikelski 2008), and in many cases we do not know why because we do not know when in the annual cycle these populations are regulated. Currently, survival is difficult to measure in most migratory animals, but as new technologies become available it should become possible to determine when and why individuals die during migration (but see Sillett and Holmes 2002 for an indirect approach). Data on when and where individuals die, gleaned from technologies that allow us to track animals across the annual cycle, may prove critical to our understanding of population demographics in migratory animals. Only with these data will we be able effectively to conserve declining populations of migratory animals.

Ultimately, understanding the correlations that exist between individual behaviour, physiology and fitness throughout the annual cycle will not be enough. Migration ecologists need to manipulate experimentally variables in one stage of the annual cycle and examine the effects on subsequent life-history events. Only then will we understand how and why animals migrate; to do so constitutes the real scientific challenge to current and future students of animal migration.

CHAPTER 5

Energy gain and use during animal migration

Nir Sapir, Patrick J. Butler, Anders Hedenström,
and Martin Wikelski

Table of contents

5.1 Introduction

Despite its widespread occurrence, long-distance migration is potentially a risky process because it commonly involves fasting periods, elevated metabolic costs, feeding on exotic diets and movement through unfamiliar environments. It is therefore not surprising that mortality rates have been estimated to be far greater during migration than during any other period of the year (Owen and Black 1991; Sillett and Holmes 2002; Strandberg *et al.* 2009, Guillemain *et al.* 2010). How can migrants prepare for, and react to, hazardous environmental circumstances encountered en route in order to alleviate the risks of migration? Animals may control their movements, such as their speed of progression and time of departure (Chapters 4 and 6). Another way to mitigate some of the perils of migration is for the animal to manage its energy budget throughout the journey. A balance of energy gain and use is needed during migration to safeguard against starvation and to minimize costs of becoming overloaded with energy stores (Chapter 4). How this is achieved is the subject of the present chapter.

Animal migration may be divided into four distinct stages; preparatory, movement, stopovers and arrival (Ramenofsky *et al.* 1999). During the preparatory stage, the animal may prepare for the expected journey using a suite of behavioural (e.g., increased food consumption), physiological (e.g., fat loading) and navigational (e.g., compass calibration) processes. After departure, the animal may divide its journey into several, and sometime numerous, movement periods, interspersed by stopovers. Stopover periods can last from several hours to several weeks, during which the animal rests, and may gain energy through feeding. An entire migration journey may take place with only a single stopover, for example Alaskan bartailed godwits (*Limosa lapponica*) stop over in the Alaskan coastal areas that are located close to, or even within, the breeding area (Gill *et al.* 2009).

During the preparatory and stopover stages, many migrating animals load their body with large fuel deposits (mainly fat but also protein) that are used throughout the journey. The ability of the animal to deposit fuel may be limited ecologically (i.e., by the abundance of food) or physiologically (i.e., by its ability to process food when food is unlimited; Lindström 1991). Physiological factors such as the transport of the fuel and its oxidation may limit the use of stored fuel during migratory movements (Jenni and Jenni-Eiermann 1998). Various environmental factors (e.g., air turbulence; Bowlin and Wikelski 2008) may affect energy expenditure during different migratory stages. Moreover, the properties of the animal, such as its mode of locomotion (e.g., passive vs. powered movement; Sapir 2009) and its size (Lindström 1991; Hedenström 1993) may govern animal energy gain and use (Chapter 4). Consequently, this chapter will deal with the following questions:

• What is the extent of fuel loading and how are its dynamics determined by different environmental factors?
• How is energy use affected by environmental factors during long-distance movement?
• How do the specific needs of the animal during distinct migratory phases affect fuel loading and energy expenditure?
• What are the methods that can be applied to answer questions related to energy gain and use by migrating animals?

We will first describe the dynamics of fuel storing and discuss factors that have been found to influence them. Second, we will describe how the migrating animal expends energy during different stages of migration. The chapter largely draws on avian examples, which reflects the dearth of studies on other taxonomic groups. However we draw parallels where possible.

5.2 Fuel storage for migration

5.2.1 Why store fuel and how?

In order to meet the high metabolic demands associated with movement between two distant geographical regions, the animal must gain energy that is later used for two purposes. First, energy is supplied to muscles to fuel their operation during movement (Pennycuick 1998). Second, energy is used to maintain body homeostasis during periods of movement when feeding cannot be conducted (e.g. Gill *et al.* 2009) or when alighting in environments scarce in food en route (e.g., Moreau 1972; Biebach *et al.* 1986). In these cases, energy stores are utilized to offset energetic costs related to the basal metabolism of the body, maintenance of the animal's body temperature under ambient temperatures outside its thermoneutral zone, foraging that involves movement, and for supporting other body systems, such as the immune system. Fat loading has been describe in many migratory taxa, including migrating insects (e.g., monarch butterfly *Danaus plexippus*; Brower *et al.* 2006), fish (e.g., European eel *Anguilla anguilla*; Svedang and Wickstrom 1997), reptiles (e.g., green turtle, *Chelonia mydas*; Kwan 1994), mammals (e.g., humpback whales *Megaptera novaeangliae*; Boyd 2004), and numerous bird species.

To measure how much fat a migrating animal is carrying and the rate of fat deposition, one can visually inspect fat deposits in some species and assess their level (e.g., visible subcutaneous fat in passerines; Sapir *et al.* 2004b), measure lean body mass of the animal in question (e.g., Piersma *et al.* 2003), or use blood measurements of fatty acids, specifically triglyceride (Jenni-Eirmann and Jenni 1994; Smith and McWilliams 2010). The advantage of the last method is that the rate of fat deposition can be estimated from a single measurement of the individual, whereas in the other two methods the animal must be captured twice. In cases where these methods cannot be applied, fat must be extracted from the animal, for example, when studying fat storage in migrating butterflies (Brower *et al.* 2006; Dudley and Srygley 2008).

5.2.2 Fuel storage in relation to migratory stages and overall migratory strategy

The extent of fuel loading can be substantial during preparation for the journey and stopover, but it has also been suggested that the rate of fuel loading may be important. Maximizing the speed of migration has been argued to be an optimal strategy for migrating birds (Alerstam and Lindström 1990) and other taxa (Hedenström 2009; Chapter 4), and this may be achieved by maximizing the rate of fuel deposition. Hence, it is not only important to load enough fuel to meet the energetic requirements of the journey, it is also advantageous to do it quickly. To store fuel, let alone at a high rate, the animal has to consume more energy than it expends during the preparatory and stopover stages (Fig. 5.1) and this may be achieved in a number of ways. First, the animal may increase its daily food intake by increasing its intake rate (Karasov and Pinshow 2000) or by extending the period devoted to feeding during the day (Kvist and Lindström 2000). Second, the animal may change its diet in order to consume food resources that are more abundant in its environment. A diet switch may also enable the animal to consume food combinations that are better suited as fuel for muscle function during the subsequent movement stage (Bairlein and Gwinner 1994; Bairlein 1998; Lepczyk *et al.* 2000). Third, the animal may assimilate the consumed food at higher

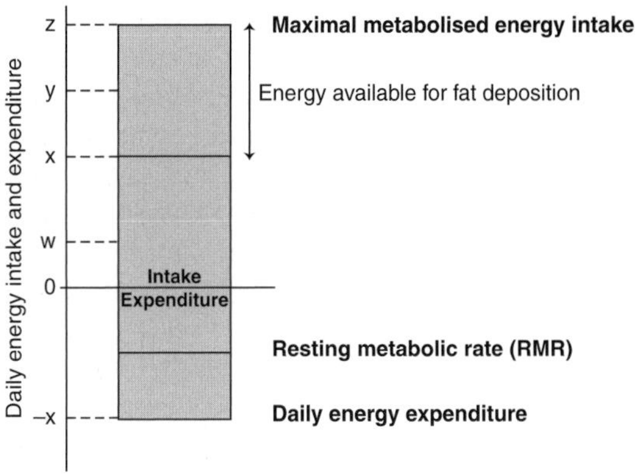

Figure 5.1 Daily energy expenditure and intake of a staging migratory animal in relation to migratory phase and ecology, following Lindström (1991). Resting metabolic rate (RMR) and the additional energy costs resulting from different behaviours of the bird other than resting comprise the daily energy expenditure with an absolute value of *x*, with a negative *x* also shown. During staging when conditions do not permit any energy intake, fat stores are used for maintenance, resulting in a decrease of fat loads. If daily energy intake equals *w*, which is defined as $0 < w < x$, then the bird must use some stored fat to compensate for its daily energy expenditure. This situation is typical of staging in poor stopover sites where food is meagre. If daily energy intake during staging equals *x*, then daily energy needs are exactly compensated by energy intake, resulting in constant fat levels. If the bird's daily energy intake equals *y*, which is defined as $x < y < z$, then energy intake is greater than energy expenditure, allowing for fat deposition. Fat deposition is limited in this case by food abundance and/or by the animal's foraging abilities. If bird daily energy intake during staging equals *z*, which is defined as the maximal metabolizable energy intake, fat is deposited at the maximal rate. Fat deposition is not limited in this case by food intake but instead by the capacity of the digestive system to process the food. Note that RMR, daily energy expenditure and maximal metabolizable energy intake are assumed to be constant over time (cf. Fig. 5.2).

efficiency through the enhancement of biochemical absorption of food in the intestine (Hume and Biebach 1996). All these changes require a highly developed digestive system that is capable of processing large amounts of food during the preparatory and stopover stages (Bairlein and Gwinner 1994; Karasov and Pinshow 2000; Gannes 2002; McWilliams *et al.* 2002; van Gils *et al.* 2006).

Attaining a highly developed digestive system may be extremely beneficial when energy gain is desirable, but this same system may become a burden during flight. This is because high body mass increases the energetic cost of transport, especially for flapping flyers that must exert a power proportional to their body mass in order to remain airborne.

Maintaining a highly developed digestive system may thus trade off with energy saving during movement. There are different solutions for this conflict, depending on the strategy of the migrant (Piersma 1998).

One solution, common in many migratory flapping birds that engage in flights of several days duration, is to decrease the digestive system during flight and to build it up again during stopovers (Weber and Hedenström 2001). However, this process takes time, usually in the order of several days (Karasov and Pinshow 2000), and therefore migrants that degrade their digestive system before takeoff to lower their cost of transport may hamper their ability to process high quantities of food during stopovers. A decrease in fat accumulation several days before takeoff has been documented in several species of staging passerines (Fransson 1998; Bayly 2006; Bayly 2007), indicating that the birds started to degrade their digestive system well before takeoff. An example of the dynamics of fat storage of a flapping flyer during stopover is illustrated in Fig. 5.2. An extreme example of this strategy is the population of bar-tailed godwits that may fly nonstop for about eight days over 11 000 km from Alaska to over-winter in New Zealand (Piersma and Gill 1998; Gill *et al.* 2009). Godwits departing from Alaska have very small digestive systems and very large fat deposits that account for 55% of their body mass (Piersma and Gill 1998).

A different solution, used by animals that feed on the move or that are involved in short duration movement, is to bear the metabolic costs of transporting an enlarged digestive system in order to maintain at least some if not all of its digestive capabilities. For example, *Catharus* thrushes migrating over the Midwestern US move at night and feed during the day (Cochran and Wikelski 2005). Presumably, these birds do not usually degrade their digestive system and deposit modest fat stores that are used for night time flights. Intermediate strategies are also likely to exist in addition to the two extremes described above.

Overall, the strategy of migrants is probably influenced by ecological factors (e.g., food abundance) and by the location of staging sites in relation to ecological barriers along the route (e.g., digestive system atrophy is likely in a warbler that

(a)

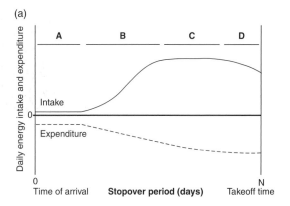

(b)

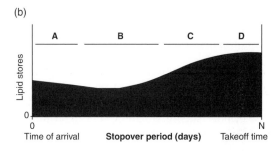

Figure 5.2 (a) Dynamics of daily energy intake and expenditure of a typical migratory bird throughout its stopover period. Daily energy intake is depicted by a solid line, while daily energy expenditure is illustrated by a hatched line. Horizontal lines are used to distinguish between stopover stages, as follows. A: Search and settlement time and the time required for building up the bird's digestive system. During this stage, feeding is substantially limited by the ability of the bird to locate food in its unfamiliar surroundings, as well as by its digestive capacity. This stage may last 1–3 days. B: Steep increase in energy intake with time due to adjustments of the digestive tract towards processing larger food quantities and better knowledge about food distribution in the stopover site. During this stage, energy expenditure increases due to intense foraging and because of an increase in fat tissue that is maintained by the body. C: High and constant energy intake, with the digestive system fully functional for processing high quantities of food. During this stage, energy expenditure further increases due to the increase in the mass of the adipose tissue that must be maintained by the body. D: Decrease in energy intake resulting from atrophy of the digestive system towards take-off for migratory flight that may last 1–3 days. During this stage, energy expenditure levels off due to a decrease in foraging activity towards the end of the stopover period. (b) Dynamics of fat stores due to changes in energy intake and expenditure of a typical migratory bird throughout its stopover period. The daily energy available for fat deposition is calculated by subtracting the daily energy expenditure from the daily energy intake (see Fig. 5.2(a)) and the actual energy (fat) deposited is this surplus energy times a conversion coefficient. A: Decrease in fat stores due to impaired food intake during search and settling and due to limited digestive capacity. B: Increase in fat stores following better foraging and better performance of the digestive tract. Note that if energy expenditure increases during this period, fat stores may increase to a lesser extent. C: High and constant increase in fat stores. D: Fat storage levels off before take-off for migratory flight.

is about to cross a wide sea where no feeding can be carried out). In addition, the type of motion may also affect fat storage strategies. For example, soaring birds that do not exert power for flapping, but rather stretch out their wings when soaring over thermal updrafts, expend much less energy than their flapping counterparts (Pennycuick 1972; Bevan *et al.*; 1995, Sapir 2009). Consequently, their fat stores are expected to be rather modest, and this has been demonstrated in white storks (*Ciconia ciconia*) that had lower fat stores during migration relative to their mid-winter stores (Berthold *et al.* 2001).

5.2.3 Ecological effects on fuel loading: food and water availability, competition and predation risk

When food is plentiful, the rate at which fuel is deposited may be constrained by the capacity of the digestive system to process and assimilate food. Under such a situation, fuel deposition is physiologically rather than ecologically constrained (Zwartz and Dirksen 1990; Kvist and Lindström 2000; Dierschke *et al.* 2003; Kvist and Lindström 2003). The capacity of the digestive system to process and assimilate food varies depending on the system's acclimation to feeding conditions (Karasov 1990). So, a staging migrant may be physiologically limited at first, but later on, when the capacity of its digestive system to process food increases, its food intake may be limited by ecological factors such as food abundance. There are only a handful of examples of birds being physiologically constrained, since ecological factors commonly limit food intake, and it is thus likely that birds rarely reach their physiological maximal fuel deposition rate in the field (Lindström 1991).

The distribution in space and time of different food types available for consumption is a key ecological factor limiting the ability of migrating animals to fuel up. For example, in the Mediterranean scrub, bird diet varies greatly depending on the season. Bird-dispersed fruits are abundant during autumn and are readily consumed by a large number of omnivorous passerine migrants that stop over in this region. Edible fruits are largely absent in this area during spring, at which time staging migrants are compelled to feed on invertebrates (Izhaki and Safriel 1985). Owing to the between-season difference in food abundance, staging migrants may be able to load fuel much faster in autumn compared with spring. However, evidence is still lacking for such a difference. Moreover, this difference may imply that autumn migrants need a certain set of physiological adjustments to facilitate rapid fuelling on large quantities of fruit, while on their northward journey, these same birds may require a different set of capabilities for detecting, handling and processing invertebrates.

In addition to food, water availability can also limit fuel loading. Water availability facilitates fuel loading in autumn migrating blackcaps (*Sylvia atricapilla*; Sapir *et al.* 2004b, Tsurim *et al.* 2008) feeding on fruits of the Mt Atlas gum tree *Pistacia atlantica*. The fruit's pulp contains only 35% water and lipids make up over 50% of its dry matter. If their water requirements are met, the birds are able to boost their fat deposition by consuming these fruits rather than insects. If not, they may deposit low amounts of fat or even lose fat. Therefore, fuel loading in blackcaps may be impeded throughout many desert and Mediterranean-type areas due to lack of drinking water. Interestingly, this effect was not found in a congeneric species, the lesser whitethroat (*S. curruca*; Sapir *et al.* 2004b).

Other biotic interactions may limit foraging and consequently fat deposition even if food is ample and water is available. Competitors may hamper feeding either by direct interference and food resources monopolization (Rappole and Warner 1976; Lindström *et al.* 1990; Carpenter *et al.* 1993) or through reduction in food availability due to exploitation (Moore and Yong 1991; Kelly *et al.* 2002). Predation risk may also reduce food intake by shortening foraging duration (Fransson and Weber 1997) and this effect may be more severe for migrants loaded with large lipid depots that are more vulnerable to predation (Cimprich and Moore 2006). Predation risk may result in preference for food-poor but relatively risk-free habitats and microhabitats (Lindström 1990; Sapir *et al.* 2004a; Cimprich *et al.* 2005), which may vary by individual depending on body mass. Under such a scenario, food-rich but dangerous habitats may be occupied by relatively lean birds, and fatter birds may sacrifice food intake for safety by inhabiting food-poor but safer sites (Ydenberg *et al.* 2002). Overall, predation risk could have a significant effect on fuel loading in staging migrants, although direct evidence for this effect is quite rare (Piersma *et al.* 2003, but see Schmaljohann and Dierschke 2005).

5.2.4 Variation by age and sex

There are many documented examples of differences in fat stores between different age groups with similar lean body mass (e.g., Ellegren 1991; Woodrey and Moore 1997; Yong *et al.* 1998). These studies found that first-year birds have lower fat reserves than those of older birds (but see Arizaga *et al.* 2008). Lower fat reserves in first-year birds may result from lower intake rates, attributed to either lower foraging success due to lack of experience, or competition with adult birds that are more successful in winning disputes over food resources. For example, Menu *et al.* (2005), who studied migratory greater snow geese (*Chen caerulescens atlantica*), reported mortality of juvenile geese under harsh environmental conditions. These conditions impaired foraging and increased thermoregulatory costs, leading to mortality of individuals that had failed to accumulate the required fat to overcome the energetic requirements of migration under severe environmental conditions. Carpenter *et al.* (1993) described the consequences of stopover territoriality on feeding in migratory rufous hummingbirds (*Selasphorus rufus*) during autumn. They found that different age and sex classes had different successes in monopolizing food resources (flowers), thereby directly affecting the birds' fuel storage. Consequently, adult females that were more successful in obtaining nectar than young females had higher fuel stores.

In general, age is a significant factor explaining inter-individual variation in fuel loading during autumn, whereas sex-related differences have been usually reported during spring. Spring-migrating males are characterized by higher fuel loads or fuel deposition rates (Hedenström and Pettersson 1986; Morris *et al.* 1996; Woodrey and Moore 1997; Yong *et al.* 1998; Lyons *et al.* 2008), and since higher rates of fuel deposition enable higher speed of migration (Alerstam and Lindström 1990; Hedenström and Alerstam 1997), males that travel at faster rates

may benefit from early arrival at breeding grounds. This may allow them to occupy high-quality territories where food can be delivered to the brood at a higher rate. Dierschke *et al.* (2005), who studied fuelling dynamics in Greenlandic/Icelandic northern wheatears (*Oenanthe oenanthe*) stopping over in a Baltic Sea island and about to cross the North Sea, reported that males had overall higher fuel loads at departure compared with females. Yet, Scandinavian northern wheatears at the same location, which is much nearer their destination than that of the Greenlandic/Icelandic birds, had lower fuel loads regardless of sex. Sex may thus affect fuel loading depending on the strategy of the animals and the position of the staging site in relation to ecological barriers en route.

5.2.5 The effect of geomagnetism on fuel loading

The extent of fuel deposition may be affected by geomagnetic fields. In the thrush nightingale (*Luscinia luscinia*), a long-distance migrant that moves between Sweden, Egypt and sub-Saharan Africa, fuel loading is low in high latitudes (Sweden) and high in low latitudes (Egypt) where the birds must store large fuel depots for crossing the 1800-km wide Sahara Desert. The upper limits of fuel loading in Sweden and Egypt are correlated with the local geomagnetic field, demonstrating that the spatial information provided by the Earth's geomagnetic field may enable the animals to overcome extensive ecological barriers that are located within their migration routes (Fransson *et al.* 2001). Similarly, the response of the robin (*Erithacus rubecula*) exposed to a simulated geomagnetic field characterizing the end point of its medium distance journey, was a decrease in fuel loading (Kullberg *et al.* 2007). These studies suggest that the local geomagnetic field has a key influence on fat storage. Consequently, studies that investigate the mechanisms responsible for spatial variation in fuel loading should consider this effect.

While not much is known about fuel loading of migratory insects, there have been a few studies focusing on butterflies. Brower *et al.* (2006) summarized extensive data on fuel loading in the monarch butterfly. Butterflies that hatch in eastern and central

N. America during summer migrate towards central Mexico during the autumn. At the beginning of their migration they deposit only small amounts of fat, while much higher lipid levels were found in migrating individuals staging in Texas and northern and central Mexico during late autumn. These lipids are used during their entire over-wintering period in Mexico in which they do not feed, and possibly also during their roughly 2000-km return migration in the following spring. A peculiarity of this migration system is that after about 2000 km of a return migration, the butterflies reproduce. Their offspring and later on their offspring's offspring continue to migrate to eastern and central N. America where the population existed in the previous summer. The entire annual cycle of this migratory population is therefore composed of several butterfly generations. This study suggested that the butterflies underwent a behavioural shift to hyperphagia (increased food consumption) at lower latitudes before the start of their extended over-wintering period. The equivalent pattern of an increase in fat deposition at low latitudes during the migratory journey in both the monarch butterfly and the thrush nightingale (Fransson et al. 2001) may suggest that a similar process controls fat loading in these two species.

5.2.6 The effect of feather moult on fat deposition

The effects of moult on migratory fuel deposition have been studied so far only very rarely. In migratory passerines, most of the adults moult after breeding and also need to migrate sometime during this period. Consequently, there may be some conflict between the timing and energy demands of the moult and of migration (Lindström et al. 1994). In most cases, migration usually starts after the moult is completed. Otherwise, moult is suspended or arrested and may be completed only after the completion of the entire migratory journey. In many waterfowl species, a complete moult of the wing feathers occurs after breeding to the extent that the birds endure an extended, several weeks long, flightless period. Waterfowl moulting is characterized by body mass reduction due to elevated energetic demands and decreased energy intake even

when food is plentiful (Portugal et al. 2007). This decrease in food intake presumably evolved to reduce exposure to predation risk when the birds' escape capabilities are severely hampered. In these species, the allocation of energy to fuel storage in preparation for migration is postponed until feather growth is completed and may depend on food availability at particular sites where the required food can be consumed during both moulting and the subsequent pre-migration period (Loonen et al. 1991).

5.2.7 The effect of fuel loading on flight performance

There are three possible effects of fuel loading on flying migrants. First, the higher energy demands of the added fat tissue must be met. Thus, the metabolic rate of the animal is expected to increase as a function of the added fuel deposits. Second, since fat is deposited subcutaneously it may change the body's streamlining and consequently cause higher drag (Chapter 4). Third, the energy cost of lifting and transporting a body of higher mass may result in higher investment being needed for the locomotory machinery of the body. For swimming migrants, the cost of increased mass is probably negligible, since they do not need to lift themselves in the water. In addition, for soaring animals that are carried by atmospheric up currents, this latter cost may be smaller and be independent of body mass (or even decrease with body mass; Hedenström 1993).

Aerodynamic theory predicts that changes in body mass have consequences on optimal flight speed. For example, higher body mass of a migrant flying by flapping flight would invariably result in higher flight speed (Pennycuick 1969; Hedenström 1993). Dudley et al. (2002), who studied diurnal flapping flights of the migratory neotropical moth (*Urania fulges*) over Lake Gatún, Panama, found no relationship between the size, mass and fat content of moths and their airspeed in individuals with a fat content of 4–38% of their body mass. By contrast, Dudley and Srygley (2008), who studied ten migratory butterfly species with fat content averaging 4.5–15.6% of their body mass in the same location, found a positive relationship between fat load and air speed in all the species with sufficiently large sample sizes, supporting the predictions made by models of optimal flapping flight speed. An interseasonal difference in flight speed was similarly attributed to body mass in Brent geese (*Branta bernicla*) migrating over southern Sweden and studied by tracking radar (Green and Alerstam 2000). This study suggested that spring migrating Brent geese flew significantly faster than autumn migrating geese, due to an average 20% increase in their body mass.

Aerodynamic theory further predicts that an animal's rate of climb during migratory flights is negatively related to body mass; therefore, animals loaded with large fuel stores will not be able to climb as fast as individuals with lower fuel stores. Hedenström and Alerstam (1992) tested this prediction by comparing climb rates of dunlins (*Calidris alpine*) in Sweden during autumn and in Mauritania during spring, with average fuel loads accounting for 17 and 50% of their body mass, respectively. Dunlins migrating during spring in Mauritania had an average rate of climb of 0.6 m s^{-1}, almost three times lower than those migrating during autumn in Sweden, with a climb rate of 1.7 m s^{-1}, consistent with the predictions of aerodynamic theory. Similarly, Green and Alerstam (2000) observed that heavier Brent geese migrating during spring showed lower climb rates than lighter birds during the autumn migration, but differences in climbing rates (0.46 and 0.62 m s^{-1}) between the seasons were quite small in this study. These findings from the field are supported by laboratory studies in which birds with different fuel levels were exposed to a predator in order to simulate a controlled escape flight. In these studies, take-off angle and take-off speed were negatively related to fuel load (Kullberg *et al.* 1996; Lind *et al.* 1999).

5.2.8 Fuel use during flight

Only rarely have researchers studied fuel use in wind tunnels during 1–12 hours long, continuous flight (Klaassen *et al.* 2000; Jenni-Eiermman *et al.* 2002; Schmidt-Wellenburg *et al.* 2008). These studies reveal that protein provides about 10% of the catabolized fuel mass during flight. Also, there was apparently no shift between fat and protein as a metabolic substrate under different levels of exercise

and the ratio between protein and fat use remained constant for at least ten hours during flight (Jenni-Eiermman *et al.* 2002). Studying fuel consumption in wild ranging birds during cross-country flights is challenging, and consequently most studies on fuel use rely on data from birds alighting at different locations relative to their departure site (e.g., Biebach 1998), or when birds unfortunately hit erected man-made structures such as lighthouses, communication towers and wind turbines (e.g. Piersma and Gill 1998). To assess fuel use during migratory flight in the field, nocturnal migrating passerines were trapped in a mountain pass and their blood metabolites were sampled. This approach allowed comparison between different species, and between migratory flight and other activities carried out by the birds during migration, such as rest and foraging during stopover. High levels of triglyceride as well as evidence for protein breakdown were found, and differences between species were quantified and partially explained by the migratory strategies of the birds (Jenni and Jenni-Eiermann 1992).

5.2.9 Physiological constraints on lipid oxidation

Animals can obtain energy from any one or from a mixture of three metabolic substrates; carbohydrates, fats or proteins. Dehydrated fats provide approximately twice as much energy as dehydrated carbohydrates or proteins but, as stored lipids are almost completely dry whereas carbohydrates and proteins are not, the difference is as much as 8–10 times on a wet mass basis (McQuire and Guglielmo 2009). Yet, both the supply and oxidation of lipids from adipose tissues to the muscles may be limited. It is believed that to overcome these potential constraints, birds have developed mechanisms to enhance the transport of fatty acids from the adipose tissue to the muscle, and to increase their aerobic capacity, possibly through the modification of enzymatic processes (Jenni and Jenni-Eirmann 1998).

Lipids are largely used by the oxidative fibres in the locomotor muscles during sustained exercise. In fish, the oxidative muscle fibres are located in two bands running the length of each side of the body

(Butler 1986), while the flight muscles of migratory birds consist mainly of oxidative fibres (Butler and Bishop 2000). However, the locomotor muscles of mammals tend to consist of mixtures of muscle fibres with the oxidative fibres being used primarily during relatively low intensity exercise and the glycolytic (anaerobic) fibres being recruited at higher intensity exercise (Armstrong and Laughlin 1985). Thus, in most species of mammals, there tends to be a decreasing reliance on lipids and an increasing use of carbohydrates (glycogen) as exercise intensity increases (see Fig. 2 in McClelland 2004). As metabolic rate is particularly high during forward flapping flight in birds and at least twice as high as that in running mammals of similar body mass (Butler 1982), it would be predicted from the mammalian data that carbohydrates would fuel migratory flight in birds, but this is not the case. In fact, migratory fish and birds have the ability to store large amounts of lipids for use by their oxidative muscle fibres during migration (Blem 1980).

The main reason why the use of lipids is restricted in most species of mammals is their inability to transport lipids sufficiently rapidly across the plasma membrane and into the mitochondria. Membrane-bound and cytosolic fatty acid binding proteins (FABPs), plus specific enzymes on the mitochondrial membranes, serve this function and migrating birds have substantially more FABPs in their flight muscles than mammals have in their locomotor muscles (McQuire and Guglielmo 2009). There is some evidence that bats use lipids as the primary fuel during flight, but whether or not they store fat for migration is not known for certain. Most species of bats for which fat stores have been determined during migration also hibernate, so it has not been possible to conclude that the lipids were accumulated specifically as fuel for migration. There may be little point in any species of mammal storing lipids for migration if it is unable to transport them sufficiently rapidly across the muscle and mitochondrial membranes. This issue is currently being investigated in migrating and non-migrating hoary bats, *Lasiurus cinereus*, (McQuire and Guglielmo 2009) but to the authors' knowledge, it has not been studied in any other species of migratory mammal.

Fatty acid binding proteins have been characterized in oxidative muscles of fish, as well as in the flight muscles of birds, and of insects, which are known to use lipids during sustained exercise, such as migration (Weber 2009). Whether or not the lipid stores of marine mammals or reptiles are used specifically for migration has not been investigated, although a number of studies seem to have incorporated the use of lipid stores during migrations to breeding areas as part of the metabolic cost of reproduction (Biuw *et al.* 2007; Southwood and Avens 2010). Examples of this are migratory cetaceans such as right whales (*Eubalaena glacialis*) and humpback whales (*Megaptera novaeangliae*) which seem to accumulate large blubber reserves during the summer when they inhabit polar seas that are rich in food resources. During winter, however, they are often found in sub-tropical waters where they breed but where relatively little feeding takes place (Boyd 2004).

5.2.10 Capital versus income breeding in migratory animals

Whether animals carry the nutrients needed for reproduction from the wintering or stopover areas to their breeding grounds (capital breeding), or whether they obtain them entirely at the breeding areas (income breeding) is a question that relates to the level of fuel carried by the migrant on arrival. Having fuel stores at the end of the journey may enable the individual to survive in the potentially harsh environment of the breeding grounds early in the season, and may help in providing for the demanding physiological processes involved in egg production and the feeding of young (Smith and Moore 2003). Yet, the amount of fat stored before breeding does not always have a positive effect on reproductive output, for example in tree swallows *Tachycineta bicolor* (Winkler and Allen 1996). Indeed, the relatively rich literature on capital vs. income breeding in Arctic waders and geese is controversial (Klaassen *et al.* 2001; Clausen *et al.* 2003; Alerstam 2006; Hedenström 2006), and the general picture is not well understood. Nevertheless, in their review, Drent *et al.* (2006) suggested that the prevailing strategy is a mixture between the two options, where fuel, acquired at stopover sites, is carried to breeding grounds and then supplemented by local resources for building up the reproductive system.

Capital breeding has also been documented in the migratory European eel, *Anguilla anguilla* (Svedang and Wickstrom 1997; van Ginneken and van den Thillart 2000; van Ginneken *et al.* 2005). Eels move to breeding grounds where they produce eggs from lipid stores that have been deposited before the start of the journey, representing an example of capital breeding in an ecological context similar to that of arctic-breeding birds.

Capital breeding considerations may have far-reaching consequences for migration strategy. For example, bar-tailed godwits that migrate during spring from New Zealand to Alaska, divide their journey into at least two different movement legs: from New Zealand to China's Yellow Sea (10 000 km), and from there to Alaska (6500 km); therefore they need to load themselves with substantial amount of lipids twice during their migration. The reasons for this bi-phasic springtime voyage are unclear. One possible explanation relates to capital breeding considerations, as a shorter pre-arrival flight from the last staging site may allow the birds to allocate the fuel deposits remaining after the termination of the journey to reproduction.

To evaluate fully the effects of different parameters on female reproductive strategy, Houston *et al.* (2007) modelled the effects of factors such as length of gestation, offspring metabolism, efficiency of energy transfer from parent to offspring, and the rates of energy intake by females with and without offspring. They suggested that the most important factor determining reproductive strategy was the cost associated with carrying accumulated energy stores, and therefore the elevated energetic demands when transporting a higher body mass during flight to springtime breeding sites may have far reaching consequences for animal life histories.

5.2.11 Conservation implications of energy gain in migratory animals

Migratory performance (e.g. overall migration speed) is determined to a large extent by the fuelling possibilities. Therefore, if suitable stopover habitats or sites are destroyed or become less profitable, the speed of migration and the time of arrival to over-wintering and breeding sites may be affected. For example, if the phenology of the emergence of insects varies due to underlying climate change, tropical wintering birds may encounter reduce fuelling rates during spring migration (Marra *et al.* 2005). In turn, this may affect the timing of arrival with consequences for the animal's fitness. Furthermore, the consequences of similar phenological changes in insect emergence that are the outcome of warming at the breeding grounds may be particularly severe for long-distance migrants. This may be because these animals cannot adjust their arrival time to coincide with the emergence of the insects due to numerous constraints operating in several geographical areas. Consequently, global warming may have severe fitness consequences for long-distance migrants (Both and Visser 2001).

Changes in the quality of individual staging points can have a dramatic effect on migratory performance. Delaware Bay is used by a population of spring migrating red knots *Calidris canutus rufa* on their northbound journey to Arctic Canada from their over-wintering area at Tierra del Fuego. The knots feed there on the eggs of spawning horseshoe crabs *Limulus polyphemus* but, due to commercial fishing for horseshoe crabs in combination with erosion of beaches, this food source has declined dramatically in recent years (Baker *et al.* 2004). The consequence is that knots cannot gain mass as fast as before in order to reach the breeding area in good condition. This knot population has declined dramatically, probably as a consequence of deterioration of this single stopover site during its spring migration (Baker *et al.* 2004). The lesson from this example is that migratory animals may be vulnerable to changes in habitats used en route during their migrations, and that conservation strategies should include the entire migration route (Chapter 11).

5.3 Energy use during migration

5.3.1 Energetic estimates of migratory movements

How animal performance and morphology affect fitness is a central question in organismal biology (Arnold 1983). Quantifying the metabolic consequences of behavioural, morphological and physiological attributes of animal migration may provide insights into their effect on the fitness of the organism. Estimation of metabolic rates during migration is, however, a challenge, since migratory organisms are difficult to track throughout their migratory journeys. Moreover, the collection of physiological data from wide-ranging animals is usually logistically limited (Cooke *et al.* 2004; Goldstein and Pinshow 2006). Currently, there is only a handful of field studies in which animal metabolism has been estimated during migratory movements.

Wikelski *et al.* (2003) estimated energy expenditure in migrating Swainson's (*Catharus ustulatus*) and hermit (*C. guttatus*) thrushes over the Midwestern US using the doubly-labelled water technique. They provided the first empirical support for the theoretical prediction of Hedenström and Alerstam (1997) that the energy expended during flight makes up only about one third of the total energy used during migration, with the other two thirds spent during stopover periods. Bowlin *et al.* (2005), and Bowlin and Wikelski (2008) tracked Swainson's thrushes (*Catharus ustulatus*) using radio-telemetry while recording their heart beat frequency (f_H) continuously during flight. The relationships between f_H and metabolic rates at rest and in flight in this species were established using respirometry chambers and doubly-labelled water techniques (Bowlin, Meijer and Wikelski, unpublished data). Bowlin and Wikelski (2008) examined the effect of several factors on f_H throughout entire migratory cross-country flights, from departure to landing. They revealed that individuals with relatively sharply pointed wingtips and higher wingload (higher ratio of body weight to wing area) were characterized by elevated f_H, suggesting that morphology and body mass have an important effect on bird energy expenditure during flight. Additional findings from this study suggest that when cruising birds encountered strong winds (regardless of wind direction) and more turbulent air, they expended more energy than during calm conditions. Mandel *et al.* (2008) studied a migratory turkey vulture (*Cathartes aura*) over North America using an f_H logger, and similarly found that the vulture's f_H increased under conditions of high vertical air speed and above rugged terrain where the structure and availability of hot air thermals may impair

their use by soaring-gliding birds, presumably forcing the birds to flap their wings.

Sapir (2009), studying migrating European bee-eaters (*Merops apiaster*) in southern Israel using radio-telemetry, found that bird flight mode had a substantial effect on bird f_H, with f_H during powered flapping flight being about 2.5 times higher than f_H during soaring–gliding flight. Moreover, the birds' f_H during soaring–gliding was similar to their resting f_H, suggesting that soaring–gliding may be particularly cheap. This finding suggests that flight mode-related differences in metabolism may be substantial and may explain the propensity of European bee-eaters to migrate using soaring–gliding, when meteorological conditions permit the use of this flight mode. The ability of these birds to time their flight in accordance with environmental conditions, permitting soaring over hot air thermals, may lead to considerable energy savings throughout their entire migratory route. Similarly, the propensity of migrating monarch butterflies to use soaring flight over flapping flight was proposed to reflect energy economy consideration (Gibo and Pallet 1979). Owing to practical limitations in estimating flight metabolism in butterflies, no empirical findings are available to test this suggestion.

5.3.2 Energetics of migrating barnacle geese

Butler *et al.* (1998) combined satellite telemetry and f_H loggers in barnacle geese (*Branta leucopsis*) migrating in autumn from Svalbard to Scotland over 2500 km. Since it has not been possible to calibrate f_H against rate of oxygen consumption ($\dot{V}O_2$) accurately, at least for the autumn migration of wild barnacle geese (Bishop *et al.* 2002), Butler *et al.* (1998) used an allometric approach based on recorded f_H and mass loss to estimate the metabolic rate of the wild barnacle geese throughout their autumn migration (Bishop and Butler 1995). Butler *et al.* (1998) found that, contrary to the conclusions of earlier studies (Owen and Gullestad 1984), the barnacle geese they studied did not fly non-stop in 30–40 h from the Arctic to southern Scotland. Instead, they flew along the Norwegian coast (Fig. 5.3), making frequent stops along the way. The total duration of the journey was between 4 and 39 days, with the longest non-stop flight being 18.6 h

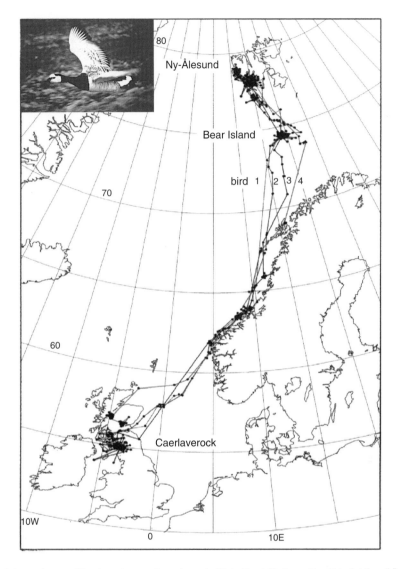

Figure 5.3 Southerly (autumn) routes of four barnacle geese, *Branta leucopsis*, fitted with satellite transmitters. Note that they all flew (and stopped) along the Norwegian coast (Butler *et al.* 1998). An individual in flight is shown in the inset (Photo: Patrick Butler).

(mean for 10 birds, 13.5 h). Despite the large variability in the total duration of the migration, the total time spent flying was reasonably consistent (range 46.5–85.2 h, mean 61 h), which means that most of the variation in total duration is due to individuals spending different lengths of time at stopover sites. The duration of stopovers was so variable because the birds presumably waited for tailwind assistance to aid their flight under the overall unfavourable (headwind) weather conditions that prevail in this

region during this time of the year. For example, one bird spent 23 days in south-western Norway before leaving the Norwegian coast for Scotland. During most of this period, strong southerly winds (mean velocity 10 m s^{-1}) prevailed.

The most intriguing data from the study of migrating barnacle geese are the f_H during migratory flight. Heart rate during the autumn migratory flight was around 300 beats min^{-1} at the beginning, gradually decreasing to a low of 225 beats min^{-1} at

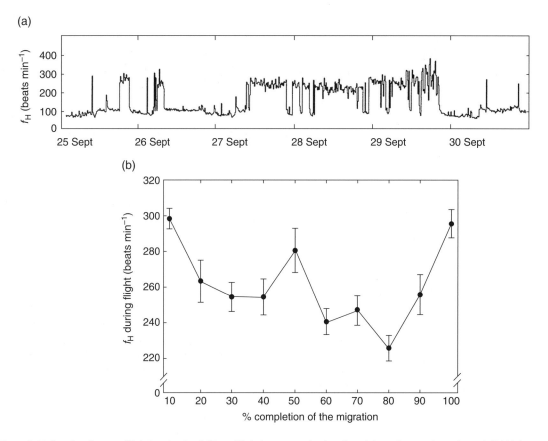

Figure 5.4 Heart beat frequency (f_H) during migration. (a) Trace of f_H during autumn migration of a male barnacle goose, *Branta leucopsis* (1.98 kg). The bird was flying when f_H was high (Butler *et al.* 1998). (b) Mean (± SEM) values of f_H while flying from ten barnacle geese as a proportion of the total flight time during migration (P. J. Butler and A. J. Woakes, unpublished data).

about 70% through the migration (Fig. 5.4). These values are lower than the lowest obtained from captive barnacle geese flying in a wind tunnel (about 350 beats min^{-1}) and within the range of the same birds running on a treadmill (about 130 beats min^{-1} to 330 beats min^{-1}; Ward *et al.* 2002). However, in the last 30% of the migratory journey, there was a steady increase in f_H during flight, which coincided with the birds leaving the Norwegian coast and heading towards Scotland. Consequently, by the end of the migration, f_H during flight was almost back to the value recorded at the beginning of the journey. One possible reason for the progressive decrease in f_H during the middle phase of the migration could have been the progressive loss of body mass (see Fig. 5.2, Part B.), but why the overall frequency was lower during this phase than at the end of migration,

and even than when running, is not so obvious. It could be that the birds were benefiting from vertical air currents as they flew along the Norwegian coast, which is possible since this species is capable of slope soaring (Butler and Woakes 1980) and could, therefore, make use of updrafts created when winds hit cliffs located on shore-lines within the bird's migratory flyway. Support for this explanation is provided by the fact that during wind-tunnel flight both f_H and $\dot{V}O_2$ fell when the trainer's presence created an updraft (Fig. 5.5).

Interestingly, at the beginning of the spring migration, f_H during flight was over 400 beats min^{-1} (P.J. Butler and A.J. Woakes, unpublished data), which is similar to that obtained for captive birds flying in a wind tunnel (Ward *et al.* 2002), although it did decrease throughout the migratory period.

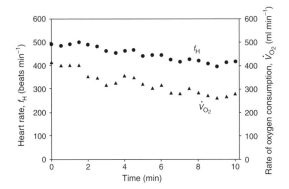

Figure 5.5 Heart rate (f_H) and rate of oxygen consumption ($\dot{V}O_2$) during a 10-minute flight in a wind tunnel by a barnacle goose, *Branta leucopsis*. During this flight, the bird flew progressively closer to the trainer (Ward *et al.* 2002).

This would suggest that during these flights the birds were not being assisted by vertical air currents. Also, as f_H during spring migratory flights and during flights of captive birds in a wind tunnel were similar, the calibration of f_H against VO_2 obtained by Ward *et al.* (2002) on captive birds is most likely to be applicable to the spring migratory flights of wild barnacle geese.

Taking an average f_H during the autumn migration of 254 beats min^{-1} and using the allometric method of Bishop and Butler (1995), gives an average VO_2 during migration of approximately 120 ml min^{-1}kg^{-1} (Butler *et al.* 1998) and mass loss data give a value of approximately 160 ml min^{-1} kg^{-1} (Bishop *et al.* 2002). These values, therefore, seem to be reasonable estimates of the energy cost of the autumn migratory flights of barnacle geese, and are 13–18 times the night time resting value for this species (Portugal *et al.* 2007). These high metabolic demands of flapping flight compared with resting metabolism are in line with allometric relationships of bird physiology and flight mechanics, suggesting that flight metabolism increases disproportionately with body mass, leading to very large power inputs in large bodied flapping flyers (Pennycuick 1969).

Another intriguing observation concerning the study of migrating barnacle geese is that, during the autumn migration, the body temperature of wild geese did not increase during the periods of flight, which could be for many hours, whereas when flying in a wind tunnel, even for only 20 minutes,

the body temperature of captive birds increased by almost 1°C, which is consistent with previous studies of birds flying in wind tunnels in temperatures as low as 0°C (Butler 1991; Torre-Bueno 1976). Reasons for this difference remain elusive. This finding and the difference in f_H mentioned above (see also Bishop *et al.* 2002) suggests that flight in wind tunnels may differ from some migratory flights to a greater extent than previously assumed.

5.3.3 Use of accelerometry to determine behaviour of migrating animals: moving towards estimation of behaviour-specific metabolic rates

Present knowledge of migratory animal energetics is still limited to the very few examples provided above, therefore currently it is not possible to give a year-round estimation of the energetics of migrating species. This means that we are unable to deduce the potential effect of variation in energy use throughout the life cycle of migrating animals on their fitness. Since different behaviours have unique metabolic rates (e.g., Bevan *et al.* 1995; Ward *et al.* 2002), a first, necessary, step in estimating the energetics of migrating animals throughout different periods of the year is to document their activity time budget. The technology of three-axial accelerometry that has been applied recently in several studies of non-migratory species (e.g., Wilson *et al.* 2006; Green *et al.* 2009; Holland *et al.* 2009) can potentially overcome the practical challenge of registering animal behaviour over extended routes throughout the entire migratory period, and even throughout the entire life cycle of a migratory animal. For example, this technology, in combination with GPS telemetry, has been recently applied to study year-round movements and activity time budgets of common cranes (*Grus grus*). Cranes are studied during over-wintering in the Hula Valley in Israel and possibly also in Ethiopia, during breeding in northern Russia, as well as during migration between these localities, allowing researchers to assess how environmental conditions affect bird behaviour throughout their annual routine. Crane flight mode and wing flap rate could be measured using this technology en route over Lebanon after a springtime takeoff from the over-wintering site in the Hula Valley (Fig. 5.6).

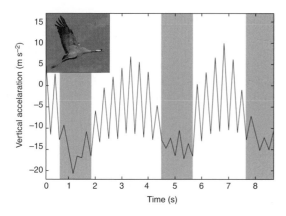

Figure 5.6 Wing flap rate registered by accelerometry of a crane (*Grus grus*) during migratory cross-country flight over Lebanon. The crane was flying 20 km away from its over-wintering area in the Hula Valley in Israel at an altitude of 800 m above the ground on 7th March 2009. Later the bird reversed its flight course and returned to the Hula Valley. Wing flap rate was registered by the body's vertical acceleration in an 18-Hz sampling rate. Crane wing flap rate is about 3.5 Hz. Shaded areas depict pauses in wing flaps (I. Shanni, N. Sapir, M. Wikelski and R. Nathan, unpublished data). An individual in flight is shown in the inset (Photo: Udi Maman).

Moving towards estimation of behaviour-specific metabolic rate and ultimately estimating the energetics of migrating animals throughout their annual cycle using accelerometry is a challenge. First, this technology could be combined with behaviour-specific measurements of metabolic rate (Green *et al.* 2009). This should be done in the laboratory and also in the field but it is extremely difficult to obtain accurate data on metabolic rate during specific activities, for example during flapping flight of birds in the field. Also, accelerometry will not be able to provide exact information on other factors known to affect animal's metabolic rate. For example, it will not always be known if an animal is feeding and/or if it is exposed to temperatures outside its thermally neutral zone. Consequently, we are not likely to have accurate values of metabolic rates for a whole range of different behaviours. Having information on the behaviour of an animal so that any measurements of metabolic rate can be related to specific behaviours (e.g., bird flight modes) would be a huge advance. We propose that a long term aim must be to use techniques such as accelerometry to provide

behavioural data to go alongside real-time estimates of metabolic rate, e.g. from measurements of heart rate. This would enable the energetics of different migratory stages and different periods of the year to be determined and thus lead to substantial improvements in our understanding of the causes, mechanisms, patterns and consequences of animal migration.

5.4 Concluding remarks

The extent and dynamics of fuel gain and its use throughout long-distance migratory routes is determined by a number of behavioural, physiological, aerodynamical and environmental factors. Some of these factors are more pronounced at certain stages of the migration and at particular points along the way (e.g., near an ecological barrier), and some trade-off with others. Consequently, the migratory animal must manage its fuel loads and energy expenditure in an optimal manner (Alerstam and Lindström 1990; Weber and Hedenström 2001).

Over the last two decades many studies, especially in birds, have documented the dynamics of fuel loading and the environmental factors affecting them, but due to practical difficulties in obtaining physiological measurements from animals that are on the move, how energy is used during migration is poorly known. There are only a few examples of studies that investigated energy expenditure at different stages of the voyage and under different environmental scenarios. Insights gained from these studies are unmatched by any other study conducted on migratory animals in the laboratory, and demonstrate the capability of such methods in resolving long-standing issues in the field of animal migration, such as the effect of loaded fuel on flapping flight performance, and the allocation of energy between stopover and movement phases during migration journeys. The combined use of techniques such as three-axis accelerometry and heart rate can help overcome the practical challenges of measuring metabolic rate in the field and relating it to specific behaviours. To explore how global climate change may affect important outcomes of animal migration, such as ecosystem services, agricultural productivity and the spread of pathogens, it is essential to further study how the environment affects animal physiology during migration and how animal physiology may influence the timing, scheduling and extent of animal migration.

Moreover, quantifying the food requirements of migratory species and securing these requirements in habitats located throughout migratory corridors may aid the conservation of migratory animals for years to come (Wikelski and Cooke 2006; Wikelski *et al.* 2007).

CHAPTER 6

Cues and decision rules in animal migration

Silke Bauer, Bart A. Nolet, Jarl Giske, Jason W. Chapman, Susanne Åkesson, Anders Hedenström, and John M. Fryxell

*Ice bars my way to cross
the Yellow River,
Snows from dark skies to climb
the T'ai-hang mountains!*

*(Hard is the journey,
Hard is the journey,
o many turnings,
And now where am I?)*

*So when a breeze breaks waves,
bringing fair weather,
I set a cloud for sails,
cross the blue oceans!*

Li Po (701–762) 'Hard is the journey'

Table of contents

6.1 Introduction

The sheer beauty and impressiveness of animal migrations have long puzzled observers and raised the questions of how these animals find their way, what initiates their migrations and how they manage to schedule their journeys at apparently the right times. There are many challenges that animals face before and during migration, and they can be grouped into two major categories namely *'where* to go?', dealing with orientation and navigation, and *'when* to go?', dealing with the timing of activities and migration schedules. As these questions are fundamentally different, we also expect different cues to provide relevant information. Furthermore, animals make decisions in relation to their physiological state and therefore, cues can be further categorized into internal and external signals.

In this chapter, we tackle these questions not in terms of *why* animals migrate (ultimate reasons), but *how* they make the right decisions before and during migration (proximate factors). Migrating animals rely on external and internal information such that they can tune their behaviour to their (changing) requirements and to the development of their seasonal environments. Here, we show that these cues, defined as 'signals or prompts for action' (*Oxford English Dictionary*), are not well understood, although they are most likely highly relevant both for advancing our fundamental understanding of migration and for increasing our capacity to manage and conserve migratory systems under the threat of environmental change.

In the following sections, we first characterize migration from different perspectives, as different migration types may require fundamentally different cues and decision rules. Thereafter, we introduce migration in the major migratory taxa that are readily observable to biologists, i.e. insects, fish, reptiles, birds and mammals, and seek to identify the cues that are used for the different steps during each of their migrations. We finish by highlighting the general lessons that can be drawn from this comparative study of cues and decision rules in migration.

6.2 Challenges in migration

6.2.1 Where to go?

Going along a specific track requires cues for positioning (navigation) and for finding the way towards the goal (orientation). Important cues for compass orientation include information from the magnetic field, the Sun and the related pattern of sky light polarization and stars, while information from, for example, landmarks and odours are used for navigation. Excellent reviews on orientation and navigation can be found in, for example, Åkesson and Hedenström (2007), and Newton (2008). However, for many migrating animals we still do not know how they find their way, which orientation and navigation abilities they have and which mechanisms they use (e.g. Alerstam 2006; Holland *et al.* 2006b).

6.2.2 When to go?

Photoperiod has been shown to be involved in the timing of activities for many species, e.g. initiating 'Zugunruhe' (restless behaviour as the pre-migration phase starts) or determining the speed of migratory progression. This may come as no surprise, as photoperiod is a reliable indicator of time of the year and thus can be a useful predictor for the phenology of resources (Fig. 6.1). Other local and short-term factors influencing timing of migration include prevailing weather conditions, e.g.

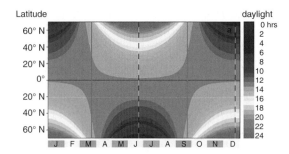

Figure 6.1 Photoperiod, i.e. day length, as a function of latitude and time of the year. Day length is here the time between dawn and dusk using civil twilight (Sun 6° below horizon). The different shades of grey indicate the day length with darker colours depicting continuous light or permanent darkness (from Bauchinger and Klaassen 2005).

temperature, wind, drought and precipitation, or water discharge in rivers, as these factors can significantly influence the costs of the travel ahead (Chapters 4 and 5).

There are also internal cues that serve as a clock or time-keeping mechanism. Additionally, physiological state and developmental stage are important cues as most migrants undergo morphological and physiological changes in preparation for migration and internal signals, e.g. hormone levels, indicate when these changes are completed (Chapter 5).

6.2.3 Seasonal and life-cycle migration

We can distinguish between two life-cycle patterns in migratory animals. In one type, which includes land reptiles, birds and mammals, most body transformations take place within the egg (reptiles and birds) or within the mother's womb (mammals), and the juveniles are only one or a few orders of magnitude smaller than the adult, and are generally well suited to the same environment as the adults. Migration in many of these animals is linked to a seasonal change in the environment and the cues involved typically predict these changes.

The alternative pattern includes organisms with complex life-cycles, such as arthropods, fish, amphibians and sea-reptiles, where animals spawn tiny eggs that develop into individuals with a body size several orders of magnitude smaller than the adults, and with a body form differing substantially from the adult form. During development, the major

changes in the body plans often entail a habitat change and therefore, internal signals are required that are linked to these developmental processes as well as cues to locate the next favourable habitat. However, these organisms may also live in seasonal environments and, thus, the timing of ontogenetic processes will also depend on the phenology of the environment (Skelly and Werner 1990).

6.2.4 Travel with or without a predefined target

The best-known type of migration is that between a few specific localities, e.g. birds between wintering and breeding sites. In many cases, however, the migration does not lead to a specific locality or even to a certain more broadly defined area: In several species of pelagic fishes, both long-distance feeding and spawning migrations need not lead to a specific target. Feeding migration is often driven by continuous local food search (Huse and Giske 1998; Nøttestad et al. 1999), while the return spawning migration combines long-distance tracking of preferred spawning sites with physiological constraints from swimming costs (Huse and Giske 1998; Slotte and Fiksen 2000). Although some large insect migrants, such as Lepidoptera (butterflies and moths) and Odonata (dragonflies) have regular, bidirectional seasonal long-distance migrations that involve movements that are directed in predictable ways but not targeted at a specific site or region (e.g. Chapman et al. 2008a, 2008b; Wikelski et al. 2006), most insect migrations do not even involve movements in consistent, seasonally preferred directions.

6.2.5 Genetic or cultural transmission of migratory behaviour

How do offspring decide where and when to migrate? Migratory behaviour can be both genetically and culturally determined. In cultural transmission, the young copy their parents' (or other group members') behaviour. Consequently, species with culturally-transmitted migratory behaviour are expected to have a social life-style, longer lifespans and (in higher vertebrates) extended parental care. Prominent examples include schooling fishes (e.g. the culturally-induced change of migration patterns and over-wintering sites in herrings; Huse *et al.* 2002), geese and swans among birds (e.g. Von Essen 1991), and large herd-living mammals such as antelopes and wildebeest.

Alternatively, migratory behaviour, e.g. routes, threshold photoperiods, or preferred directions, can be genetically transmitted when there are no parents, peers, or elders—be it due to high mortality, short adult life-span, solitary life-style, absence of parental care or separation of age classes, e.g. age classes or generations have different requirements or constraints (differential migration). Examples of genetic transmission of migratory behaviour include some birds (e.g. the majority of small passerines and the European Cuckoo *Cuculus canorus*), all insect migrants and sea turtles (hatchlings complete their migration alone, and have to return to the same beach to breed when reaching sexual maturity many years later).

Besides the general insights, how (part of) migratory behaviour is transmitted is also highly relevant with regard to global and local environmental changes. A first review of the effects of environmental change by Sutherland (1998), concentrating on birds only, showed that none of the species with culturally determined migration routes had sub-optimal routes, i.e. longer than necessary, while approximately half of the species with genetically transmitted routes had become sub-optimal. There is thus a risk that environmental changes may occur faster than natural selection, particularly for long-lived and less fecund life forms. Whether these findings also apply to other taxa has yet to be shown. Exceptions to this general pattern appear to be zooplankton and insects—with their short generation times and high reproductive rates, many insect pests, for example, are able to adapt to changing conditions rapidly.

6.3 Cues in the different phases of migration

Migration can be divided into a few major steps—preparation, departure, on the way, and termination—a cycle that might be repeated if migration is suspended at intermittent stopover sites. Each of these steps potentially requires specific cues and decision rules as their demands on the animal's physiology and behaviour differ. Similarities might exist across taxonomic groups in how animals deal with each of these steps but differences may also be expected depending on the specific way of migrating or their particular environment.

6.3.1 Migration in plankton

The annual or seasonal migrations in plankton probably include a higher number of migrants than any other group (e.g. 10^{15} individuals of Antarctic krill *Euphausia superba*). As an example we present the much-studied *Calanus finmarchius*, which is among the most abundant species of marine calanoid copepods. These North Atlantic copepods reach an adult body size of a few millimetres, and spend most of the year in a survival mode in deep waters. Although the exact depth varies with local conditions and the state of the individual, and may range from a few hundred metres to >1000 m (Kaartvedt 1996), it is vital that they descend deeper than the winter mixing zone to avoid passive retransport to the surface layers during winter. The minimum energetic cost during over-wintering occurs where the organism is buoyant, so the individual variation in over-wintering depth probably comes from variation in storage tissue in the form of wax esters (Heath *et al.* 2004). Only a small fraction of the wax esters produced in the preceding feeding season are consumed during over-wintering (which is also sometimes called hibernation, diapause, dormancy or resting stage, Hirche 1996). Most is saved for conversion to eggs in or near surface waters in spring.

Preparation and departure: These copepods undergo a series of moults during their life, with six nined naupliar stages followed by five copepodid stages before adulthood. Overwintering is usually restricted to the fifth copepodid stage (C5). Since the maximum efficiency in converting food to storage occurs in the C3–C5 stages, the eggs of the over-wintering adults must hatch in time to grow and develop through the naupliar stages in time for C3–C5 to hit the spring-peak in phytoplankton production. Thus, ascent from deep waters must be timed well in advance of the peak. Depending on the food conditions in the surface waters, the copepods may produce one or several generations during spring and summer. Only the last of these generations will descend to the diapause depth. This migration pattern is therefore not genetically hard-wired, but also depends on one or more environmental signals.

On the way and termination: There are still several plausible suggestions for cues involved in the seasonal migrations for plankton in general and the species *C. finmarchicus* in particular. The matter is further complicated as this latter species lives in a very diverse range of environments in the North Atlantic Ocean and adjacent seas, and cues that are reliable in one area may not be so in another. Therefore, Hind *et al.* (2000) modelled the seasonal dynamics of the species in four different areas: the North Sea, the Norwegian Sea, the Iceland Shelf and in a northern Norwegian fjord, to test which set of cues would produce viable populations in all of these areas. They found only one set of cues that produced realistic population dynamics in all areas. This set consisted of four cues: (i) an external signal; (ii) an inherited threshold for the downwards migration; (iii) a physical characteristic of the over-wintering depth for the organism (buoyancy); and (iv) an internal cue for timing of ascent. If ambient food concentration is above the inherited threshold value for environmental food concentration, C4 copepodids develop towards adults and another generation in surface waters. If food levels are below the threshold, they sink after moulting to C5. Having reached the over-wintering depth, they continue to develop at a constant rate, but slower than for surface dwelling organisms. The cue for ascent to the surface is that the organism has completed 80% of the development of the C5 stage. However, one should bear in mind that this is only an ultimate test (population dynamics modelling) of the proximate mechanisms—neither the physiological nor developmental mechanisms are understood so far.

6.3.2 Migration in insects

Although migration occurs in all major insect orders, the actual migrations may often go unnoticed due to the small size of most insects, and the tendency of many species to migrate at great heights above the ground. However, the utilization of fast air currents allows many species to cover enormous distances (hundreds or even thousands of kilometres), often within just a few days and the consequences of these invisible large-scale insect movements may be highly conspicuous wherever they terminate. Some insect migrations are highly noticeable; among the most impressive of natural phenomena are the mass migrations in enormous cohesive swarms of a few species (e.g. the desert locusts *Schistocerca gregaria*, the dragonfly *Aeshna bonariensis*, and the monarch butterfly *Danaus plexippus*), which rival the largest flocks and herds of migratory birds and mammals in terms of biomass, and far exceed them in total numbers (Holland *et al.* 2006b).

Insect migrants typically do not make round-trip journeys, where the same individuals return to their natal area, nor do most species carry out bi-directional seasonal movements between separate breeding and wintering grounds. Instead, successive generations engage in windborne displacements through the landscape, most likely in an attempt to locate transient and patchily distributed favourable habitats. The majority of insect migrants take advantage of fast windborne dispersal and fly at altitudes of from several tens of metres up to a few kilometres above the ground. Relatively few species migrate predominantly within their flight boundary layer (FBL), i.e. the narrow layer of the atmosphere closest to the ground within which their airspeed exceeds the wind speed (Taylor 1974)—this is mostly restricted to large, day-flying species, such as butterflies and dragonflies (e.g. Dudley and Srygley 2008).

In most species, migration is restricted to the adult—winged—life stages and to a single brief time window of just a few days, due to the short adult life-span and further because migration typically takes place in the brief period of sexual immaturity immediately following metamorphosis from the immature stage to the adult (aka oogenesis-flight syndrome, Johnson 1969).

Preparation: The development of full-sized wings and associated musculature is obviously the most important preparation and many species, e.g. aphids, have the ability to produce offspring with varying levels of flight capability in response to environmental conditions, e.g. decreasing plant nutritional quality, and increased crowding. Exceptions to this general pattern exist in longer-lived species such as the monarch butterfly *Danaus plexippus*, which builds up substantial fuel reserves by foraging as adults, and tops these reserves up during intermittent stopover episodes.

The juvenile hormone and its esterase mediate a range of correlated factors associated with migration, e.g. timing of reproductive maturation, fuel deposition, development of larger wings and wing-muscles, and increased flight capability.

Departure: Owing to the very short window for migration in most species, opportunities to choose the departure time are rather limited and mainly concern questions of whether to migrate on a particular occasion and at what time of the day to migrate.

For time of the day, two basic options exist—diurnal migrants take advantage of the higher air temperatures and greater illumination (presumably facilitating orientation), while nocturnal migrants benefit from the absence of convective up-draughts and down-draughts and thus can control their altitude to a much greater extent than day-flying insects, taking advantage of warm, fast-moving, unidirectional air currents (Wood *et al.* 2006; Chapman *et al.* 2008b).

The decision whether to initiate migration on any particular occasion varies between species. Many insect migrants will not take off when wind speeds at ground level are too fast (more than a few m/s), as they cannot control their flight direction immediately after take-off (e.g. green lacewings *Chrysoperla carnea*, Chapman *et al.* 2006). However, as the migration window of most species generally lasts for just a short period (e.g. two nights in lacewings), they are unable to migrate if confronted with extended periods of strong winds, or are forced to do so in unfavourable conditions.

More complex decision rules are required for species that need to move in a particular direction, e.g. south in the autumn to escape northern hemisphere winter conditions. Some species are able to gauge the presence of favourable high-altitude tailwinds, facilitating southerly displacement in the autumn. An example for this is the potato leafhopper *Empoasca fabae*—a small insect that is entirely dependent on windborne displacement to escape deteriorating winter conditions in northern regions of the US by migrating to its diapause site in the southern US. Autumn migrants initiate their flights in response to falling barometric pressure, which is indicative of the passage of weather fronts that are followed by persistent northerly air flows, thus facilitating long-range transport of the leafhoppers to the south (Shields and Testa 1999).

Green darner dragonflies have a number of simple decision rules that guide their autumn migrations along the eastern seaboard of North America in a favourable, southerly direction (Wikelski *et al.* 2006). They initiate migratory flights on days following two preceding nights of dropping temperatures, which are highly likely to be associated with persistent northerly air flows, and then simply fly in the downwind direction while avoiding being carried over large water bodies (and thus out to sea). Red Admiral butterflies also choose cold northerly tailwinds for their return migrations from Scandinavia—they fly at high altitudes when fast-moving winds from the north predominate, but low down when migrating in headwinds (Mikkola 2003).

On the way: Many insect migrants are too slow-flying compared with the speed of the air currents to influence the direction and speed of their movement. The most efficient strategy then is simply to fly downwind if the migrants are able to perceive the direction of the current (either through visual assessment of the direction of movement relative to the ground, or via some wind-related mechanism). There is considerable evidence that many high-altitude migrants are capable of aligning their headings in a more-or-less downwind direction (e.g. Reynolds and Riley 1997), and given that winds blow in favourable directions, displacement distances will be considerably longer than if the insects flew across or against the wind (e.g. Wood *et al.* 2006).

Migrants that fly predominantly within their flight boundary layer (FBL; Taylor 1974) can control their direction of movement irrespective of the wind direction. This is the case for butterflies, which are powerful fliers and can maintain migration speeds of 5 or 6 m/s for several hours a day, and for many consecutive days. To guide their migrations in seasonally-favourable directions, these butterflies must have a compass mechanism. A well-known example is the monarch, whose eastern North America population undergoes an annual autumn migration of up to 3500 km from the late-summer breeding grounds in eastern Canada and North-Eastern United States to the communal wintering site in central Mexico. But how do the monarchs orient their flight headings in the correct direction? Work by Mouritsen and Frost (2002) has demonstrated that autumn-generation monarchs have a preferred migratory heading towards the southwest, and that during sunny conditions they use a time-compensated solar compass to select and maintain this heading. In spring, successive generations of monarchs move progressively northwards through North America and presumably use the same orientation mechanism, but in reverse, to guide their migrations.

Migratory tracks of day-flying FBL insect migrants (butterflies over the Panama Canal) occur in predictable seasonal directions (from the Atlantic wet forest to the Pacific dry forest at the onset of the rainy season), and in at least two species (*A. statira* and *P. argante*) these preferred directions are maintained by reference to a time-compensated solar compass (Oliveira *et al*. 1998), i.e. use of visual landmarks on the horizon to compensate for crosswind drift away from their preferred migration directions (Srygley and Dudley 2008). Measurements of wind speed and air speed also indicated that these butterflies adjusted their air speed in relation to wind speed and their endogenous lipid reserves, so that they maximized their migratory distance per unit of fuel (Dudley and Srygley 2008; Srygley and Dudley 2008).

Chapman *et al*. (2008a, 2008b) have demonstrated that high-flying migrants, hundreds of metres above their FBL, can also influence their displacement direction even though wind speeds far exceed their own air speed. The moth *Autographa gamma* is able to select flight headings that partially compensate for crosswind drift away from its preferred seasonal migration directions, thus maximizing the distance travelled while influencing its migration direction in a seasonally-advantageous manner. Using a combination of altitude selection to fly in the fastest winds, and taking up advantageous headings, they can cover up to 600 km in seasonally adaptive directions during a single night's flight (Chapman *et al*. 2008a, 2008b).

Termination: The very act of migration slowly reduces the inhibition of responsiveness to 'appetitive' cues that is typical of migratory flight (Dingle and Drake 2007), and thus migratory behaviour itself slowly promotes its own termination. The vast majority of insect migrants only undertake one, or at most a few, bouts of migratory flight, and so the factors that bring about termination of a single bout of flight are often the same as those that bring about the termination of the whole migratory phase. These include depletion of fuel reserves, changes in photoperiod (e.g. nocturnal insects rarely migrate into daytime, and diurnal species rarely carry on into night-time: e.g. Chapman *et al*. 2004; Reynolds *et al*. 2008), and changes in temperature (e.g. migrations of nocturnal insects are often terminated due to a drop in temperature as the night progresses (Wood *et al*. 2006)). If the habitat after the termination of the initial migratory bout is suitable, then that will usually signal the end of the migratory phase, otherwise migration may continue for another bout. In some species, the flight muscles are autolysed and converted to increased egg mass after migration, i.e. they become effectively flightless.

6.3.3 Migration in fish

Although there are about 30000 species of teleost fish, only a small fraction of them are currently known to be migratory. However, these few species are the dominant marine species in terms of biomass and numbers, and most of the world's fish catches are based on them. Many types of migration exist in fishes, e.g. from freshwater natal areas to the sea (or vice versa), and between feeding and breeding grounds in the sea. Here we illustrate some characteristics of fish migration by introducing two prominent examples, namely life-stage migration in salmon and feeding migration in herring.

6.3.3.1 Life-stage migration in Atlantic salmon
As predation pressure is considerable in the estuary and beyond, schooling behaviour is advantageous and therefore so is size similarity amongst smolt. This is achieved through growth control by the parr,

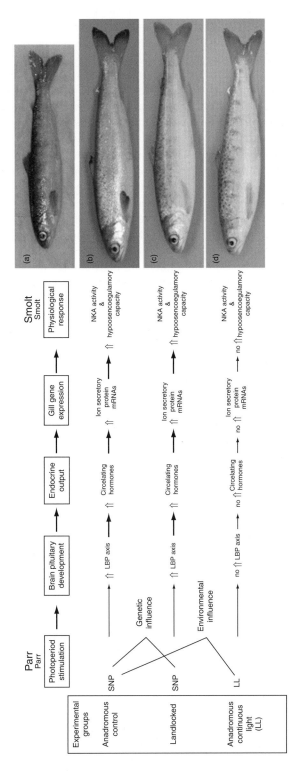

Figure 6.2 The sequence of events occurring during the parr–smolt transformation (smoltification) in salmon leading to hypo-osmoregulatory development, or not. Here three experimental groups of Atlantic salmon are presented: anadromous control, parr in February (A) and smolt in May (B) reared under simulated natural photoperiod (SNP); landlocked in May (C) reared under SNP; and anadromous control in May reared under continuous constant light (D) to demonstrate the importance of the light–brain–pituitary axis (LBP) early in smoltification on the downstream endocrine output, gill gene expression and hypo-osmoregulatory capacity. The degree of LBP development is reflected through all downstream processes including physiological development and Na$^+$, K$^+$-ATPase (NKA) activity (Ebbesson *et al.* 2007; Nilsen *et al.* 2007; Stefansson *et al.* 2007). Reproduced with permisson.

which is the life stage of young fish in the river before smoltification. Smoltification is the process in salmonid parr of preparing for their downstream, seaward migration and includes a suite of physiological, morphological, biochemical and behavioural changes (Fig. 6.2). Under very benign growth conditions, smoltification can happen during the first year of life, but in northern populations this may take up to seven years. The smoltification decision is based on internal stimuli but the relationship between the parr's body condition and the initiation of smoltification is poorly understood (Stefansson et al. 2008). According to Thorpe (1977), a bimodality in size appears in the second (or later) autumn. Only parr larger than 7.5–8.5 cm fork length eventually leave the river as smolt the coming spring. Once the decision on smoltification is made, the parr changes into a fast growth mode, where the growth rate may be 4–5 times higher than before. If the parr decides to wait, it even goes into anorexia during the winter (Stefansson et al. 2008).

The next external stimulus is the change in day length in the following spring, probably combined with exceeding a temperature threshold. This leads to growth of the brain and the pituitary gland, leading afterwards to the release of a series of endocrine hormones. This, in turn, activates gill genes and initiates the physiological processes leading to the smolt stage (Stefansson et al. 2008). The final downstream migration is triggered by a combination of light regime, temperature and river discharge (Hoar 1988), leading to simultaneous mass migrations into the estuary. The smolt will usually remain there for some months before migrating into the open ocean, usually as solitary individuals.

How can the salmon find its way back to its native river, small or large, hundreds of kilometres away and 1–4 years later? Homing to the river is also driven by a combination of internal and external factors. Several hypotheses have been suggested including a pheromone trail left by out-migrating fish, counter-current swimming, navigation by stars, and geomagnetism (Lohmann et al. 2008). It is quite clear that salmon utilize smell and learned cues in the later homing phase. Magnetic crystals have been found in the lateral line sensory system of salmon (Moore et al. 1990) and in the olfactory

lamellae of trout (Walker et al. 1997), which they might use for long-distance navigation. Therefore, they probably use a combination of geomagnetic information (for long-distance directional migration) and smell and imprinting cues (for choosing the correct river and stretch of it).

Preparation in parr: While food abundance is the driving force for the seawards migration in all size classes, the change from hypo- to hyper-salinity requires a major transformation of the metabolism and, additionally, changes in behaviour and skin pigmentation as the young salmon transforms from a bottom-dwelling territorial parr into an open-water schooling smolt. As this decision is taken long before the actual migration, cues for preparations come from both internal and external sources—once a threshold body condition (size) is reached, day-length initiates the onset of body and metabolism changes.

Departure in smolt: After having completed all body changes, the smolt often waits for the autumn river discharge to depart and go downriver seawards.

On the way smolt: Seawards, the smolt remains for some time at the estuary to become imprinted and then leave for the ocean in groups, where they become solitary again and mainly follow food. On their way back, they probably initially use some sort of magnetic field orientation, and gradually change to olfactorial orientation when they are near the home-river (e.g. Healey and Groot 1987).

Termination in returning adults: Once they have arrived in their target area in the natal river, migration is suspended.

6.3.3.2 Feeding migration in pelagic fish: an undefined target

In several species of pelagic fish, long-distance migrations may be directional rather than to a specific localizable target, e.g. mackerel Scomber scombrus and blue whiting Micromesistius poutassou migrate northwards from spawning areas around the British Isles into feeding areas in the Norwegian Sea. Thereby, they benefit both from the later spring and summer further north, and also from the gradual increase in daylight-hours in the northern summer. Both factors contribute to prolonged high feeding and growth rates (Nøttestad *et al.* 1999). Similarly, herring *Clupea harengus* migrate westwards from the mild Atlantic waters off the coast of Norway towards the colder waters in the west, with delayed spring production (Varpe *et al.* 2005). For some decades, the whole adult population of Norwegian spring spawning herring has been over-wintering in the deep Tysfjord in northern Norway. Spawning migration in spring is southwards along the coast of Norway. The further south the eggs are spawned, the better the prospects for larval growth and survival. However, as migration is energetically costly, there is a trade-off between fecundity and migration distance such that small individuals migrate shorter distances and larger individuals longer, i.e. further south (Slotte and Fiksen 2000). Hence, the likely cues are a combination of physiological state (spawning migration in herring) and a seasonal signal such as day length (end of northwards or westwards feeding migration).

Very little, if anything, is know about how fish find their way and make decisions. Geomagnetism has been proposed for long-distance navigation (e.g. Lohmann *et al.* 2008). Many of the species also migrate in large schools, which may act as cooperative units for food searching (Clark and Mangel 1986) and decision-making (Huse *et al.* 2002). Thus schooling acts both to reduce predation risk, and to increase the chance of being on the right track for future food resources. During the feeding migration, models indicate that a long-distance direction finder may not be needed as the fish simply follow the seasonal development of the food, which will automatically lead them to profitable places (Huse and Giske 1998). However, models also indicate that a separate 'homing motive' is needed for the return migration, during which local gradients in food or temperature may not be helpful (Huse and Giske 1998). Unfortunately, it is not known whether the decision to return is based on some seasonal signal or the state of the organism, or both.

6.3.4   Turtle migration

Long-lived sea turtles regularly commute between two completely different environments, the open ocean for foraging and sandy shores for egg-laying. Some sea turtles, e.g. the leatherback turtle (*Dermochelys coriacea*), spend several years foraging in pelagic habitat (e.g. Hays *et al.* 2004) and accumulate body stores, which they later use for mating and producing eggs. These two habitats are usually separated by vast areas of unsuitable habitat and, consequently, migrations are often long.

Others, e.g. herbivorous green turtles (*Chelonia mydas*), also lay eggs on sandy beaches but forage as adults on sea grass and algae along shallow coastal areas (e.g. Mortimer and Carr 1987, Bjorndal 1997). One of the longest distance migrations is performed by Ascension Island green turtles, migrating between breeding sites at Ascension Island and foraging areas along the Brazilian and Uruguayan coasts (Carr 1984; Papi *et al.* 2000; Luschi *et al.* 2001).

It is well known that sensory information is important for navigation by hatchling sea turtles when they depart to the sea (Lohmann and Lohmann 1996; Lohmann *et al.* 2008), but it is less well understood what information is used by the adults when returning to breed (Luschi *et al.* 2001; Åkesson *et al.* 2003). Even less is known about the migratory behaviour and information used by sub-adult sea turtles (Godley *et al.* 2003) and how the transition takes place from the genetically programmed guidance of the hatchlings into the migration programme guiding the sub-adults and adults later in life (Åkesson *et al.* 2003). Most likely, the turtles use partly genetically encoded behaviours, but also learn to incorporate a number of cues into their navigational toolbox (Åkesson *et al.* 2003).

Preparation: Many sea turtles need several years to recover from a major migration and egg-laying event and during this time they store fat as fuel. For example, female Ascension green turtles migrate to the island to lay eggs, where they do not forage at all for 5–6 months. Apparently they exhaust most of their reserves during the event, such that their recovery and preparation for the next migration and breeding bout requires approximately 3–4 years (Carr 1984).

Departure: Hatchlings: When the hatchlings in a clutch escape from the nest, they first climb to the surface of the sand during the day and await the night. At that stage they are stimulated by their nestmates' movements such that all siblings leave at night in a synchronized fashion. Once in the water, they mix with other hatchlings and depart on their independent migratory journeys. When they depart to open sea, their movement is both active and passive, i.e. partly swimming and partly drifting with the currents. The timing of departure relative to the season very much depends on the timing of egg-laying, which again depends on the foraging conditions encountered in the wintering areas (Godley *et al.* 2001).

On the way: Sea turtles have been shown to use a number of different cues to orientate and navigate during migration (e.g. Lohmann and Lohmann 1996, Åkesson *et al.* 2003). Loggerhead turtle hatchlings (*Caretta caretta*) respond to light when leaving the sand and moving along the beach; later they have been shown to swim against the waves to leave the shore and, once they are in more open water, they probably use magnetic field information for navigation as has been shown in experiments manipulating the magnetic field (Lohmann *et al.* 1999, 2001). Studies on adult green turtles have tried to identify cues used during migration, but also when searching for the breeding island after displacement. It was found that successfully homing turtles responded to local information, suggesting they are using information carried with the wind from the island (Papi *et al.* 2000; Luschi *et al.* 2001; Åkesson *et al.* 2003).

Termination: For all sea turtles, breeding migration ends as soon as they have reached the breeding grounds. During foraging migrations, differences exist between the pelagic species, such as the leatherback turtle, that can be considered to be constantly moving and exploring the open ocean environment (Hays *et al.* 2004), and coastal foragers, such as the green turtle, which forage along shallow coastal sea-grass beds.

6.3.5 Bird migration

The classic bird migration is the biannual migration between breeding and wintering grounds. The breeding grounds are suitable for nesting and hatchling/fledgling survival, whereas the wintering grounds are more suitable for post-fledgling and adult survival. Because birds are able to fly, they can travel long distances relatively cheaply and quickly (Chapter 4), e.g. the longest non-stop migratory flight recorded is that by bar-tailed godwits (*Limosa lapponica*) crossing the Pacific Ocean from Alaska to New Zealand, a distance of more than 10 000 km (Gill *et al.* 2009).

Exceptions to this are moult and facultative migrations. In moult migrations, birds appear to migrate to predator-free areas where they can safely shed their flight feathers. In facultative migrations, birds only migrate long distances when food is sparse, e.g. many finches (Newton 2006). At an extreme end of the spectrum are birds that are nomadic, like the grey teal (*Anas gracilis*) looking for ephemeral water and food sources in a desert landscape in Australia (Roshier *et al.* 2008).

Two main flight modes exist—flapping and soaring—each having particular consequences: Flapping flight is very costly but can be used under a wide range of weather and topographic conditions, whereas for soaring, thermals or wind are needed (Chapter 4).

The majority of birds cannot feed while flying, and in many cases the total travel distance exceeds the maximum flight distance. Thus, the birds need stopover sites where they can replenish their reserves. A good example is tundra swans (*Cygnus columbianus*), which migrate 4000–5500 km (Nolet 2006), whereas their maximum recorded non-stop flight is 2850 km (Petrie and Wilcox 2003). These swans mainly refuel on energy-rich, below-ground parts of macrophytes in shallow lakes and wetlands along the route (Beekman *et al.* 1991).

Preparation: Before actually embarking on migration, most birds partly change the composition of their bodies, e.g. increase flight muscles at the expense of leg muscles, atrophy digestive and metabolic organs (Piersma and Gill 1998; Biebach 1998; van Gils *et al.* 2008; Bauchinger and McWilliams 2009) and accumulate body stores. Photoperiod is an important external signal for preparations; it has been shown to initiate 'Zugunruhe', i.e. migratory restlessness in many migratory passerines (Gwinner 1990), but also many geese, swans and waders start accumulating body stores, altering their digestive system and building up flight muscles from a particular day length onwards. The specific value of day length at which these transformations are started is under strong genetic control, as evidenced by hybridization, parent–offspring comparisons and effects of changing selection pressures (Newton 2008). Birds kept under constant day length for up to several years still showed a circannual rhythm with the right sequence of annual events (migratory fat deposition and restlessness, gonad development, and moult) suggesting that getting into the migratory state is under internal control (Gwinner 1977). But these cycles tend to drift and be either shorter or (most often) longer than a calendar year. This internal control is most rigid in long-distance migrants that are normally confronted with most variation in day length.

Thus, under natural conditions the exact timing of events is most likely determined by a combination of internal and external factors such that the internal system is adjusted by seasonal changes in photoperiod, as has been shown with experiments with extra light or shorter than annual cycles (Newton 2008).

Departure: The exact timing of migratory departure is fine-tuned by secondary factors like temperature, wind, rain and food supplies (Newton 2008). Birds have been shown to choose favourable flight conditions and preferably leave on days with tailwinds and no rain. In the Swainson's thrush (*Catharus ustulatus*), departure decisions are best predicted by both a high daily temperature (>20°C) and low wind speeds (<10 km/h) at the time of presumed take-off. If one of these conditions is not met, the individual will not take off. However, such apparently strict rules also lead to serious errors, e.g. individuals take off at low local winds, and yet ascend into air streams that will push them backwards against their flight direction (Cochran and Wikelski 2005).

One means by which birds may forecast improving weather conditions before they actually occur has been hypothesized to be sensing air pressure changes (Newton 2008, Keeton 1980). In facultative migrants, departure may also be delayed until weather conditions for refuelling deteriorate (Newton 2008; Gilyazov and Sparks 2002).

The decision to depart from a stopover site is probably based on rather simple behavioural rules. Passerines that lose or increase fuel stores at a high rate leave a site quickly, whereas the intermediate birds stage the longest (Schaub *et al.* 2008). Geese use a mixture of endogenous and external cues, with the endogenous cues having a stronger effect as the season progresses (Bauer *et al.* 2008, Duriez *et al.* 2009).

On the way: Birds have been shown to use several cues to guide them in the right direction on long-distance migrations. The combination of cues may be essential for correct navigation as directional cues change with place (e.g. the magnetic compass; Wiltschko and Wiltschko 1972) and time (e.g. the sun compass; Kramer 1959). Birds use a combination of cues for recalibration. For instance, recent experiments suggest that birds use cues from the setting Sun to re-calibrate their magnetic compass before migrating at night (Cochran *et al.* 2004). At sunrise or sunset, birds can use skylight polarization, especially visible close to the horizon, as a compass (Able 1982; Muheim *et al.* 2006). Uniquely to birds, star patterns that indicate the axis of rotation of the night sky have been hypothesized to be a directional tool (Newton 2008). However, laboratory-based experiments on stellar orientation in birds could also be explained by the fact that individuals have the rule to use the single brightest and non-moving light as the main orientation cue.

If birds do not compensate for the change in local time when travelling across longitudes, a sun compass would lead them along routes similar to great circle routes (Alerstam and Pettersson 1991). In contrast, if they compensate and reset their internal clocks regularly while crossing longitudes, they would follow a constant rhumbline route (Fig. 6.3). Birds following one (Alerstam *et al.* 2001) or the other (Green *et al.* 2002) have been found.

The direction of migration is also under endogenous control (Gwinner and Wiltschko 1980; Helbig 1991). The direction is reversed when the return migration starts (Newton 2008). Together, the inherent period and direction of migration result in naive migrants being able to stop in the right area. Juvenile starlings that were trapped on migration and displaced by aeroplane to Switzerland, east of the usual wintering area of the population, continued their migration in the same direction and over the same distance that they would otherwise have flown and ended up in southern France or Spain (Perdeck 1967). In contrast, the trapped adult star-

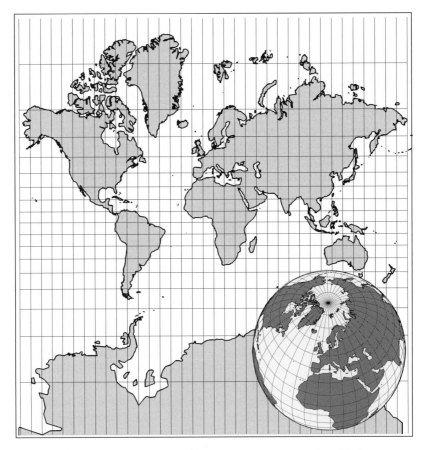

Figure 6.3 Two extremes: a global view of the Earth from space (insert; orthographic projection centred in the Wadden Sea 54 °N, 8.5 °E), and a Mercator projection flattening and stretching the globe. On the Mercator projection all straight lines are constant geographic bearings (loxodromes, rhumblines), whereas straight lines through the centre of the orthographic projection represent great circles (orthodromes). Note that the scale at the Equator is similar on both maps (from Gudmundsson *and* Alerstam 1998). Great circle routes are thus the shortest distance between any two points on the Earth's surface whereas rhumbline routes may be easier for navigation as they require no re-adjustments of headings but always cross meridians at the same angle. The difference in distance between both routes is small (<1%) at high latitudes and distances of less than 30° longitude but increases significantly thereafter. For instance, for travelling along 50° latitude and across 180° longitude, the rhumbline route is 45% longer than a great circle route.

lings were found in their traditional wintering area in northern France and England, so they must have used goal orientation. Interestingly, the juveniles returned to their new wintering area in the subsequent years, showing they also switched to goal orientation in later life. Similarly, white-crowned sparrows (*Zonotrichia sp.*) were caught in Washington while migrating from Alaska to California, and were flown to the US East coast. From there, adults headed back towards Californian wintering grounds, whereas juveniles headed south, presumably in an innate direction (Thorup *et al.* 2007).

Some migrations require changes in direction or migratory steps along the way, e.g. to avoid inhospitable environments. Birds from populations that change direction during migration show a corresponding change in direction in orientation cages as the season progresses, indicating that this is also under genetic control (e.g. Gwinner 1977; Helbig 1991). In other cases, local conditions serve as cues. For instance, pied flycatchers *Ficedula hypoleuca* changed their directional preference only when confronted with the magnetic conditions where they normally change direction, and not when mag-

Figure 6.4 Adult dark-bellied Brent geese (*Branta b. bernicla*) arriving at their still largely ice-covered breeding grounds in Taimyr, northern Siberia. In experienced birds, the decision to stop migrating is influenced by cues indicating that a familiar nesting locality has been reached (Photo: Andries Datema, Alterra Wageningen-UR).

netic conditions were kept constant (Wiltschko and Wiltschko 2003). Thrush nightingales *Luscinia luscinia* that were captured in southern Sweden at the start of their first autumn migration were exposed either to the local magnetic field or to an artificial magnetic field typical of northern Egypt, where they are thought to prepare for crossing the Sahara (Fransson *et al*. 2001). The latter group responded by accelerating fat deposition, suggesting that there is a built-in genetic response to local conditions.

Replenishment of the fat store itself may act as a cue to continue migration: in several experiments it was demonstrated that migratory restlessness and inclination to leave were higher in fatter than leaner individuals. Most of these studies were performed at localities where the birds were preparing for a major crossing (Newton 2008).

Termination: When tested under identical conditions in the lab, the duration of migratory restlessness is longer in long- than in short-distance migrants, even within species (Berthold and Querner 1981). Cross-breeding experiments showed that this is an inherited trait (see also Berthold 1999 for an experiment with hybrids of redstarts). In birds from the same species, those wintering furthest away from the breeding grounds show a tendency to start spring migration earlier (King and Mewaldt 1981).

Juvenile blue-winged teal *Anas discors* caught in the autumn and held captive for a while, migrated less far than normal after release at the same site (Bellrose 1958). This shows that the decision to stop is at least partly under genetic control. However, in adult birds the opposite was found, with the migratory restlessness continuing longer than normal when held captive for a while during spring migration (Newton 2008). Also, when held at the breeding location, indigo buntings *Passerina cyanea* did not migrate after release in the spring, whereas the control birds that were displaced 1000 km to the south did (Sniegowski *et al*. 1988). The same was true for white storks *Ciconia ciconia* reared in captivity and released in a reintroduction programme (Fiedler 2003). In experienced birds, the decision to stop is therefore apparently influenced by cues indicating that the familiar locality has been reached (Fig. 6.4).

6.3.6 Migration in mammals

6.3.6.1 *Migration in bats*
Even though bats are mammals, their ability to fly makes them more like birds in terms of their opportunities for and ecology of migration, but relatively little is known about their migration biology in

comparison with birds. This is probably because temperate bats have adopted hibernation as their main strategy for surviving periods of resource depression in a seasonal environment. Yet, in the family Vespertilionidae, migration occurs in 23 out of 316 species classified (7%), and has evolved in 15 genera with apparently little phylogenetic inertia (Bisson *et al.* 2009). The distances of bat migrations are shorter than for birds, with maximum migration distances around 2–3000 km (Hutterer *et al.* 2005). In temperate bats, long-distance migration occurs mainly in species that use trees for roost sites (Fleming and Eby 2003), but in these species migration is combined with hibernation at the wintering site. Migration also occurs in tropical species but movement distances are generally rather short and, in most cases, are driven by the phenology of fruiting trees (Fleming and Eby 2003). Differential migration is common in bats, with females migrating further north than males to raise their young, presumed to be due to higher resource needs for raising the young (Fleming and Eby 2003). As in birds, partial migration also occurs, i.e. part of the population migrates and the other part is resident.

Preparation. Bats accumulate fat deposits before hibernation (e.g. Kunz *et al.* 1998), and therefore fat is probably the main fuel used during migration (McGuire and Guglielmo 2009).

Departure. Bats most likely depart during the early night hours, similarly to nocturnally migrating birds. Departure conditions are little studied, but is seems as if migration activity is highest when wind speeds are low (Petersons 2004).

On the way. It has recently been shown that bats possess a magnetic sense (Holland *et al.* 2006a), and it therefore seems likely that this is involved in orientation during migration. Otherwise, next to nothing is known about orientation and navigation in bats, although recent evidence suggests that the greater mouse-eared bat *Myotis myotis* calibrates a magnetic compass with sunset cues (Holland *et al.* 2010). Some frugivorous species seem to track the phenology of their main food source on migration (Fleming and Eby 2003).

Long-distance migration generally consists of several cycles of fuelling followed by migratory flight. It remains to be shown whether bats follow this model, but there are some indications that they do stop over for fuelling (Petersons 2004). Since bats are mainly nocturnal they must divide their active period (the night) between foraging and migratory flight during migration. Because migrating bats' rate of energy consumption is typically much higher than that of fuel accumulation, it is expected that the proportion of time spent on stopovers should be much longer than that spent on migratory flights (Hedenström 2009). One way of saving energy on migration is to use torpor during periods of fuelling, i.e. lowering the body temperature during daytime roosting and thereby increasing the net rate of fuel accumulation (and hence overall migration speed).

In flight, there are alternative 'optimal' flight speeds predicted from flight mechanical theory (Hedenström 2009), with the maximum range speed being the best option for minimizing energy cost per unit distance. A comparison between foraging and commuting flights in *Pipistrellus kuhlii* showed that these bats select flight speed according to this prediction. Overall, bats seem to fly at slower speeds than birds of similar sizes (Hedenström *et al.* 2009).

The overall migration speed includes time for fuelling and flight (Hedenström *et al.* 2009), and is predicted to be about 46 km/day on the basis of fuel accumulation rate, energy consumption during flight and flight speed. Ringing recoveries of Nathusius's bat *P. nathusii* showed a migration speed of 47 km/day (Petersons 2004), which is comparable to that of short–medium distance migrating birds. Flight altitudes of bats on migration appear to be rather low (Ahlén *et al.* 2009), although free-tailed bats *Tadarida brasiliensis* may reach altitudes of about 3000 m when foraging (Williams *et al.* 1973).

Termination. Birds have an inherited migration programme that determines when to cease migration, but whether bats have a similar mechanism is not known.

6.3.6.2 *Migration in large herbivores*

Seasonal nomadism and migration have been documented numerous times in terrestrial mammalian herbivores and occur on every continent (Fryxell and Sinclair 1986). There are three types of situation in which herbivore nomadism or migration are common, perhaps even typical. The first situation is species inhabiting montane environments, such as elk, mule deer; red deer, and montane ecotype caribou (Albon and Langvatn 1992; Brown 1992; Horne et al. 2007; Hebblewhite et al. 2008). Migration is relatively common in herbivore species inhabiting open savannah or tundra environments, such as tundra ecotype caribou in North America, wildebeest, zebra and Thomson's gazelles in the Serengeti and Tarangire ecosystems of Tanzania, white-eared kob and tiang in the Boma ecosystem of Sudan, and Mongolian gazelles (Pennycuick 1975; Inglis 1976; Fryxell and Sinclair 1986; Durant et al. 1988; Fryxell et al. 2004; Boone et al. 2006; Mueller et al. 2007; Holdo et al. 2009). Finally, seasonal migration by ungulates also occurs in temperate regions subject to severe climatic variability, such as woodland caribou, pronghorn antelope, saiga antelope, white-tailed deer or mule deer (Rautenstrauch and Krausman 1989; Nelson 1998; Johnson et al. 2002; Ferguson and Elkie 2004; Berger et al. 2006; Sawyer et al. 2009; Singh et al. 2010).

Seasonal onset of vegetation growth is strongly temperature-dependent in montane ecosystems in temperate to arctic regions. As a consequence, snow melt occurs later at high elevations and vegetation growth is retarded relative to lower elevations (Pettorelli et al. 2005). It is common for terrestrial herbivores to exhibit seasonal shifts in accordance with seasonal green-up (Albon and Langvatn 1992; Horne et al. 2007; Hebblewhite et al. 2008; Berger et al. 2006; Sawyer et al. 2009). In low-lying forbs and grasses, structural compounds, such as lignin and cellulose, are incorporated more and more into stem and leaf tissues as the plant grows taller, reducing digestibility and lengthening the processing time in herbivore digestive tracts (van Soest 1982). As a consequence, optimal rates of nutrient intake can often be best achieved by feeding on relatively immature ramets (McNaughton 1984; Hobbs and Swift 1988; Fryxell and Sinclair 1988; Illius and Gordon 1992). By appropriate timing of migration up the elevation gradient, herbivores are able to access young vegetation and maintain optimal nutrient intake over a prolonged period. Over the course of the winter, animals usually retreat to low-lying areas, which are less exposed to severe climatic conditions and have residual plant standing crop from the growing season. A similar pattern is

seen in tundra systems, with animals retreating to woodland margins during winter, but venturing far out on to the tundra during the brief growing season.

In savannah environments, animals usually follow rainfall gradients, from the more arid rangelands used during the brief growing season to higher rainfall areas used during the driest part of the year (Pennycuick 1975; Fryxell and Sinclair 1988; Mueller et al. 2007). As nutrient quality is often inversely related to annual rainfall levels (Bremen 1983), migrants are able to access young vegetation at an optimal growth stage during the growing season in the arid areas, while retreating to high rainfall areas when arid lands dry out. Nomadism is characteristic when rainfall is unpredictable in space (Fryxell et al. 2004; Mueller et al. 2007, 2008), though this is often superimposed on a relatively dependable migratory pattern at coarser temporal and spatial scales (Wilmshurst et al. 1999; Boone et al. 2006; Holdo et al. 2009).

In temperate regions with extreme seasonal variation in climate, it is common to see migration from summer home ranges, presumably chosen primarily to obtain food and reduce predation risk, to winter ranges with less snow cover or improved shelter from snow and wind (Rautenstrauch and Krausman 1989; Nelson 1998; Johnson et al. 2002; Ferguson and Elkie 2004; Berger et al. 2006; Sawyer et al. 2009, Singh et al. 2010).

It seems likely that there is at least some learned or cultural component to migration behaviour in terrestrial herbivores, though this question has received relatively little attention in the ungulate literature. The cultural conjecture is based on well-documented examples of altered migration routes, adoption of migration by previously resident animals and even cessation of migration within a single generation (Nelson 1998). Longitudinal studies clearly suggest that partial migration is typical of northern white-tailed deer, with young individuals typically mimicking the migratory behaviour of their mothers, but capable of shifting to different strategies (resident or mixed) later in life (Nelson 1998). Similarly, elk in Banff National Park were largely migratory before the 1990s (Woods 1991). Re-invasion of wolves into areas in the Bow Valley from which they had been extirpated led to a dramatic change in space use patterns by elk over the course of 10 years (Hebblewhite *et al*. 2005), with most individuals concentrating year-round near towns that provided security from predation as well as improved nutrient intake on a year-round basis (McKenzie 2001; Hebblewhite *et al*. 2005, 2008).

Preparation and departure. Because large herbivores are highly mobile, feed while they travel, and have relatively slight energetic costs of movement relative to other taxa, there is little indication of extensive physiological preparation for seasonal movements. Cues for the initiation of migration and nomadism are thought to include seasonal changes in temperature, precipitation, and water quality (Pennycuick 1975; Rautenstrauch and Krausman 1989; Albon and Langvatn 1992; Nelson 1998; Wolanksi and Gereta 2001; Mahoney and Schaefer 2002; Boone *et al*. 2006; Gereta *et al*. 2009), although evidence is largely anecdotal. For example, Serengeti wildebeest have been seen to reverse direction and return to previously vacated areas when temporary periods of drought interrupt the usual onset of the rainy season (Pennycuick 1975). This suggests that rainfall is a key variable for this species, but it is hard to disentangle that from other putative causal variables (young vegetation abundance, water quality) that co-vary with rainfall.

On the way. Once underway, it is not clear what proximate cues migratory herbivores use to guide their movements. Some species have specific migration routes that are travelled year after year, such as pronghorn antelope in the mountains of Wyoming, Idaho, and Montana (Berger 2004, 2006), mule deer in Wyoming (Sawyer *et al*. 2009), and montane caribou in Alaska (Horne *et al*. 2007). In each case, specific individuals travel the same corridors as they shift from winter to summer ranges. Anthropogenic habitat changes that create bottlenecks in such migration corridors are a source of considerable conservation concern, because there are clear examples of migrants being negatively affected by anthropogenic barriers to movement (Williamson et al. 1988; Mahoney and Schaefer 2002; Berger *et al*. 2006; Ito *et al*. 2005, Chapter 11).

Migration routes in other systems seem much less repeatable within individuals from year to year (Wilmshurst *et al*. 1999; Thirgood *et al*. 2004; Boone *et al*. 2006), suggesting cues may be regional in nature. For example, movement trajectories of Serengeti wildebeest and Thomson's gazelles can be fairly precisely predicted in coupled map lattice models on the basis of local rainfall, grass biomass and soil nutrient levels (Fryxell *et al*. 2004; Holdo *et al*. 2009), but only when animals are capable of choosing new locations to move to within ranges of the order of 100s of km². Smaller zones of perception would probably lead individuals to concentrate in areas of local fitness peaks, thereby disrupting the migration that is repeatedly observed at a coarser spatial scale. In a particular area, residents can prefer different habitats from migrants (M. Hebblewhite, pers. comm.), suggesting either that migrants and non-migrants differ in their selective constraints or that migrants are unable to choose the best local resources because of lack of familiarity with the area.

6.4   Discussion and Integration

In this chapter, we have considered the cues that are used in several phases of migration across taxonomic groups. Although naturally many differences appear due to the specifics of each species' migration, considerable similarities appear to exist in the cues involved in the different phases of migration (Table 6.1).

In all species, preparations for migration involve entrainment to time of the year, as all environments are seasonal to some degree, thus particular times are more suitable for particular activities. Indeed, even at very low levels of seasonality, animals should migrate in order to make use of the varying levels of food in different areas (Barta *et al*. 2008). Therefore, the occurrence of photoperiod as a cue in almost all taxa is not surprising.

However, as migration is a daunting activity in the life-cycle or annual cycle, it also requires bodily changes, such as the accumulation of energy stores, the build-up of the locomotion apparatus—often at the expense of the digestive and/or reproductive system, and the transformation of a freshwater- to a seawater-adapted life-form or the achievement of a particular developmental stage. Whenever these changes are accomplished, an internal cue is produced indicating that the animal is ready to depart.

Table 6.1 Summary of the cues identified for the four major steps of migration, in all major migratory taxa

Cues for	Preparation	Departure	On the way	Termination
Plankton: *Calanus finmarchicus*	Undergo all naupliar stages and reach 5th copepodid stage. Storage of wax esters	Descending migration: Food level below threshold. Ascending migration: C5 stage to 80% developed	Continue until buoyancy depth and below mixing depth	Descend: arrival in buoyancy depth; Ascend: arrival in surface waters
Insects	Photoperiod, Crowding during immature stages, Habitat deterioration (food, predation or parasitism)	Favourable flying conditions, e.g. tailwinds	Time-compensated sun compass	Migration reduces inhibition to appetitive cues (in mig. bout); depletion of fuel reserves, changes in photoperiod or temperature
Fish	Reach minimum body size; for some: physiological adaptation to new environment	Salmon: Autumn river discharge	Local food search, or long-distance spawning location tracking	Arrival in locations favourable for spawning
Turtles	Light regime, internal status, and migratory restlessness	Favourable departure conditions, e.g. night, with currents	Visual information (bright skylight), direction of waves, geomagnetic field, wind	Arrival on specific target location, e.g. feeding or wintering grounds
Birds	Photoperiod, Build-up flight apparatus, Reduction digestive system	Favourable flight conditions (wind, rain, air pressure). Fuelling rate and body stores. Cumulative temperature or related proxy	Sun compass, magnetic field, skylight polarization, star pattern. Direction under hormonal control, sometimes responses to local conditions	Arrival on specific target location, e.g. breeding or wintering grounds. Naïve birds have an inherent migratory period
Mammals: Bats	Accumulate fat deposits (torpor during fuelling periods)	Early night hours, low wind speeds	Not much known, probably use magnetic field calibrated by direction of sunset	Unknown
Large mammals	No particular (physiological) preparations	Seasonal changes in temperature, precipitation and water quality but evidence anecdotal	Not much known; may follow gradients.	Unknown

For the actual departure, another external cue is often involved, which is usually related to travel conditions, e.g. wind, precipitation, temperature. Thus, animals prefer to depart during periods of favourable conditions, for instance, flying animals wait for tailwinds in their preferred directions; swimming animals use river discharge or sea-currents.

On the way, orientation and navigation determine the migration route taken but they may also be involved in indicating when migration is to be terminated. Animals heading for a specific location need to recognize this location, which is an option only for experienced animals, whereas naive individuals (e.g. first-time migrants) need to have a genetic programme that signals when to stop. Alternatively, migrations without clear endpoints, e.g. between feeding locations, may involve physiological cues for the termination of migration. Here again, internal signals play a greater role as they indicate when a threshold state is reached, e.g. sufficient body reserves have been accumulated for a subsequent breeding attempt.

Although we can make very rough generalizations such as these, we need to realize that currently we know the full set of cues and decision rules used throughout their annual cycle for hardly any species. For most species, we don't have any idea which cues and decision rules, orientation and navigation mechanisms they use during (specific parts of) their migration. However, such knowledge is all the more urgently required in the face of human-induced environmental changes. These changes affect the size and quality of habitats as well as the distances that separate suitable environments. Furthermore, climatic conditions are changing, but to complicate matters some areas on the globe are expected to be affected much more than others. For migratory animals, such changes pose particular challenges as they visit multiple, distant sites during their annual or life cycles—often even in different ecosystems. If we are to predict the consequences of such changes for migratory animals, we need to close the gaps in our current knowledge and gain a thorough understanding of the cues and decision rules used during migration as well as animals' orientation and navigation mechanisms.

To this end, we need to overcome the considerable bias both in the species and taxa studied and also in the type of questions asked and the approaches used. Birds are by far the best-studied taxon at present, followed by the economically relevant fish species, while comparatively little is known for the other taxa.

Most studies on navigation, orientation and decision rules have so far been conducted in captivity. Although such studies can provide important first indicators of the processes involved in natural migration, the relative importance of different cues can only be established in complex environments. Hence, it will be essential to study migratory decisions of wild, naturally migrating individuals (Wikelski *et al.* 2007).

Traditionally, migrations of animals (in particular, birds) have been identified using recoveries and resightings of marked individuals. Especially for larger birds such as swans and geese, individual marking, e.g. with neck-rings, has been possible, allowing detailed observations along their routes. More recently, the advent of increased communication possibilities and technological advances have led to significant progress in, for example, the development of satellite transmitters, geolocators, or the miniaturization of existing devices such that the movement of individuals of smaller and/or clandestine species can be followed in great detail (e.g. discovery of migratory routes in turtles, Hays 2008).

Data obtained with these devices can provide insights into the *individual* level of decision-making involved in the different steps during migration and thus provide mechanistic rather than phenomenological insights (Chapter 8). Such individual movement data can be analysed across taxa (www.movebank.org). Furthermore, these tracks can be integrated with detailed geographical and dynamic meteorological information allowing the identification of both the internal and external (environmental) determinants of migration decisions.

Another avenue for further advances in our understanding is improved integration of theoretical and empirical efforts (Bauer *et al.* 2009; Chapter 8). Significant progress in science has often been achieved when theoretical developments have inspired new experiments or when startling empirical findings have inspired the development of new theories. Despite pioneering efforts (e.g. Alerstam and Lindström 1990), the interaction between theoreticians and empiricists has been too limited to date in the study of animal migration. Several modelling approaches exist, ranging from simple optimality models (e.g. Alerstam and Hedenström 1998), dynamic optimisation models (e.g. Houston and McNamara 1999), game-theoretic models (e.g. Kokko 1999), individual-based models (e.g. Pettifor *et al.* 2000) to models based on evolutionary methods (genetic algorithms and neural network models, e.g. Huse *et al.* 1999). Again, the use of these models has been highly biased, with birds being the most studied taxon with the widest variety of theoretical approaches used.

Methods for the identification of cues and decision rules are numerous and include (but are not restricted to) translocation/displacement experiments (e.g. Luschi *et al.* 2001), cross-breeding experiments (Helbig 1991) and a combination of theoretical and empirical approaches (confronting models with data), e.g. simulation models (e.g. Duriez *et al.* 2009) and proportional hazards models (e.g. Bauer *et al.* 2008).

We believe that much could be learned by overcoming taxonomic borders and integrating theoretical and empirical efforts—particularly in our rapidly changing world that challenges migratory animals with 'large-scale experiments'; this will give us important new insights and advance our understanding of migration.

PART 3

Migration in time and space

Female saiga antelope (*Saiga tatarica*) with calf.
Picture: Rory McCann (www.rmillustrations.com)

CHAPTER 7

Uncertainty and predictability: the niches of migrants and nomads

Niclas Jonzén, Endre Knudsen, Robert D. Holt, and
Bernt-Erik Sæther

Table of contents

7.1 Introduction

Environmental conditions vary across space, fluctuate over time, and ultimately define the area that can be inhabited by an organism at a given time. The ecological requirements of individuals may also fluctuate seasonally, for instance with specific requirements for successful reproduction. Movements in response to the environment can be seen as a general strategy for dealing with such variability, but can take on a variety of forms. For instance, birds may undertake local movements at scales up to a few kilometres over the course of a year to long-distance intercontinental migrations over distances exceeding 10^4 km. We will here contrast two patterns differing in the regularity of movements—*migration*, here defined as seasonally recurring and predictable movements of individuals along a geographic or environmental gradient, and *nomadism*, at seasonal time scales involving less regular movements in response to environmental fluctuations, and typically also characterized by between-year variability in the geographic location of reproductive events.

At a population level, migration and nomadism resemble, respectively, advective and diffusive processes, although both involve directed species- and state-dependent individual movement decisions, some of which are independent of the environment. Nevertheless, we can conceptualize the two patterns as being close to the end points of a continuum of advection–diffusion processes, with the diffusion coefficient being smaller (relative to the advection coefficient) for migrants than for nomads. Understanding the variability between species in movement decisions is therefore important, not only for understanding the diversity of movement patterns, but for predicting the patterns and consequences of population spread.

Understanding movement decisions requires insight into ecological niches and their dynamics. While acknowledging the role of predator avoidance and competition as important for the evolution and maintenance of many migratory systems (Fryxell and Sinclair 1988b), we will follow the Grinnellian/Whittakerian tradition of focusing on environmental requirements and resource variability. We argue that much of the diversity observed in seasonal movement patterns by migrants and nomads in many taxa can be understood and predicted by spatiotemporal variation in the resources needed for survival and reproduction, as modulated (and sometimes driven) by density-dependence within and across species. We do this by discussing spatial and temporal niche dimensions and their functional significance for animal movements, within a context of life history evolution, annual cycle organization and the proximate control of movement decisions. Many of our examples will be for birds, since this class of animals is particularly well-studied and illustrates well the diversity of movement patterns, many of which are intermediate between migration and nomadism, and many of which remain enigmatic after centuries of research interest.

7.2 Seasonality, environments and the evolution of migration and nomadism

The factors governing the evolution of migration are not completely understood (Chapter 2). Early work on the evolution of migration and partial migration, where only a fraction of a population migrates, emphasized the importance of environmental stochasticity and competition (e.g. Cohen 1967; Alerstam and Enckell 1979). Even though later studies have argued that environmental stochasticity is not necessary in order to generate partial migration (Kaitala *et al.* 1993) and instead emphasized the role of frequency- and density-dependent processes (Lundberg 1987; Kaitala *et al.* 1993; Taylor and Norris 2007), it is generally believed that migration is more likely to evolve in species that are dependent on seasonal, largely predictable, and heavily fluctuating food resources (e.g. Fryxell *et al.* 2004; Boyle and Conway 2007; Chapter 3). For the large-scale bird migration systems of the northern hemisphere, there has been much focus on whether seasonal migration originated in birds breeding in northern habitats (shifting wintering areas southwards due to climatic cooling) or southern habitats (taking advantage of emerging breeding sites further north). However, this debate has largely failed explicitly to incorporate the fossil record and fluctuations in climatic seasonality over geological time scales (Louchart 2008), and the repeated occurrence and loss of migratory behaviour within taxa at the level of modern bird families (e.g. Helbig 2003). For instance, sedentary oceanic species of duck such as the Hawaiian Duck (*Anas wyvilliana*) and Laysan Duck (*Anas laysanensis*) have presumably all derived from widespread migratory Palearctic waterfowl species. More surprisingly, migratory behaviour can re-emerge from sedentary ancestors. The 'threshold' model of expression of migratory behaviour in partial migrants (Pulido 2007) suggests that a latent tendency toward migratory behaviour can persist in avian gene pools, which opens the possibility of evolving towards obligate migration patterns if the environment changes or the species disperses into more seasonal habitat (Salewski and Bruderer 2007; Chapter 2).

Resource dynamics vary between environments, and the relationship between organisms and their resources is not consistent across taxa. An exciting hypothesis is that habitat or resource associations might present evolutionary precursors for long-distance migration. Levey and Stiles (1992) showed that Neotropical–Nearctic migration tended to develop within families of frugivorous or nectarivorous birds of edge, canopy or open habitats—i.e. performing altitudinal or local movements within the tropics—that are therefore predisposed to tracking variable resources across space and time. This was largely confirmed in a phylogenetic comparative analysis of Austral migration in the Neotropics (Chesser and Levey 1998).

The evolutionary origins of nomadism are even less known. Nomadism is characterized by a lack of the regularity in spatiotemporal movements that we normally ascribe to migration, and this has been linked to the characteristics and resource dynamics of ecosystems where it is commonly found. At any given site, nomadic species may appear in large numbers, then not be seen for many years. The typical environment where we find nomads is a low-productivity region where the resources are highly variable and unpredictable in both time and space (Davies 1984; Dean 2004). Deserts and semi-deserts are the most frequently cited examples of such environments. Productivity in these ecosystems is clearly affected by rainfall, but it is much less clear which components of abiotic and biotic environmental variability are most important for promoting a nomadic life-style (Wiens 1991; Dean 2004; Newton 2008). Also, not all species in these environments are nomadic and it has been debated to what extent nomadism is best predicted by diet or other factors (Allen and Saunders 2002; Woinarski 2006; Allen and Saunders 2006).

Nomadism is not restricted to desert areas but is also found in raptors and owls tracking rodent abundance in northern Europe, North America and along the Siberian tundra. Even though the climate is clearly seasonal at these latitudes, the resource environment is unpredictable. Many rodent populations display quasi-cycles but the synchrony levels off with distance, and there is spatiotemporal variation in rodent density that has favoured nomadic behaviour in rodent predators, rather than regular migration or a sedentary lifestyle. Another interesting dynamic resource is tree-fruit crops in boreal forests, whose densities vary greatly from year to year, with peak years practically always followed by years of poor production. These resources are exploited by seed- and/or fruit-eating birds often classified as irruptive migrants. When resources are abundant they can be resident but in some years they undertake long-distance movement far beyond their normal range. The minimum distance they have to move is set by the spatial correlation in resources (Koenig and Knops 1998). In terms of regularity, irruptive migrants fall somewhere between regular migrants and nomads, but are closer to the latter.

Since nomadism is not as clearly defined as migration, the classification of a species as nomadic or not could simply reflect the amount of knowledge we have on movement patterns. Frequently, a lack of observed regularity for a given species has resulted in movements being classified as nomadic. However, a closer look at the movements of eastern Australian landbird species, using a combination of large-scale survey databases, suggested a number of distinct movement patterns, many of which can be seen as a nomadic movements superimposed on more regular north–south movements (Griffioen and Clarke 2002). Furthermore, partial migration may in fact be quite widespread among Australian landbirds (Chan 2001), resulting in a mixture of migratory and non-migratory populations. Variability in migratory and nomadic movements between individuals in a population, and between years for an individual may further obscure the distinction between migration and nomadism (Mueller and Fagan 2008).

To sum up, neither regular migration nor nomadic movements seem to be strongly evolutionarily constrained, and movement strategies are variably expressed both across and within ecosystems. Many organisms show movement patterns that challenge any attempt to classify species as being strictly either migrants or nomads (Cheke and Tratalos 2007), and there exists a continuum of movement strategies that cannot be easily categorized. A major challenge, therefore, is to understand how variability in contemporary ecological niche characteristics can predict the diversity of large-scale movement patterns, given contrasting life histories.

7.3 Annual cycles and the control of movement decisions

From the perspective of life history evolution, movement is not an isolated phenomenon, but has evolved in concert with the other biological events in the annual cycle. Therefore, it is not surprising that we find differences between regular migrants and nomads not only in the mode of large-scale movement, but also in many other respects. For long-distance migration to work, individuals need adaptive linkage of endogenous rhythms to predictable environmental states (Newton 2008). Nomadism, on the other hand, can be seen as an adaptation to variability, and the lack of predictable changes in resource abundance demands more spatial and temporal flexibility in the organization of the annual cycle. Hence, movement decisions also need to be less rigidly controlled by the endogenous mechanisms ultimately controlling the annual cycle. This decoupling increases the responsiveness of nomads to environmental cues, both in terms of movement trajectories (Nathan *et al.* 2008) and the possibility of environmental input triggering transitions between life history stages such as movement and breeding phases (Dawson 2008). Hence, one can also expect larger variability between individuals and populations with respect to the life history stages that are expressed at any time. The less rigid endogenous control of nomads also implies that different states can more easily mature and/or wane simultaneously, and more easily be expressed at the same time (Wingfield 2008). Conversely, migrants may be more time-constrained due to prolonged migratory periods and the need to complete associated life history stages, such as pre-migratory fattening and moult in birds (Chapter 4).

It seems reasonable to argue that the reduced flexibility in annual cycle organization found in migrants compared with nomads might render them more sensitive to temporal resource dynamics, and thereby also more prone to cascading seasonal cycle effects. Understanding the balance between ultimate mechanisms controlling annual cycles and the proximate environmental control of movement decisions might therefore be important for understanding the evolution of large-scale movement patterns such as nomadism and migration.

This balance can sometimes be addressed in terms of trade-offs. All mobile organisms face the trade-off between staying or leaving (Andersson 1980; Dean *et al* 2009), and the difference between migrants and nomads is mainly in the extent to which annual cycles are influenced by environmental input. However, even for the 'evolutionarily fixed' annual cycle of many migratory species, movements within the non-breeding area can greatly affect survival. Resident species also face such a trade-off; they may for instance need to disperse in order to escape sudden inhospitable conditions. Post-reproductive and juvenile dispersal is a common feature of many animals, and frequently combines with migratory or nomadic movements, yet the two are rarely considered jointly (see, e.g., Winkler 2005). Nevertheless, interesting new perspectives are likely to arise from considering variability in movement strategies both in terms of factors promoting dispersal on the one hand and factors shaping the dispersal event on the other hand.

7.4 Habitat selection and niche dimensions

Migratory and nomadic movements can be considered in terms of habitat selection across space and time. Much focus has been on habitat preferences per se and the link with food resources, such as the need to move in order to track specific food resources or disperse in search of habitat of sufficient quality to ensure survival or reproduction. However, many species are generalists, so the link between habitat and food resources may be rather weak. Furthermore, food resources are only one of several habitat requirements, and movements will be constrained by landscape-specific patterns and individual requirements.

Even for species having a rather specialized diet during the reproductive phase, it is quite common to switch to a more generalist diet during migration. Most passerine birds breeding at northern and temperate latitudes are mainly insectivorous during the breeding season, but rather omnivorous for the rest of the annual cycle. Some species, such as migratory *Sylvia* warblers, switch to a diet consisting mainly of fruits prior to and during migration (Chapter 5). Interestingly, non-migratory *Sylvia* warblers are omnivorous or insectivorous year-round in the Mediterranean region (Jordano 1987). The utilization of fruits is morphologically constrained in these species, but it is also a rather predictable and stable food source in seasonal environments. Many nomadic or invasive species also rely on widely available food resources, such as seeds or fruit. On the other hand, fluctuating abundances of rodents seem to be the primary driver for the nomadic movements of many specialist predators such as boreal and Arctic owls and raptors. The diversity of alternative food sources may be important for determining movement strategies in these species; for instance, a north–south gradient from nomadism through partial migration to residency was found for European populations of Tengmalm's Owl (*Aegolius funereus*; Korpimäki 1986).

Hence, while fluctuating food abundance seems important for triggering movement in the first place, diet may or may not constrain the actual migratory or nomadic movements. There are certainly cases where species track food resources closely, such as birds following swarms of army ants in order to feed on flushed insect prey. But, typically, migratory movements and the exploratory or anticipatory movements of nomads force organisms into new kinds of habitat where food resources are likely to be different. An important issue is therefore to what extent animals prefer and search for certain habitat types when they are on the move. In other words, to what extent do migratory and nomadic animals occupy similar niches throughout the year? Much early work on large-scale bird migration systems hypothesized that species were adapted to certain habitat types and kept these habitat associations year-round, although increased competition with residents on wintering grounds might force migrants to shift or broaden their niche, e.g., by utilizing alternative habitats. Empirical results do not, however, clearly support such generalizations; although many species in a broad sense occupy similar niches throughout the year (Salewski and Jones 2006), microhabitat associations may be species-specific or even state-specific. As an example of the latter, American Redstarts (*Setophaga ruticilla*) breed in moist deciduous forests with abundant shrubs, but survival and breeding output is affected by the quality of the wintering habitat. Competition on Jamaican wintering grounds leads to habitat segregation, with dominant individuals (adult males) occupying high-quality mangrove and moist forest and less dominant individuals (females) ending up in drier

scrub habitats of lower quality (Studds and Marra 2005). In general the idea of niche-following versus niche-switching is an unexplored frontier in understanding the diversity of migratory systems (Nakazawa *et al.* 2004).

While there has been some interest in linking the diversity of movement types to variability in habitat characteristics, predictive success has been limited. For instance, while nomadic movements in birds are frequently linked to desert habitats, there is an interesting difference between deserts in the northern and southern hemispheres. In the former, migratory bird species are more common than year-round nomadic species, whereas the opposite pattern is found in the latter (Dean 2004). In Australia almost half of the desert bird species are nomadic and the general explanation is the lower and more erratic primary and secondary production in the Australian deserts compared with, for example, North American deserts, where nomadism is rare (Wiens 1991). Dean (2004) noted a general trend for nomadic species to occur in more open habitats, but in general it remains an open question what aspects of a habitat are important.

Our focus in the remainder of the chapter will therefore be on general niche dimensions common to the diverse habitats and regions where we find migrants and nomads. Habitat structure and resource variability in space and time are such dimensions. A central feature of both migratory and nomadic lifestyles is the movement of individuals across landscapes and the need for integrating information en route in order to determine movement paths, search out suitable locations for foraging or breeding and otherwise allocate time or resources (Chapter 6). Such integration is constrained by the movement modes of organisms; whereas many migratory species follow a comparatively fixed schedule of alternating movement and foraging/resting, nomads typically have a much more flexible schedule. Conversely, the distribution of habitats and resources across a landscape determines the feasibility of various movement modes. In other words, we may need to bring in elements of landscape ecology.

An explicit consideration of landscape heterogeneity brings in complexity, but also interesting new perspectives. Some overarching notions of landscape ecology are always useful to keep in mind, such as the scale dependency of ecological patterns and processes, and the distinction between the scale of observation and biologically relevant scales. It can for instance be helpful to recognize that the definition of migration is actually scale dependent; even a perfectly regular migrant with strong site fidelity might show random movement patterns at short time scales, and some migratory birds, such as the White Stork (*Ciconia ciconia*) and swifts (Aves:Apodidae), may be largely nomadic on the wintering grounds. More generally, nomadic behaviour is constrained by seasonal fluctuations in home ranges, and these vary across species as well as habitats and food resources. During migration, scale dependent patterns of habitat preferences may be evident. For instance, Deppe and Rotenberry (2008) found species-specific preferences for broad vegetation types at a Yucatan stopover site, as well as a response to vegetation structure within these vegetation types.

Other general issues of relevance for animals searching for resources are the hierarchical nature of patchiness and the tendency for characteristic temporal scales of biotic variability to increase with their spatial scales. Hence, free-ranging predators may search for high-density prey patches in a hierarchical manner—a classical example being pelagic seabirds seeking out highly productive areas such as frontal zones, where large concentrations of forage fish can be found, and within this area using a range of cues, such as the presence of other seabirds, for locating individual fish schools within which they hunt for individual fish. Variability in abiotic structuring variables such as temperature often show such 'reddened' spectra (variability increasing with increasing observation scale). On the other hand, when considering their functional significance, spatial and temporal scales may sometimes need to be decoupled. Whereas temperatures typically vary smoothly (and hence somewhat predictably) both over time and space, rainfall is rather unpredictable in arid regions of the world, but may induce spatially predictable patterns in resource abundance. A high spatial correlation may for instance occur if a sudden rainfall moves quickly over a large area, thereby synchronizing the growth conditions. The synchrony may soon break down

depending on local characteristics. Nomads that are moving around in the landscape may either decrease or reinforce the patchiness of resources depending on how easy it is for them to spot the good patches.

Finally, we note that the functional significance of niche dimensions should be considered at the appropriate level of response; while one or a few food resources may be integrating the environment at scales appropriate for understanding the movements of a single species, landscape patterns may in fact be more successful in integrating the environment at scales relevant for explaining community composition in space and time (Pavey and Nano 2009). As a first approximation, we will, however, consider resource dynamics at the level most proximate to animal movement. In the following two sections we will discuss the functional significance of temporal dynamics and spatial heterogeneity of resources for migratory and nomadic animals.

7.5   Temporal dynamics of resources

7.5.1   Seasonality

Most resources are dynamic, with local resource abundance fluctuating over time. These fluctuations occur across different time scales and the magnitude of the fluctuations may vary with time scale. For many organisms, the amount of food available is rather predictable from one day to the next. For others, there can be enormous variation and little predictability even at very small time scales. Such lack of predictability can be due to infrequent pulses of food resources. For instance, the Giant Red Velvet Mite (*Dinothrombium pandorae*) in the Mojave Desert of North America lives in burrows underground without feeding, sometimes for years on end (Tevis and Newell 1962). After very unpredictable heavy rains, swarms of flying ants and termites appear, and the mites emerge, to hurriedly scurry over the desert floor eating their fill during a few hours of frantic activity. During the brief resource pulse the mites mate; once the pulse wanes, they return to their burrow, to wait again for the next chance at feeding and reproduction. In the study cited above, only ten such emergences occurred in a 4-year period.

In many cases, large-scale movements can be found when the resources in a landscape vary on such a long time scale that animals that cannot simply hunker down locally (like the Giant Red Velvet Mite) need to travel far beyond their home ranges to survive and/or reproduce. Within a year there may be seasonal variation in resource abundance driven by periodic climate forcing, a phenomenon often exemplified by the arctic, boreal and temperate areas of the Northern Hemisphere, where latitude is assumed to reflect differences between summer and winter productivity (Dingle *et al*. 2000). Almost all of the variance in the proportion of migratory birds is explained by latitude in both North America and Europe (Newton and Dale 1996a,b) and a high proportion (78%) is explained in butterflies along the Australian east coast (Dingle *et al*. 2000). The seasonality in temperate areas is mainly driven by changes in temperature (Fig. 7.1(a),(c)), whereas in Australia, being the driest continent, seasonality (if present) is often, but not exclusively, due to variation in precipitation. In many tropical regions, seasonality is pronounced in precipitation, leading to regular altitudinal and latitudinal migrations, e.g. in frugivorous birds in the Neotropics, and ungulates in the savannas of east Africa (Holdo *et al*. 2009b). The

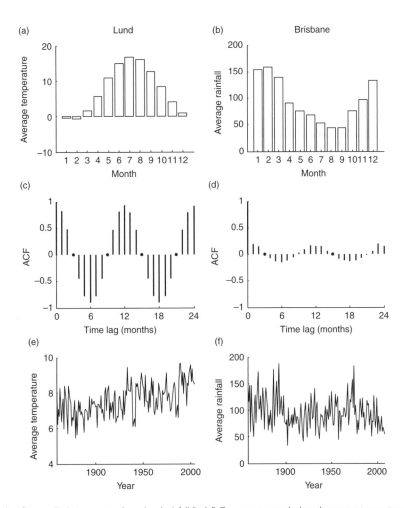

Figure 7.1 Examples of seasonality in temperature (a, c, e) and rainfall (b, d, f). The upper top panels show the average temperature for each month (1 = January, 2 = February…) in Lund (55°43′N, 13°09′E) and the average rainfall in Brisbane (27°29′S, 153°02′E). The mid panels show the autocorrelation function (ACF) and in the bottom panels the time series of average annual temperature and rainfall are plotted. The data from Lund are from 1859 to 2004 (SMHI) and the Brisbane data were collected at the Brisbane Regional Office during 1859–2007 (Rainman software; Clewett *et al*. 2003).

seasonal pattern of rainfall may be obvious in a given year but due to inter-annual variation it is less predictable than are the seasonal shifts in temperature in temperate Europe (Fig. 7.1). The striking seasonality in the temperate areas of the Northern Hemisphere is very predictable, in the sense that winter is always followed by spring. The regularity of such seasonal variation is most probably a necessity for the successful adjustment of morphology, physiology and behaviour needed for an individual to implement a long-distance migratory annual life cycle. However, the exact timing of events, such as the last frost night, bud burst, etc., varies between years, so many details of migration can vary due to behavioural and other modes of plasticity.

In all oceans, vast basin-wide fronts can separate low-productivity sub-tropical gyres and high-productivity temperate gyres (Polovina *et al.* 2001). These fronts can provide transient bands of high resources for mobile predators such as Albacore Tuna (*Thunnus alalunga*) and Loggerhead Turtle (*Caretta caretta*). The positions of these fronts can vary seasonally by up to 1000 km, and also vary more erratically between years due to factors such as El Nino events. Movements of these consumers are thus comparable to nomadic movements in terrestrial environments. Other marine taxa have very regular seasonal migrations; e.g., grey whales (*Eschrichtius robustus*) in the eastern Pacific and high-latitude dolphins (Mammalia:Delphinidae). *En passant*, we note that the term 'migration' is sometimes used by aquatic ecologists to refer to diel movements of individuals (e.g., zooplankton in the water column), or to historical dispersal events in biogeography. Although these certainly can represent long-distance movements, their relevance is mainly at either very short or very long time scales, and are different (particularly the latter) from our focus here.

7.5.2   Stochasticity

Environmental stochasticity is random variation that affects the whole or parts of the population in a similar way (Lande *et al.* 2003), and adds a stochastic element to the regularity of seasonal environments (Fig. 7.1(e),(f)). On an inter-annual time scale, resource dynamics are stochastic, for example in the sense that the onset and length of the growing season varies unpredictably between years, as does

productivity and the total amount of resources available. An example of stochastic seasonality is given by the occurrence of insect larvae, vital food resources for many bird nestlings (e.g. Visser *et al.* 2006). There is strong seasonality based on the annually recurrent period every spring, set by the phenology of plants, when resource availability for growth and reproduction is optimal for herbivorous insects. However, since plant phenology varies from year to year depending on environmental conditions, the timing of the optimal period for the herbivore also varies annually (e.g. Asch and Visser 2007). Hence, individual insect predators trying to optimize the timing of migration returning from the wintering grounds have to make this decision in the face of uncertainty. More generally, individuals may have to rely on environmental cues to trigger the transition from one life history stage to another, as well as to adjust to local conditions (Ramenofsky and Wingfield 2007).

In contrast to the strongly seasonal environments of temperate areas, many arid regions are characterized by unpredictable and highly variable rainfall with low or no correlation between monthly and even daily rainfall (Box 7.1, Figs 7.2–7.4). All organisms inhabiting this environment have to adapt to the tension between drought and flood, boom and bust. In stochastic and unpredictable environments anticipatory movement is not feasible and organisms benefit from any adaptations that make it possible to respond rapidly to changing conditions, and for mobile organisms, to glean information about long-distance rainfall events (Shine and Brown 2008). Some taxa may be able to detect spatially patchy pulses of production over long distances, for instance by sensing the position of rain clouds. One advantage of group living is that if a few individuals are able to pick up appropriate cues, other individuals can exploit this information. Holdo *et al.* (2009b) have argued that a combination of both effects may underlie observed population fluctuations of migratory wildebeest (*Connochaetes taurinus*) in the Serengeti.

Since a lack of temporal (and spatial, see below) predictability of resources is unlikely to favour the evolution of migration (see Chapter 2) many organisms have instead adopted a nomadic life style in unpredictable and fluctuating environments

Box 7.1. Exploring resource dynamics: Australian rainfall and bird movements

In order to understand the evolution and maintenance of large-scale movement patterns it can be helpful to explore the spatiotemporal dynamics of the resources making up the niche of a species. In particular, it is important to understand how predictability, variability and other statistical properties vary across space and time. Long time series of data on resource dynamics are rarely available and it is often difficult to find good enough surrogates (see Boyle and Conway 2007), particularly at the large geographic scales required for comparative analyses of movement patterns. However, in arid regions rainfall is a driver of plant growth, which also affects insect density and water availability in general. Therefore we use rainfall as a proxy in an exploratory exposition of temporal dynamics and spatial variability of resources important for understanding the diversity of movement patterns in Australian landbirds.

Australia is the second driest continent on Earth, only surpassed by Antarctica. Most of the interior is extremely arid, with more or less seasonal rainfall along the coasts, most notably along the easternmost (east of the Great Dividing Range) and northernmost parts of the continent (Fig. 7.2). Such seasonal predictability is reflected in positive (although weak) autocorrelation at time scales of a few months and a characteristic sign-switching pattern in the autocorrelation function (Fig. 7.1(d)).

Alternatively, the issue can be explored in terms of variance spectra, quantifying how the variance of the time series varies with observation scale. Wavelet analysis (e.g., Percival and Walden 2000) has become a popular tool for such analyses, not least since it allows a time series to be expressed as a function of both time and scale (Fig. 7.3). The seasonality in rainfall along the eastern and northern coasts of Australia can be seen as

(a)

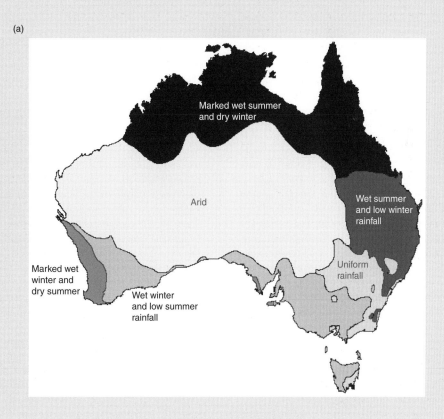

continues

Box 7.1 (*continued*)

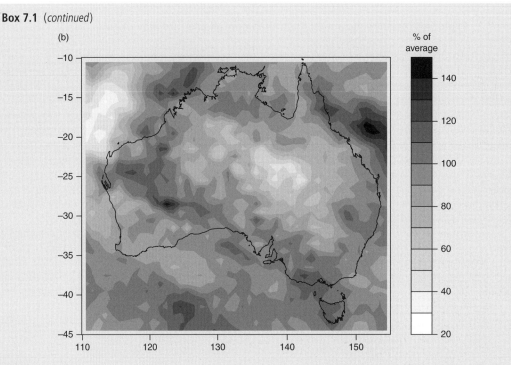

Figure 7.2   (a) Climatic zones of Australia, defined by seasonal rainfall patterns (redrawn from a map from the Bureau of Meteorology, Australian Government; http://www.bom.gov.au). (b) Example of spatial heterogeneity in rainfall that is assumed to produce spatial variability in vegetation growth and resource abundance: showing rainfall in February 2004–January 2005, relative to the yearly average (data as in Fig. 7.4), by latitude and longitude.

a peak in the wavelet variance spectra at approximately semi-annual time scales, whereas rainfall in the arid interior appears completely random and unpredictable (Fig. 7.3(a)). A lack of predictability does not, however, mean that there is no structure (Fig. 7.3(b)); clusters of rainfall events are still evident at observation scales ranging from weeks to months. While periods of high rainfall are most clearly distinguishable at temporal scales of months in the more seasonal environments, temporal contrast in conditions in the arid environment is due to sudden rainfall events (Fig. 7.3(c)). Hence, no overall 'reward' is expected for a periodic life cycle temporally matching the rainfall pattern, which contrasts with the potential costs of temporal mismatch in seasonal environments.

Movement patterns of birds within the Australian continent are not well known, but a combined analysis of large-scale survey databases suggested migratory or nomadic movements for up to 36% of 407 eastern Australian landbird species, and a number of distinct movement patterns (Griffioen and Clarke 2002). Close to

40% of these species were classified as performing local or unclear movements. More regular large-scale movement patterns mainly occurred within or between regions of seasonal and relatively abundant rainfall—i.e. between Tasmania and southeast Australia, and along the sub-tropical eastern coast and towards the seasonal northern coastal and inland areas (Fig. 7.2). Only 20% of the non-sedentary species showed movement toward the drier and less seasonal inland areas. In western Australia, migration is observed in a number of species breeding in the south-western parts of the continent, where rainfall is seasonal. Recent changes in the timing of migration appear to be associated with changes in precipitation rather than with changes in temperature (Chambers 2008).

It is interesting to note that the infrequent rainfalls of interior Australia not only contribute to high temporal variability (Fig. 7.4(b)). Apart from some higher-altitude areas, the temporal predictability in rainfall is also lower in the interior than along the east coast and in northern Australia, both at a day-to-day scale (Fig. 7.4(c))

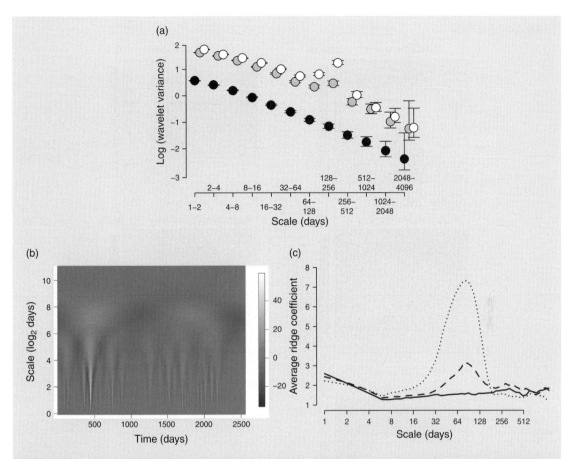

Figure 7.3 Time-scale decomposition of Australian temporal rainfall patterns using wavelet analysis. Data are daily precipitation values from a desert site (Marree; 27°29'S, 138°03'E; 1885–1994; black circles/solid line), a humid sub-tropical site (Brisbane; 27°29'S, 153°02'E; 1887–1994; grey circles/dashed line) and a tropical wet-dry site (Darwin; 12°25'S, 130°52'E; 1872–1996; open circles/dotted line). (a) Estimated wavelet variance based on the maximum overlap discrete wavelet transform, along with 95% C.I. The graphs show how the variance of the time series is partitioned into intervals of observation scale. At all sites, short-term variability dominates, but whereas the linear decrease with increasing scale (on a log–log scale) observed for the desert site indicates a pattern of white noise, the peaks in wavelet variance at semi-annual time scales show a superimposed pattern of seasonality, being stronger for the tropical compared with the sub-tropical site. (b) Time-scale representation of wavelet coefficients from a continuous wavelet transform ('Mexican hat' wavelet) of data from Marree, 1988–94. A lighter colour indicates larger coefficients, i.e. more precipitation. Note how the representation of the time series changes with observation scale; at small scales the detailed day-to-day variability is evident, whereas at large scales we only see coarse fluctuations. (c) Averages of wavelet coefficients along the maximum ridges (lightly coloured in Fig. 7.3(b)), as a function of observation scale. Since the time series here were standardized to zero mean and unit variance before applying the continuous wavelet transform, the three lines indicate how sites differ with respect to the scale dependence of the temporal contrast in rainfall (i.e. the difference between high rainfall events and 'average' conditions).

relevant to en route movement decisions, and at a time scale of weeks (Fig. 7.4(d)), relevant to the timing of life history events such as breeding, migration, staging and wintering. A general north–south gradient in predictability can also be seen for much of the continent, corresponding to the overall north–south axis of movement observed for many species. On the other hand, there is overall weak positive spatial autocorrelation in rainfall over much of the interior, southern and western parts of the continent (Fig. 7.4(e)). This is in contrast to the situation along the east coast (no overall autocorrelation) and in the north (autocorrelation due to

continues

Box 7.1 (*continued*)

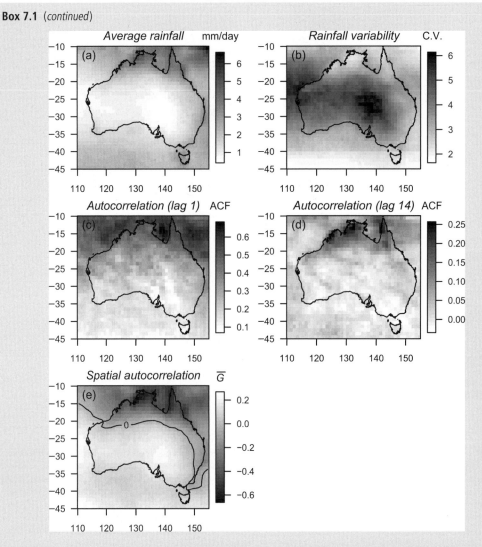

Figure 7.4 Some simple attributes of continental-wide rainfall patterns, potentially useful for understanding niche dynamics, shown by latitude and longitude. (a) Average rainfall, (b) overall temporal variability (coefficient of variation), (c–d) temporal predictability (autocorrelation function at lags 1 and 14 days), (e) time-averaged spatial predictability (local spatial autocorrelation) as estimated by the local G statistic for local rainfall anomalies (deviation from the temporal mean, calculated separately for each cell). The magnitude of the G statistic determines the strength of autocorrelation, and the sign indicates whether autocorrelation is mainly due to spatial clustering of low precipitation values (negative G) or high values (positive G). The analysis was based on GPCP 1-degree gridded daily precipitation data for the period 1997–2007 (http://www.gewex.org/gpcp.html).

spatial clustering of droughts), and could facilitate local movements and nomadic lifestyles. Most of interior Australia is grassland and bushland, and rainfall from moving weather systems could synchronize vegetation productivity at large spatial scales, thereby increasing landscape connectivity and enhancing the detectability and use of high-productive patches for wide-ranging species.

(Newton 2008). A classic example of nomadism is the life history of the Banded Stilt (*Cladorhynchus leucocephalus*), a wader bird that survives along the Australian coast but only breeds when the drought breaks and the rain pours down in the inland desert lakes thousands of kilometres away from the coast (Robinson and Minton 1989). The rain transforms the dry and hostile salt pan lakes into a haven for the banded stilts and large numbers of brine shrimps (*Parartemia* spp.) are produced. The high concentration of brine shrimps provides the resources needed for their energetically demanding breeding, similar to the insect peak crucial for successful reproduction of insectivorous birds in temperate areas (Visser *et al.* 2006). In Australia, nectarivorous birds are also known to undertake regional-scale movements in response to variation in the distribution of flowering plants (e.g. Keast 1968). It has been suggested that, in general, nectar (and probably fruit) is a far more dynamic resource than terrestrial invertebrates and hence demands a more mobile lifestyle for its consumers (Woinarsky 2006). An important challenge is to understand how organisms glean appropriate cues from the environment about the spatial position of patchy and fleeting resources.

7.6 Spatial heterogeneity of resources

Resources are not evenly distributed in space and spatial heterogeneity at the landscape level makes it possible for individuals to compensate for temporal variability by adaptive movement towards transiently rich patches (e.g. Fryxell *et al.* 2005). It is important to make a distinction between spatial variance, i.e. the frequency distribution of resource density, and spatial pattern per se, i.e. the spatial distribution of resources. These are different aspects of a heterogeneous environment, and their temporal dynamics may differ (Klaassen and Nolet 2008). If there is high spatial variance there exists a potential for adaptive movement to take advantage of local food abundance (e.g. Fryxell *et al.* 2004, 2005). The spatial pattern, on the other hand, affects landscape connectivity and the spatial scale of resource heterogeneity, and therefore the possibility to exploit the resources in the first place. If patches are too small or too hard to reach, they may be unusable by a forager of a given body size.

Similar to the temporal dimension of resource dynamics, there are fundamental differences between the spatial properties of environments where we typically find migrants and nomads. In the temperate areas where there is strong seasonal variation in temperature there may also be a strong spatial correlation in some environmental conditions. For instance, a snapshot of the spring temperatures in Europe would give a gradual increase from north to south. As time goes by a wave-like pattern of increasing temperature moves from south to north, which allows, for example, migratory butterflies and birds to track the arrival of spring conditions, like a surfer riding the crest of a wave.

The situation is fundamentally different in, for example, the wet–dry tropics and the arid zone of inland Australia where day-temperatures are less variable but rainfall fluctuates, often greatly (Fig. 7.2(a)). Rainfall is often patchy and one may find areas experiencing drought close to areas having normal conditions (Fig. 7.2(b)). For example, clear spatial patterns in monthly rainfall means only become evident in rainfall data for the Wet Tropics region of north-eastern Australia when very long time scales are considered (Hancock and Hutchinson 2006). In fact, weather stations separated by only 1 km can record very different rainfall patterns (Shine and Brown 2008), while the spatial correlation of annual rainfall may be rather high on a larger scale.

Both undirected movements in response to local conditions and directed movements, possibly in response to previous prospecting or long-range correlations in weather variables, may be of importance in explaining the sometimes remarkable ability of birds to arrive at distant sites close to the timing of rainfall events (e.g., Roshier *et al.* 2008). Organisms can respond to spatial structure across a range of scales, but the question of which spatial scales are relevant when evaluating the abiotic environment in which strategies such as migration and nomadism evolve, is likely to depend on landscape-level constraints as perceived by a given species. For instance, species differ in habitat requirements as well as in perceptual and movement capabilities. In general, there is a smallest spatial scale ('grain') of environmental heterogeneity to which an organism is sensitive, and movement capability, sedentary behaviour or home range behaviour such as territoriality or homing may limit the response to landscape structure at large scales, hence determining a spatial extent of sensitivity. Food or other habitat requirements may be structured by the environment in a way allowing the identification of characteristic scales of resource variability (e.g. Bradshaw and Spies 1992). Such variability and pattern can be translated both into distribution patterns and movement attributes of moving organisms.

Positive spatial correlation in temporally fluctuating resources not only adds an element of predictability for animals making movement decisions, but also affects the area needed for survival of species tracking resources more locally. In the Serengeti National Park Thomson's gazelles (*Gazella thomsoni thomsoni*) have been shown to make adaptive movements on a landscape scale to track rain-driven and spatiotemporally fluctuating resources (Fryxell *et al*. 2004). Modelling work for this system has suggested that the gazelles require unrestricted access to rather large areas to guarantee long-term population persistence, and the area needed increases when resources are positively spatially correlated (Fryxell *et al*. 2005). Hence when resource abundance is low and positively correlated across vast areas, individuals are forced to move long distances, especially if there is temporal autocorrelation as well, for example, in seasonal environments. But when should an individual leave an area, where should it go, and what route should be followed? What are the sources of information available in the environment?

7.7 Spatiotemporal dynamics: information and uncertainty

Both temporal and spatial dimensions are needed to understand the selective pressures for movement. Mueller and Fagan (2008) suggest that resource gradients can vary across four axes: (i) resource abundance, (ii) spatial configuration, (iii) temporal variability of resource locations, and (iv) temporal predictability of resources. These axes are not orthogonal; for example, temporal variability is not independent of abundance, and temporal predictability depends on both temporal and

spatial variability as well as spatial configuration of resource patches. Hence, temporal and spatial attributes of the resource landscape are linked. For instance, high spatial predictability is often accompanied by high temporal predictability. The linkage may be strong, as in the case of temperature gradients experienced by migratory birds at intermediate latitudes during spring migration, or weak, as in the case of patchy and unpredictable rainfall in arid areas. In the former case, spatial and temporal attributes may yield much of the same information to an individual moving in response to a resource gradient, while in the latter case, spatial and temporal attributes to a larger extent yield independent information, and the exact nature of their interaction becomes important. Also, spatial and temporal information can be weighted differently—for instance, temporal dynamics are likely to be more important for sedentary species and decisions regarding the timing of breeding, whereas spatial variability could be regarded as more important for movement decisions in free-ranging organisms.

As a simplification, we will in the following paragraphs consider how individuals and their movement strategies are challenged by the temporal predictability of the conditions at the current site, conditions en route, and the conditions at the destination. The environmental conditions and their predictability will affect when to move, what route to follow, and when to stop. This analysis takes a deeper look at the predictability components of the decisions that are more generally reviewed in Chapter 6, and also considers the decisions made by nomads.

7.7.1 When to move?

The timing of movement is an important aspect of a survival strategy in spatiotemporally fluctuating environments. The role of resource dynamics in initiating movement has been studied in detail from a foraging perspective (e.g., Stephens and Krebs 1986), but the importance of resource abundance, its variability, and its predictability for the initiation of migration have received much less attention. It has been suggested that an important component is pre-emption, i.e. to leave an area before habitat quality has declined too much (Dingle and Drake

2007). Pre-emption is typical of seasonal breeders inhabiting and adapted to strongly seasonal and predictable environments; such species are often obligate migrants. These organisms are adapted to escape in time before it gets rough and/or to arrive in time elsewhere when conditions there are benign. For such species the timing of migration is often 'hard-wired', under endogenous control. This contrasts with facultative migrants that respond more directly to prevailing conditions rather than showing anticipatory movements (Newton 2008). In this group, we find aseasonal breeders that often have a nomadic life style.

For migratory birds, it has been argued that pre-emption cannot rely on proximate cues, and that selection ought to act on the migrant's responses to endogenous rhythms or cues that reliably predict habitat deterioration (Newton 2008). An example of the latter is photoperiod, but one could also think of abiotic factors such as rainfall. However, in practice rainfall is likely to vary too much from year to year to be reliable as a cue for anticipatory movements. Nevertheless, environmental variation is likely to modulate the timing and speed of migration. For instance, the earlier spring arrival of migratory birds at their breeding grounds is often interpreted as a response to climate changes (Jonzén et al. 2007b).

Environmental processes can either affect the migration process through their effects on food abundance and hence the scheduling of movement and stopover, or they may act primarily as cues. In a thought-provoking paper, Saino and Ambrosini (2008) show that meteorological conditions in Europe during the breeding seasons of many birds covary with those in the winter quarters south of the Sahara desert. They further suggested that migrants may therefore be able to predict meteorological conditions at the beginning of the breeding season and adjust migration schedules accordingly. Saino and Ambrosini do however note that the ability to use this climatic information as a cue depends on the balance between the fitness benefits it provides and the potentially dramatic costs experienced by early arriving individuals in the years when the cue fails to provide reliable information. In more general terms we can ask the following:

7.7.2   When should information be used?

A key aspect to consider is the reliability of information, i.e. the probability that the information is correct. Whether or not a given cue should be used when the true state of nature is uncertain is a decision theoretical problem and the best tactic is to make a decision that maximizes the expected payoff. When individuals do not control the reliability of information they should be more discriminating as the benefit of correct information declines, as the cost of misinformation increases and when acquisition costs are greater (Koops 2004). In the face of uncertainty, collective decision-making strategies regarding the timing and directionality of movement can be adopted, because taking the average of individual strategies within a group or aggregation may lead to lower error than that inherent in each individual's strategy (Simons 2004; Hancock and Milner-Gulland 2006).

For nomads (and nomadic movements of migrants outside the breeding season), the timing of movements is much more flexible and likely to be more directly related to deterioration of resources for breeding or foraging. Movements can either reflect local resource tracking or simply a choice to disperse. In either case, there are trade-offs between staying with or leaving a resource (Dean et al. 2009), so in the typical environment of unpredictable and scarce resource pulses, movements are likely to be strongly influenced by local resource dynamics and patch-leaving criteria, combined with the gleaning of information in order to detect resources from a distance.

7.7.3   What route to follow?

By having an inherent time–distance–direction programme synchronized with photoperiod and combined with external cues, for example magnetic conditions, migrants can, in principle, successfully travel across the globe even in their first year (Chapter 6). This is an example of memory-based movement *sensu* Mueller and Fagan (2008), who suggest that all active animal movements can be assigned to one of three different classes; non-oriented mechanisms, oriented mechanisms, and memory-based mechanisms. In their definition, memory mechanisms include genetic inheritance, individual experience, and information from communication between conspecifics. Non-oriented movement refers to movement where sensory stimuli at the current location affect the movement parameters, whereas oriented movement relies on perceptual cues in areas other than the current position. In practice, long-distance migrants may use a combination of mechanisms, including compasses that may be cross-calibrated or integrated for direction finding (Åkesson and Hedenström 2007). There are clearly some systems that rely heavily upon endogenous control, such as the famous migration of the Monarch butterfly (*Danaus plexippus*), where there is reproduction and mortality en route, so that several generations elapse within each annual migratory cycle.

It is, however, important to be aware that strong endogenous control of migration does not mean a lack of variability between individuals, seasons and years. Even in regular long-distance migrants such as Eleonora's Falcon (*Falco eleonorae*), individuals of the same population can follow different routes in a given year (Gschweng *et al.* 2008). Environmental factors such as wind assistance vary between years, and global wind patterns are generally regarded as an important driver of the loop migrations seen in many seabirds and passerine birds migrating between North and South America (Newton 2008; Chapter 4). We need to study the statistical properties of the resource environment to appreciate fully the processes shaping migration patterns and strategies. For instance, inexperienced individuals may have to include external spatial cues to make the preparation for crossing ecological barriers, as shown for the Garden Warbler (*Sylvia borin*) before crossing the Sahara desert (Fransson *et al.* 2008). The statistical properties of the environment may also be influential on the outcome of migration since spatial correlation of environmental conditions en route allows migrants to fine-tune their migration schedule. Hence, it also becomes important to understand how migration is constrained by movement modes, time constraints and their interactions within the annual cycle. For example, climate change effects on the timing of migration for birds are generally more consistent for the more time-constrained spring migration (when there is strong selection to arrive in time), compared with the more 'relaxed' autumn migration (Lehikoinen *et al.* 2004). Theoretical considerations suggest that ways of integrating environmental change when adjusting to climate change may be rather different for species migrating quickly and with few stopovers, compared with species progressing more slowly and gradually across the flyway (Hedenström *et al.* 2007).

Compared with the overwhelming literature on migration ecology, less is known about the life of nomads (but see Dean 2004). Little is known about the detailed movements of nomadic birds and their proximate control, since they typically inhabit the most sparsely populated regions of the world. Owing to the lack of predictable changes in resource abundance, fixed orientation mechanisms are not useful, and the organisms may suffer from low information availability. The ability to track resources varies across species; highly mobile organisms (e.g., swifts; Aves:Apodidae) can potentially track ephemeral local resource pulses by sensing rainfall events from great distances. However, all nomads can improve their chances of finding resources by adapting their movement to the statistical properties of their environment (Sims *et al.* 2008). Given the energetic costs of large-scale searching, behavioural features favouring an intermittent locomotion that promotes efficient search patterns may represent a real adaptive advantage (Bartumeus and Levin 2008). To our knowledge it is not known if, for example, nomadic desert birds are using 'Lévy-like' movements (where movement lengths are drawn from a distribution characterized by many short movements and fewer longer movements) similar to those found in foraging movements of many organisms. However, the strength of the empirical evidence for biological Lévy flights has been questioned (e.g. Edwards *et al.* 2007).

7.7.4   When to stop?

Birds often switch from an obligate mode to a facultative mode and start responding to local conditions by the end of the migration (Helms 1963). This makes perfect sense, because as the migrants approach benign areas by the end of the trip they should start searching for suitable habitat. Alternatively, conditions at high-latitude or high-altitude breeding grounds may still not be suitable, and birds may need to rely on nearby or distant staging sites for survival or build-up of energy resources for breeding (Chapter 5). It has been speculated that migrants have an endogenous response to the expected external cues that would indicate that they have arrived in the winter quarters (Newton 2008). Such a mechanism seems to trigger fat deposition and directional changes, as suggested by results from experiments on Pied Flycatchers (*Ficedula hypoleuca*) held in captivity (Wiltschko and Wiltschko 2003).

The weaker endogenous control of nomadic movements enables flexibility in breeding locations and presumably a quicker transition from movement to breeding phases. Still, there may be a need to fine-tune both movements and breeding locations. This is seen in the Red-Billed Quelea (*Quelea quelea*), being intermediate between migratory and nomadic, since it deposits fats and performs directional long-range movements, but shows great variability in breeding locations. Rainfall drives their movement patterns, but since these granivorous birds might starve if they wait until rainfall germinates the seeds, they move ahead of the rains, and when the rain comes they fly back to areas where the previous rainfall has already resulted in new seeds (Cheke and Tratalos 2007). In this way they can manage three yearly breeding events.

Finally, it should be noted that environmental information is not always used. Cultural imprinting may sometimes be strong and result in surprising migration patterns. For instance, a moose (*Alces alces*) population in central Norway migrates from low-lying summer areas through large favourable areas, to end up in poor winter areas situated at higher altitudes (Andersen 1991). According to archaeological records this is a traditional migration route that still persists even though habitat quality has deteriorated.

7.8   Intrinsic control of mass movements

We have in this chapter discussed resource dynamics as an extrinsic driver of movement patterns, but the role of intrinsic factors for modifying and driving movement patterns should also be mentioned.

Some of these may themselves be sufficient for explaining dramatic mass movement patterns resembling those of migrants and nomads. Density-dependent feedback may lead to self-reinforcing spatial flows; for example, travelling waves have been suggested for a range of phenomena, such as the propagation of feeding fronts of sea urchins (Abraham 2007).

A clear class of such intrinsically generated movements can be found in insects in arid and semi-arid environments. For instance, mass migration has arisen independently multiple times in grasshoppers (Orthoptera:Acrididae). There are typically dramatic shifts between two phenotypes in locusts, one that is solitary, and the other that is gregarious. The transition between these phases is believed to be driven by local interactions, such as the 'dilution effect' of grouping behaviour on mortality from predators. Once local groups are sufficiently abundant, they can engage in intense exploitation and interference competition, which leads to movement over large distances. This modality of nomadism is thus at least in part driven by the density-dependent impacts of locusts on their own environment (Simpson and Sword 2008). Climatic variation may trigger the migration, but it develops its own momentum due to the locusts being able to escape limitation by predation, overexploit local resources, and in the course of so doing, gather enough resources to keep moving as a band of destruction across vast landscapes. Abstractly, this example seems to involve an Allee effect (Courchamp *et al.* 2008), in that locally, locusts can escape predation, i.e. reduce mortality, by aggregation. It is not clear if this is a necessary ingredient of such intrinsically driven migration syndromes.

7.9   Conclusions and perspective

Whereas migration can be seen as an adaptation to avoiding severe conditions and exploiting predictable spatiotemporal variation in resource abundance, nomadism can be seen as an adaptation to variability rather than severity. These two strategies can be seen as the end points of a continuum of movement strategies, and in a given species regular migration and nomadism may occur in different seasons or in different areas of the range. In order to understand this diversity, we may need to advance our understanding of niche dimensions for migrants and nomads beyond a focus on seasonality, habitat and resource associations per se. Landscape attributes and statistical properties of the resource environment are general features that provide a key to understanding and predicting the relative merits of large-scale movement strategies in different environments. These may need to be considered in a context of life history evolution; a pattern of obligate, regular migration is likely to be largely controlled by endogenous processes and characterized by rigidity, whereas nomadism is more under external control and can be seen as flexible rather than rigid (Newton 2008).

Explicitly linking the statistical properties of movement patterns to specific internal traits and/or behaviours has been identified as a major challenge in movement research (Nathan *et al.* 2008). In this chapter we have focused on the statistical properties of resource dynamics motivating movement in the first place, and how they may link to individual movement decisions such as when, how and where to move—i.e., important components of the emerging paradigm of movement ecology (Nathan *et al.* 2008). Spatial and temporal variability and predictability of resources are poorly explored dimensions of the niches for migrants and nomads, and empirical analyses of space–time data of resource abundance (or any index thereof; Box 7.1) may be helpful for linking resource dynamics to the spatial distribution of movement strategies. Simulation studies based on mechanistic movement models coupled with dynamic resource maps would be an alternative approach (see Chapter 8 for a discussion of these alternatives). Climate change offers an opportunity to study variation and adaptation of migration patterns in time and space, including the timing of migration (e.g. Jonzén *et al.* 2007a). A key challenge is to move from purely theoretical, exploratory or correlative studies to making quantitative predictions that can be tested using available data. This may require further efforts to identify the functional importance of various components of resource variability.

Habitat and resource requirements do, however, vary over the course of annual and life cycles, and cannot always be considered separately from an animal's life history stage. Phenotypic differentiation between life history stages is a central issue

here; when annual environmental variability is low, few and relatively undifferentiated stages are needed, whereas high variability may favour more and differentiated life history stages, including switching between mobile and sedentary phases (Wingfield 2008). On the other hand, migration or nomadism may permit species to track conditions to which they are already well-adapted, hence weakening evolutionary pressures to shift, for instance in tolerance of climatic extremes or in the utilization of alternative resources. Also, switching between multiple movement modes such as free-ranging and home-ranging behaviour, searching and resource tracking, may in fact be a general feature of animal movement not requiring special adaptations (e.g., Fryxell *et al.* 2008).

Conversely, the ability to utilize resources may be constrained by the movement modes and perceptual capabilities of organisms. Here lie some of the most intriguing and fascinating questions. For instance: How do nomadic species know when and where to go? How can nomads find suitable localities in vast desert areas? From how far away can they detect suitable conditions? The explicit treatment of the basic mechanistic components of animal movement emphasized by the movement ecology paradigm (Nathan *et al.* 2008) is likely to facilitate our understanding of the interaction between spatiotemporally fluctuating environments and adaptive behaviour, including the large-scale movements of organisms. This may in fact require bringing together elements from diverse fields, such as sensory ecology, communication networks, animal signalling and the use of cues. In a broad sense this is also a question of extending our notion of niche dimensions and resource utilization; different species will have different perceptual constraints, and sensory capabilities will be physically constrained by the environment (Dusenberry 2001). Factors affecting movement and sensory capabilities might in fact covary with resource abundance and variability; for example one could expect low productivity and variable environments also to be open and structurally simple.

In a wider context, a deeper understanding of the niches of migrants and nomads will increase our understanding of the factors structuring animal communities and assemblages. Most of our under-standing of, for instance, the structure of bird communities (Wiens 1989), builds on theory developed in a static framework emphasizing species interactions. However, long-distance animal movements render such interactions weaker and more variable, and the ability to escape predation or competition may in fact be a driving force for migration (Fryxell and Sinclair 1988b).

Both migration and nomadism can have ecosystem-level effects (Chapter 9): directly by, for example, migrants' exploitation of resources, or indirectly by their mere spatiotemporal distribution that can act as a cue for other species tracking seasonal changes. The seasonality of environments where we typically find migrants becomes reinforced by the seasonal movement of organisms. In temporally less predictable environments, nomads that are moving around in the landscape may either decrease or reinforce the patchiness of resources, depending on how easy it is to spot the good patches.

Finally, because mistakes in movement do occur, at the level of the species, migration and nomadism may permit exposure of individuals to novel conditions that would simply not be encountered by more sedentary lifestyles, and so more broadly contribute to species diversification. The avifauna of the Lesser Antilles, for instance, tends to be dominated by clades comprising migratory and nomadic taxa characteristic of North America (warblers, flycatchers, finches, mimids, doves, and pigeons), and are devoid of sedentary, non-migratory taxa from South America (e.g., antbirds, ovenbirds, woodcreepers). Some of these taxa (e.g., the Brown and the Grey Trembler; *Cinclocerthia ruficaudia* and *C. gutturalis*) have evolved to become quite specialized on the islands they occupy.

In this chapter we have explored the extremes of the continuum of movement patterns exemplified by migrants and nomads, and considered the importance of spatial and temporal heterogeneity and uncertainty, and the scale at which these operate, in shaping movement strategies. In so doing we have demonstrated the strength of the niche concept in framing questions surrounding the mechanisms underlying migration.

CHAPTER 8

Migration quantified: constructing models and linking them with data

Luca Börger, Jason Matthiopoulos, Ricardo M. Holdo, Juan M. Morales, Iain Couzin, and Edward McCauley

In many cases, the daily movements of animals represent in miniature movements similar to migration, and require similar mechanisms of operation.

Woodbury (1941)

(…) we now probably see more clearly than ever before the intimate relation existing between the animals and the conditions which influence their migrations.

Adams (1918)

Table of contents

8.1   Introduction

'Why did the warbler on my summer place in New Hampshire start his southward migration on the night of the 25th of August?' (Mayr 1961). Questions about the motivation and orientation capabilities of migratory animals have fascinated natural historians since Aristotle and Pliny the Elder. More recently, faced with the challenges of conserving migratory species (Bolger *et al.* 2008; Wilcove 2008), other problems have gained in importance, such as mapping migratory routes (Sawyer *et al.* 2009; Strandberg *et al.* 2009), understanding the role of individual and environmental drivers of migration patterns (Alerstam 2006; Bolger *et al.* 2008), or estimating the size of trans-continental migratory populations (Hahn *et al.* 2009). Our overall aim in this chapter is to discuss how models, combined with modern data sources and statistical methods, can be used to test different hypotheses about the causes of migration. In Sections 8.2–8.4 we structure our presentation around these three essential components (models, data, and inference) and, in Section 8.5, we illustrate their linkages by means of several case studies.

8.2   Identifying potential causes of migration: translating hypotheses into models

The drivers of migration may be investigated by translating contrasting, verbal hypotheses into animal movement models. These models must then be fitted to movement data (see Section 8.3) and the quality of fit evaluated with formal inferential methods (see Section 8.4). A prerequisite for all of this is a definition of migration. There is, however, a striking lack of consensus about what constitutes a migration and which biological processes give rise to it (Newton 2008; Chapter 1). For example, Baker (1978) used migration as a shorthand for any type of animal movement, Fryxell (1991) defined it as the

seasonal movements of animals tracking changing resource distributions, Hansson and Hylander (2009) interpreted it as periodic vertical diel movements of invertebrate zooplankton between surface and deep waters, and Holland *et al.* (2006b) as one-way movements of single insect generations that may, over multiple generations, lead to a return of the entire population to its point of origin. Several authors have debated whether migration should be defined solely on the basis of movement patterns or whether it should also include behavioural and physiological processes, such as we considered in Theme 2 of this book (Dingle 1996; Dingle and Drake 2007; Hobson and Norris 2008). In Subsections 8.2.1 and 8.2.2, we argue that, to approach migration quantitatively, it is necessary to place it in the wider context of animal movement models (conceptual as well as mathematical). In Subsections 8.2.3–8.2.6 we discuss the ecological mechanisms that could spontaneously have given rise to migration-like patterns of space use.

8.2.1   Mathematical formulations for migration

Dingle (1996) remarked on the difficulty of clearly separating migration from other, often equally poorly defined, types of movement such as nomadism, ranging, dispersal or home-ranging (see Chapter 7). Such classification difficulties usually indicate that the phenomena of interest are part of a continuum. This need not preclude a discrete classification of movement as long as we are prepared for the occurrence of non-archetypal movement patterns. For example, there is a continuum between the extremes of sedentarism and nomadism, along which lie intermediate types of movement that share characteristics of both (see also Mueller and Fagan 2008). In fact, many authors have remarked upon the intimate connections between sedentarism, dispersal, and migration (Bruderer and Salewski 2008; Mayr and Meise 1930). Here, as in

Chapter 7, we propose that by including nomadism, a better conceptual understanding of the dynamics and patterns of animal movements can be obtained.

An additional difficulty is that some movement patterns (such as ranging) are simpler than others (such as migration). Often, this is because animal movement patterns are the expression of several different behaviours corresponding to different life history priorities (see also Nathan *et al*. 2008). It may therefore be helpful to think about *compound* movement patterns comprising two or more *elementary* types of movement (e.g. Morales *et al*. 2004). Once again, this is a constructive conceptualization, as long as we are prepared for the fact that transitions between elementary modes of movement might often be gradual rather than clear-cut.

We can classify elementary types of movement from a time series of spatial locations by asking: what is the minimum set of features that a movement model needs to include in order to replicate the observed movement geometry? This broad question can be posed more precisely in three parts: (i) Can the patterns of movement be replicated with reference solely to the animal's present or past positions? (ii) Do we also need to involve information on the environment, or on individual behavioural state and physical condition? (iii) Do we need to involve information on the position of conspecifics?

The modelling literature offers mathematical formulations in response to each of these questions. For example, if movement is purely self-referential we can use models such as the simple random walk (Turchin 1998), the correlated random walk (Turchin 1998) or the Lévy walk (Bartumeus *et al*. 2005; Edwards 2008). The effect of the environment can be incorporated in different ways depending on whether it acts through local conditions/gradients (biased random walk, Turchin 1998) or some global, fixed points of reference (Ornstein–Uhlenbeck process, see Blackwell 1997, 2003). Finally, if we consider conspecifics as part of an animal's extended environment, the same models can be used if we allow the environmental conditions/gradients/points of reference to be dynamic and interactive (Haydon *et al*. 2008; Moorcroft and Lewis 2006). For example, a territorial animal exploiting the resources within its domain in competition with its neighbours might be modelled by an Ornstein–Uhlenbeck process in which the territorial centroid drifts according to the ebb and flow of conspecific pressure. Or, by using differential game theory, the behaviour of territorial individuals balancing territory expansion against the risk of encountering hostile neighbours, conditional on the distance from the centroid, can be modelled in a spatially explicit way (Hamelin and Lewis 2010). There are statistical difficulties in separating the effects of internal, environmental and conspecific influences from a set of data (e.g. Benhamou 2006). We discuss these further in Subsection 8.4.2.

Having described the elementary types of movement, it is necessary to decide how these should be combined to form a compound movement process. Here, we need to ask: (i) What is the time scale over which an elementary type of movement occurs? (ii) What is the probability that a given type of elementary movement will succeed another? and (iii) Is there any long term pattern (e.g. periodicity) in how different elementary types of movement occur in relation to each other?

The lifetime tracks of migratory animals are probably as complicated as any movement trajectory can be; for part of the year, migratory animals appear to be nomadic and for other parts they appear to be home-ranging. Their movement can be affected by multiple environmental gradients, fixed reference points and the movement of conspecifics. Different types of movement succeed each other with stochastic regularity, which may or may not be seasonal. The modeller's task is to construct an appropriate mathematical formulation that is capable of reproducing this richness of pattern thus facilitating understanding of (and making testable predictions regarding) the underlying processes. The task of statistical inference is to identify which elementary types of movement are found in the data, how each of them is specified parametrically and how they succeed each other through time. Each of these tasks can usefully be guided by ecological first principles. Finally, the specific question asked will determine the spatiotemporal scale of analysis—for example, for many questions regarding migration, small-scale temporal variation in movements might be ignored.

8.2.2   Ecological mechanisms for migration

Active movement allows animals to exploit spatially separated resources and reduce the risk of mortality. Passive movement is in addition shaped by environmental driving forces. In either case, migration is unlikely to occur independently of an animal's environment (Adams 1918; Nathan *et al.* 2008; Schick *et al.* 2008). A full, mechanistic understanding of migration requires knowledge of: (i) an animal's life history priorities (survival, growth, fecundity) and how these vary with age and season, (ii) the set of behaviours (exploring, foraging, commuting, mating, avoiding risk) that are used to satisfy these priorities, (iii) the spatial distribution of resources or conditions (food, risk, cover, cues, conspecifics) that may affect movement and (iv) the cognitive abilities of the animal (memory, perception, communication, intention) that enable it to acquire and exploit information from its environment. These processes contribute to the state of an animal; its membership of a particular life-history stage, its energetic condition and spatial position. Although it might not always be necessary to include all these components in order to model migration, it is useful to enquire how they can spontaneously lead to migratory movements. However there is a caveat when applying these insights to genetically hard-wired movement behaviours; evolved behaviours may become maladaptive over time, making it hard to understand their existence from observation of the animals' current environment. It might be interesting for future research to explore if/when, and in which systems, hard-wired behaviour can be identified in the data. For example, if there is a component of movement that remains fixed year after year despite inter-annual environmental variation, or a movement component that is conserved within genetically-distinct populations or individuals. Understanding the flexibility of migration behaviour to environmental variation is also of applied interest under climate change scenarios.

8.2.3   Migration emerging from individual-environment interactions

It has been postulated for some time that migration-like patterns can emerge in several different ways

from the dynamic interaction of the individual with its environment (e.g. Adams 1918; Mayr and Meise 1930). Apart from the historical interest in how some migratory patterns might have evolved, these interactions allow us to model environmentally driven migration from first principles. There is a certain conservation urgency in understanding such environmentally determined migratory patterns because they are the most likely to be disrupted by environmental change (Chapter 11).

In many cases, the environment may be solely responsible for driving migratory movements. If animals track desirable conditions and environmental conditions vary along smooth geographical gradients, then seasonality in temperature, rainfall and the availability of resources may lead to regular movement of animals between locales (Chapter 7). In an environment where two resources are spatially separated, an animal may be forced to undertake long-range movement because its resource requirements are not satisfied by the local habitat. As the need for one resource becomes satisfied, the animal may return to its original position. This may give rise to regular sorties, as seems to be the case in plankton migration (Hansson and Hylander 2009). Whether this should be classified as migration is unclear (but see Woodbury 1941 for an early suggestion): Dingle's (1996) definition of migration requires that during a migratory move, animals should not be responsive to resources that would otherwise elicit a reaction. We think that this is perhaps too restrictive. For example, movements of long-ranging capital breeders such as elephant seals (Le Boeuf *et al.* 2000) are the result of extreme separation between breeding sites and seasonal foraging grounds. Their annual movements are therefore characterized by a regularity and spatiotemporal extent that are more often associated with migration than central-place foraging. Yet, it is likely that they would remain close to their breeding grounds throughout the year if they could feed at a sufficient rate.

Perhaps less obvious are migratory patterns that arise from two-way interactions between the animals and their environment. A nomadic animal living in a fragmented environment may, by chance, be captured within a small subset of a habitat patch network. This is especially likely if the geometry of the network prevents the animal from perceiving, or makes it unwilling to move within, the wider network. Such movement may even be periodic under certain conditions. For example, if the animal is solely responsible for depleting the resources in a two-patch system it will regularly alternate between patches. Seasonal variation in environmental conditions could cause the same patterns, especially at the edges of species ranges (Mayr and Meise 1930). Alternatively, an energy-intake maximizer with good knowledge of the position of food patches but incomplete knowledge of their current value, may alternate between exploitation and exploration. Hence, even if the individual is not the primary cause of patch depletion it could nevertheless present regular attendance patterns. This could conceivably give rise to patterns such as the traplining behaviour observed in some hummingbirds and bumble bees (Ohashi *et al.* 2008; Ohashi and Thomson 2005). Although the limited spatiotemporal scale of the above examples appears to be stretching the definition of migration, these are the same essential mechanisms that underlie mass migrations (see also Woodbury 1941) and may also explain white shark migration in the eastern Pacific (Jorgensen *et al.* 2010).

8.2.4 Mass migrations: the emergence of collective movements

The interaction between several individuals has the potential to amplify the mechanisms discussed above. For example, several consumers acting together are more likely to deplete local resources. When such groups are found in environments containing few patches, it is possible that consumers will move in synchrony, particularly in the presence of conspecific attraction, e.g. as a predator-avoidance strategy. Conversely, conspecific avoidance can also play a role—recent research has shown that coordinated insect mass migration can be generated by cannibalistic local interactions between individuals (escape and pursuit responses, Bazazi *et al.* 2008; Simpson *et al.* 2006) but that insect density must be above a critical density for coordinated motion to result (Buhl *et al.* 2006). In general, group dynamics can yield migratory movement through the effects of local interactions, even without forcing

from environmental drivers. Through a combination of laboratory experiments and individual-based modelling, recent research has obtained a detailed understanding of the mechanistic basis for the onset and maintenance of such mass movements in locusts, as we describe in detail in the case study in Section 8.5.1.

8.2.5 Interactions between different types of conspecifics: partial and differential migration and group leadership

Partial migration occurs when only one part of a population migrates while the other remains sedentary. Differential migration refers to populations whose members follow different migration strategies (thus, partial migration could be defined as a special case of differential migration, yet this is counter to common usage). Populations are not made of identical individuals and some partial migrations are age-, stage- or sex-dependent. For example, Daphnia of different sizes are able to assess the threat level of ultraviolet radiation and predation risk and respond by adjusting their depth distribution, resulting in size-dependent differences in diel vertical migration (Hansson and Hylander 2009).

Differences in life history priorities and energetic requirements between different population components may interact in unpredictable ways with local population density and the spatial distribution of resource abundance. For example, increases in red deer (*Cervus elaphus*) density in the North Block of the Isle of Rum after cessation of culling (Clutton-Brock *et al.* 1982) led to an increase of male emigration rates, not so for female emigration rates, with virtually no resident males in high female density areas by the late 1990s (Coulson *et al.* 2004); males migrated to these high female density areas only during the rut. The underlying cause is probably forage availability mediated through the effect of sex-biased body-size differences; the smaller female red deer are better able to utilize the short-cropped high-quality vegetation on the greens, which presumably has led the males to search for areas with lower grazing pressure (Conradt *et al.* 1999). A similar example of differential migration by sex has been shown recently in Great bustards (*Otis tarda*) in central Spain (Palacín *et al.* 2009).

Furthermore, differential informational states among individuals can play an important role in migratory processes (for a similar suggestion concerning the importance of information in animal movements see Clobert *et al.* 2009). In some cases more experienced (typically older) individuals, dominant individuals, or those with most to lose by not migrating may lead groups. Couzin *et al.* (2005) developed an individual-based model to explore such leadership behaviour. They assumed that social individuals have a conflict of interests when making movement decisions—to remain with group members or to move in a goal-oriented (desired) direction of travel. When conflicts between these desires arise, individuals must reconcile their behavioural tendencies. In the context of migration, a desired direction could represent the memory (or a genetically encoded representation) of a migratory route, or a direction based on an individual's assessment of a local environmental gradient. This model revealed that leadership can spontaneously emerge in groups without requiring signalling or individual recognition and that only a few such 'informed' individuals can accurately guide a large number of others. Furthermore as group (or population) size gets larger, the proportion of 'informed' individuals required for effective navigation diminishes rapidly, such that only a very small percentage of individuals (<5%) need actually be actively migrating and yet collective performance is very similar to the maximal possible migration accuracy (such as if all individuals are actively migrating).

If detecting environmental gradients is costly— energetically or cognitively, or through time lost in other activities such as feeding or vigilance—then one could envisage a scenario where relatively few individuals are actually navigating and a large proportion of others are using a cheaper strategy of responding to social cues. This is what appears to occur in honey bee colony dispersal; in inter-nest movements of social insects such as honey bees, where acquiring information about where to go is known to be costly, less than 5% of the colony obtain information and guide the whole swarm (Seeley 1985). Thus, in migrating species, individuals may accept such costs if they aid their kin by leading the group, or even among unrelated organisms if the risk of not migrating is sufficiently high, setting the scene for frequency-dependence in migratory strategy.

A further aspect of aggregation among individuals during migration is that individuals may benefit from the directional assessment of others. If each individual makes their own, error-prone, assessment, but then (in addition) tends to align with the direction of motion of others, environmental noise can be dampened and long-range noisy gradients more easily assessed. This has been termed the 'many-wrongs principle' (Simons 2004; Hancock and Milner-Gulland 2006; Codling *et al.* 2007).

Animal groups may be capable of even more sophisticated collective assessment of noisy environments. For example Torney *et al.* (2009) used an individual-based model to demonstrate that groups can climb environmental gradients with minimal cognitive or sensing capacity even if, at the level of an individual, an explicit assessment of the gradient *cannot* be made. If grouping individuals employ a context-dependent strategy such that social interactions (alignment with and/or attraction towards others nearby) become increasingly dominant when the local environment is decreasing in quality, and they ignore others when the local environment is increasing in quality, then groups can climb gradients that are impossible for individuals in isolation. Groups can therefore display an awareness of their environment not present at the individual level.

Understanding the relative contributions of individual interactions, genetically hard-wired behaviours, and environmental variation in determining synchronization of migration over large spatial areas, for example in birds (Nowakowski *et al.* 2005), as well as temporal variation between years, is an important topic in migration studies. The ideas outlined above will be a useful guide to tackling these questions. Information transfer need not be limited to conspecifics. Instead, inter-specific dynamics may be very important for the emergence of migratory patterns. Escape-and-pursuit responses are more prevalent in multispecies interactions. A predator population tracking a migrating prey is the most trivial way in which inter-specific interactions could result in species migration through predator avoidance. Similarly, information transfer can also occur between individuals of different species which might also lead to the emergence of, or influence the dynamics of, migratory patterns.

8.3 Observing the process of migration: different types of data and their uses

Traditionally, migration was the most challenging movement behaviour to observe because of the large spatial scales involved. Technical advances are now revolutionizing our ability to collect detailed movement data over large spatial scales (Green *et al.* 2009; Moil *et al.* 2007; Ropert-Coudert and Wilson 2005; Rutz and Hays 2009; Wikelski *et al.* 2007). Depending on the research objectives and the characteristics of the study system, many different methods can be used to observe migration (Section 8.3.1). After collecting appropriate movement data, several issues relating to data quality need to be addressed (Section 8.3.2), such as location error or imputation of missing data. Traditionally, these operations were separated from statistical modelling, but modern analysis methods incorporate the treatment of sampling issues within model fitting. Finally, we briefly address the importance of good-quality environmental data for understanding the emergence of migratory patterns (Section 8.3.3).

8.3.1 Types of movement data

There are three main types of data on movement; transect data, trajectory data and mark-recapture data. Transect data (Buckland *et al.* 2001) can be collected from various platforms (visual or acoustic, ship, aerial or shore-based) and by various methods (strip-, line-, point-transects, cue-counting). Depending on the type of transect and platform used, surveys can cover large areas and because the regions of observation are subject to survey design they can theoretically achieve representative spatial coverage. However, this is not easy in practice and the challenge for survey design is to reduce bias (Buckland *et al.* 2001; Buckland *et al.* 2004) caused by, for example, coastline geometry and imprecision due to variation in detection probability (Strindberg and Buckland 2004). Examples and discussions of the advantages and limitations of transect data for migration stud-

ies (especially, absence of individual-based information) are provided in the case studies (Sections 8.5.2 and 8.5.3).

Trajectory data or movement path data are obtained using telemetry methods (Bowlin *et al.* 2005; Cooke *et al.* 2004; Fedak *et al.* 2002; Kenward 2001; Millspaugh and Marzluff 2001), ranging from visual tracking and radio-telemetry to geolocation and real-time satellite telemetry. Thanks also to technical developments (Reynolds and Riley 2002; Rutz and Hays 2009; Wikelski *et al.* 2007), these are the only data with which to answer more detailed questions such as the location and characteristics of migratory routes, the type of movement modes employed during migration (including stopovers) or the balance between active and passive movements (e.g. influence of ocean currents), as a detailed record of the actual movements over large spatial scales is required. Two examples of the use of trajectory data in migration studies (Bunnefeld *et al.* in review; Fieberg and Del Giudice 2008) are discussed in Section 8.4.1.

Data obtained through mark-recapture methods (Amstrup *et al.* 2005; Turchin 1998), such as trapping, photo-ID, transponder tags and ringing, have features from both transects and trajectory data. Hence, although getting more than one observation for some animals is possible in some experimental set-ups, the number of repeated observations is never as high as with telemetry. The spatial resolution of the observations is usually error-free because the recapture locations are known, but spatial coverage is generally poor. Representation of the environment at first capture/sighting is within the observer's control but the recapture/resighting probability depends on where individuals go. Furthermore, the number of recapture/resighting stations are usually fewer than the observation points in a transect survey. Mark-recapture methods were originally developed to study bird migration; as early as 1800, the naturalist John James Audubon attached strings to the legs of migratory birds, to successfully demonstrate that the same individuals tended to return the following spring (which he indeed could verify). A recent example of the use of mark-recapture data in migration studies (Ambrosini *et al.* 2009) is discussed in Section 8.4.1.

An emerging type of data for migration studies comes from methods involving stable isotopes, strontium isotopes and trace elements contained in feathers, blood, fish otoliths or other tissues (Barnett-Johnson *et al.* 2005; Brattstrom *et al.* 2008; Hobson and Wassenaar 2008; Kaimal *et al.* 2009; Norris *et al.* 2005). These are integrative or cumulative data, because information about position is compounded on to the organism's chemical make-up, and thus they might be classified as 'indirect data on position' (genetic markers can also be considered as part of this group). The advantage of these techniques is that information like natal origin, overwintering location, and even overwintering habitat use (Norris *et al.* 2005), can be quickly obtained for each individual without the need for any tracking or banding device. However, the spatial resolution is generally coarse and strictly determined by the available information on the spatial geochemistry of the chosen marker(s), which may limit the applicability of the techniques in certain areas (Lank *et al.* 2007). A statistical problem is that without the knowledge of an individual's previous life history a large part of the natural variation in tissue markers cannot be controlled for (Brattstrom *et al.* 2008). This problem may be partly circumvented by combining techniques, e.g. trace elements and stable isotopes (see Kaimal *et al.* 2009 and references therein). A more general limitation of these techniques for detailed movement studies is that, usually, only one location record is obtained for a particular individual (but see Cherel *et al.* 2009).

8.3.2 Preparing the data for analysis

Animal location data are never sampled without positional error (D'Eon and Delparte 2005; McKelvey and Noon 2001; Witey *et al.* 2001) and, depending on the study objectives, even small location errors can bias the inferences obtained from the data. For example, Hurford (2009) showed that GPS collar measurement errors give rise to a systematic bias in the distribution of turning angles for successive positions less than 20 m apart; a stationary animal is mostly incorrectly measured as turning backwards by 180° or moving forwards towards a fixed point in space. Consequently, measurement

error must be considered as a possible cause of 180° turning angles.

Various techniques are available to deal with these problems. For example, sparsely or unevenly sampled locations may be interpolated, using simple linear interpolation or more flexible curvilinear methods depending on the study objectives and characteristics (Tremblay *et al.* 2006). Unlikely locations may be filtered out prior to analysis based on the expected distribution of movement speeds and turn angles, or the habitat of the position (e.g. land for a fish).

Using state-space models (SSMs; Patterson *et al.* 2008, see also Section 8.4) all these operations can be carried out as part of model fitting. SSMs are hierarchical time-series models, usually estimated by Bayesian techniques. They comprise two probabilistic components, a process model that refers to the biological mechanism and an observation model that describes how observations of the system are obtained. Thus, SSMs explicitly model the data collection process and use information about data quality, as well as other, ancillary data (e.g. sea-surface temperature or geographic boundaries), to make inferences about an animal's true position (Jonsen *et al.* 2005; Patterson *et al.* 2008; Schick *et al.* 2008). However, SSMs are technically and computationally demanding, so prompting the development of various ad hoc methods (e.g. Tremblay *et al.* 2009), that focus primarily on data-filtering.

We have so far focused on the effects of location errors and missing data on trajectory and mark-recapture data. Transect data are affected by other types of error, such as the imperfect detectability of animals from the transect. These problems can be dealt with separately using distance methods. Alternatively, their treatment may be embedded in the model fitting process (e.g. Royle and Dorazio 2008).

8.3.3 Environmental covariates

Good-quality environmental data are essential for modelling individual-environment interactions and thus for understanding the emergence of migratory patterns. This topic is considered in more detail in Chapter 7, but it is important to remember that, just like movement data, environmental covariate data may require lengthy preparation prior to modelling

(Aarts *et al.* 2008; Haining 2003; Hunsaker *et al.* 2001).

8.4 Inferential approaches to migration: linking models to data

There are three main questions we can ask in our analyses: (i) Are the observed patterns of movement migratory? (detecting migration; Section 8.4.1); (ii) What are the quantitative properties of the migratory patterns? (estimation; Section 8.4.2) and (iii) What are the underlying mechanisms (model choice; Section 8.4.3)?

8.4.1 Identifying migratory patterns

We address here the challenge of devising methods for identifying migratory patterns and the underlying movement modes, as a prelude to modelling. A central theme of current movement theory is the recognition of the fundamental spatiotemporal scaling of the movement of organisms (Nathan *et al.* 2008; Chapter 7). Scales extend from a single movement step, to entire movement paths (defined as temporal sequences of location data), to the lifetime track of an organism, to the multigenerational tracks of populations. Based on the assumption that the structure of a movement path is a reflection of the basic processes that produced it, the first step for obtaining a mechanistic understanding of movement is to identify the basic functional units of the lifetime track, called movement phases (Nathan *et al.* 2008).

Most populations have a characteristic spatial structure (e.g. colonies, territories, metapopulation structure). An interesting question for migratory species is whether these spatial relationships between individuals stay consistent between areas. In other words, do individuals of one colony or population migrate to the same areas? This has been termed migratory connectivity and has been studied extensively in birds (Webster *et al.* 2002). Recently Ambrosini *et al.* (2009) proposed a new method to quantify migratory connectivity: Mantel correlation coefficients were used to test if the observed connectivity deviated markedly from a random distribution of individuals and clustering methods were used to analyse whether observed patterns of connectivity were produced by aggregation of individuals after the migration or as the result of a direct transference of spatial relationships across migration areas. Application of these methods to a large dataset of ringing recoveries of barn swallows (*Hirundo rustica* L.) migrating from Western Palearctic breeding areas to sub-Saharan winter locations showed that, for this species, migration connects specific breeding and wintering areas. These results were also in accordance with information from studies based on different methods. Consequently, this approach might be used to quantify migratory connectivity in other animal taxa.

Thanks to the current revolution in biotelemetry technologies we can start mapping the migration routes of many taxa. Several questions arise from such data: Why do individuals choose some particular routes? Why do migratory routes of a species converge in specific geographical areas (e.g. Strandberg *et al.* 2009)? To address these questions it is important to map all feasible migration routes. Pinto & Keitt (2009) proposed an extension of the least-cost approach, based on a GIS implementation of graph theory methods. Using simulations, the authors showed that, in landscapes with randomly distributed habitat, multiple movement routes converged in the same areas. Conversely, once the distribution of favourable habitat became more clustered, movement corridors became less redundant and bottlenecks could be identified. The authors exemplified the approach by showing the effects on movement routes of simulated habitat destruction on a real landscape. Hence, this approach might be employed to map the distribution of alternative migration routes and compare it with observed movement data.

Finally, how can we detect migratory patterns in the first place? Ad hoc criteria are generally used to determine migration patterns, such as the onset of migration or migration distance. A further complication is that generally the available data rely on periodic observations of animals (see Section 8.3.1). Thus the resulting response times, like the onset of migration, are known to occur only within a certain time interval, e.g. a week (such data are called interval-censored data, as opposed to known-event times). Fieberg and Del Giudice (2008) recently addressed this issue. The authors compared non-parametric and parametric methods for exploring migration data, using white-tailed deer autumn migration as their study system, and suggested a new parametric model that also accounts for facultative migrants. Furthermore the authors illustrated methods for analysing environmental effects (climate) on the timing of migration and suggested that it may be necessary to account for interval-censoring.

A different approach to identifying and quantifying migration patterns has recently been developed by Bunnefeld *et al.* (in review), based on non-linear mixed effects modelling of net squared displacement (NSD) patterns. The idea is that different movement behaviours may generate different space use patterns and an informative measure of these patterns is given by Net Squared Displacement statistics (Turchin 1998). NSD can be used to identify distinct movement phases, like searching vs. homing behaviour (Börger *et al.* 2008; Moorcroft and Lewis 2006). For example, Fryxell *et al.* (2008) characterized different movement phases of dispersing elk using generalized additive models to analyse net squared displacement patterns, and Kölzsch and Blasius (2008) analysed the characteristics of stork migration by inspecting mean-squared displacement patterns. Börger *et al.* (in review) developed a novel approach, derived from animal movement theory, to characterize and quantify the movement patterns of dispersers by means of non-linear hierarchical modelling of net-squared displacement patterns. Building on this approach Bunnefeld *et al.* (in review) developed a model to identify migratory individuals and to quantify the timing, duration, and distance of migration, all within the same approach. The authors exemplified the approach using data from two populations of GPS-collared migratory moose (*Alces alces*) in northern Scandinavia, showing that individual differences accounted for a significant part of the variation in the distance of migration but not in the timing or duration and that the model had good explanatory and predictive power between individuals and populations.

8.4.2   Estimation

Once the data have been collected, prepared for analysis and migratory patterns have been detected, what is the best way to obtain inferences regarding the underlying ecological drivers? Inferential modelling approaches may be split into Bayesian vs. frequentist (Cox 2006; Gelman and Hill 2007), dynamic vs. static, probabilistic vs. algorithmic/data mining (Breiman 2001; Hastie *et al.* 2001), mechanistic vs. statistical (for a discussion see Bolker 2008; Hilborn and Mangel 1997). In practice, the results of different approaches will often be similar (e.g. Spiegelhalter and Rice 2009) and the improvements in computer hardware and statistical software facilitate the application of multiple methods to a given problem. Importantly, using multiple methods is especially powerful if the choice is guided by biological knowledge.

Notwithstanding the inferential method employed, however, there are statistical difficulties in separating the effects of internal, environmental and conspecific influences from a set of data. For example, the state–space modelling approach is quite appealing for dissecting movement trajectories into different movement strategies (Patterson *et al.* 2008; Schick *et al.* 2008) but parameters on speed and sinuosity of movement could play off against each other to give similar overall displacements; problems may arise due to the interplay between observation error in telemetry data and movement parameters; or behavioural and movement parameters may conflict such that, in a model that allows multiple behaviours, it might become impossible to separate between a switch in behaviours and random heterogeneity in movement paths. This is a general problem in statistical analysis, called the inverse or non-identifiability problem, which arises every time there are multiple (or even infinite) combinations of parameter values that produce equivalent model fits. Thus, the problem cannot be solved by increasing sample size but additional independent information is necessary. Consequently, we urge researchers to consider carefully which data will be necessary to investigate specific hypotheses—even the most detailed GPS data might not be sufficient but will need to be integrated with independent data on the state or behaviour of individuals. Thanks to the revolution in bio-logging techniques (Green *et al.* 2009; Moil *et al.* 2007; Ropert-Coudert and Wilson 2005; Rutz and Hays 2009; Wikelski *et al.* 2007) such requirements are increasingly easy to meet.

8.4.3   Model choice

Model criticism, model understanding and confidence building are fundamental components of modern statistical modelling (Gelman and Hill 2007) and take considerably more time to complete than model fitting itself. Especially for more complex models, this may involve various steps such as debugging, software validation, checking the convergence of numerical algorithms and checking the fit of a model to data using computer intensive randomization approaches.

Different candidate models can be fitted to the data and model selection approaches can be used to choose between them or come up with a confidence set of models (Burnham and Anderson 2002; Murtaugh 2009). For example, Morales et al. (2004) fitted several movement models to reintroduced elk in Ontario and used the Deviance Information Criteria (DIC) to weight the merits of alternative models, concluding that individuals alternated between 'encamped' and 'exploratory' movement modes. The encamped mode was characterized by short daily displacements and large turning angles while the exploratory movements included large displacements and small turning angles. For some of the animals, a very similar fit to the data was obtained when step length in exploratory movement had mode zero and a long tail or when the distribution had a large positive mode (the Weibull distribution used for step length has zero mode and a long tail for shape <1, and positive mode for shape >1). That is, some very small displacements could be included as part of the exploratory movement by having a shape parameter of less than one, which contradicted the idea of exploratory movements being large. Morales et al. (2004) then decided to constrain the shape of the step length distribution using a strong prior for this parameter and looked at the posterior predicted distribution of autocorrelation in step length as another tool for model comparison. Movement data from collared elk showed strong autocorrelation patterns in step length as they alternated between movement modes and only the 'constrained' model was able to reproduce this feature of the data, providing good justification for the use of a strong prior in the shape parameter of the exploratory movements. Note that the autocorrelation pattern was not modelled directly but instead was an emergent property of the movement process. In this example, Morales et al. (2004) could discriminate between models that provided very similar fit to the data in terms of DIC but that had different performance when accounting for other interesting properties of the data. Research on model selection procedures is currently very active (e.g. Kuha 2004; Raffalovich et al. 2008; Sauerbrei et al. 2008; Ward 2008) and we anticipate further developments in this area.

8.5   Case studies

In previous sections we have discussed the three essential components for testing hypotheses about the causes of migration behaviour—models, data and inference. Here, we illustrate the links between them using three case studies, each taking a different modelling approach to the problem of migration (individual-based, population-level and a combination of the two).

8.5.1   An individual-based approach: understanding the mechanisms of mass migration in the desert locust

The basis of many of the individual-based (Lagrangian) techniques that are currently used to understand animal motion, such as random walks (Berg 1983) and bio-social interaction forces among individuals (Couzin and Krause 2003), is in statistical physics (statistical thermodynamics/mechanics), where the aim is to relate the properties of individual atoms or molecules to the large-scale (macroscopic, or bulk) properties of materials, liquids and gases. Recently, there has been growing interest in applying similar mathematical techniques to study biological processes in which a large number of individuals interact to produce collective behaviour. Notwithstanding the increased complexity of organisms compared with atoms and molecules, there is an important difference in living systems—the constituents are not at the mercy of thermal noise, but rather act to propel themselves. Buhl et al. (2006) and Yates et al. (2009) have employed this approach to study the onset and maintenance of collective motion in locust swarms.

Insects such as the desert locust (*Schistocerca gregaria*) form some of the largest migrating groups found in nature (up to 10^9 individuals). Approximately one fifth of the Earth's land surface is susceptible to invasion by this species. Locusts, counter to popular expectation, are shy, green cryptic grasshoppers whose natural tendency is to move little and to avoid one another. It is only when forced to feed together under conditions of resource limitation (which is particularly likely when sudden drought follows a period of high food availability) that locusts change behaviour and become attracted to one another. Locusts in the gregarious state are then prone to forming mobile aggregations that can extend over tens of kilometres and have a devastating impact on local communities through their consumption of crops. This process of behavioural 'gregarization' occurs within a few hours (Simpson *et al.* 2001) and results from a combination of sight and smell of conspecifics, or tactile contact to the hind legs. Recently it was discovered that release of the neurochemical serotonin is both necessary and sufficient for such a transformation (Anstey *et al.* 2009). The full process of gregarization involves a complex cascade of physiological changes including black body patterning and shape changes that occur over a longer period (especially across moults). Wingless juvenile insects—termed nymphs or 'hoppers'—are more prone to such aggregation since their motion is largely restricted to the plane and indeed large aggregates of nymphs (called 'hopper bands') invariably precede the flying swarms of adults.

Locust gregarization is thus an exquisite example of phenotypic plasticity. But what drives the migration of locusts once they are gregarious? Buhl *et al.* (2006) built an annulus arena in which locusts could move under controlled laboratory conditions (Fig. 8.1). This represented (in reasonable space) a typical subsection of the much larger swarms that form in nature. Such never-ending 'periodic boundary' conditions are frequently used in simulations and experiments for physical systems (as in particle accelerators). In this environment the density of insects can be controlled and its relation to migratory behaviour studied.

In order to quantify the individual and group-level motion, computer vision software was developed

that could track the motion of up to several hundred insects simultaneously and in real time (Fig. 8.1, top panel). Thus individual positions and motion were known, from which macroscopic properties such as the degree of collective motion exhibited by the population could be derived. This latter 'instantaneous alignment' is analogous to 'order parameters' in physics and was close to 1 if all individuals were moving in the clockwise direction, close to −1 if they were moving counter-clockwise and close to zero if an equal number were moving in each direction at the same time (see Fig. 8.1, lower panels). While this is a deliberately simplified version of reality, it allowed the detailed analysis of the essential features of the onset of collective motion.

In conjunction with the experiments, to enable a deeper theoretical understanding for this specific system, the authors used an individual-based modelling approach in which individuals were considered similar to interacting particles in physical systems; specifically, a 'self-propelled particle' (SPP) model in which particles (i.e. individuals) exhibited a stochastic propensity to align their direction of travel with their neighbours. This had the advantage of being a mathematically tractable minimal model of the dynamics exhibited by the real system. Using this approach Buhl *et al.* (2006) revealed that both gregarious hoppers and the self-propelled particles in the model spontaneously formed mobile bands when the local population density exceeded a critical value. Below this density, the insects behaved much like particles in a gas, with interactions between individuals in close proximity, but no long-range order or collective motion at the group level. As insect density increased, however, a local behavioural tendency (over a scale in the order of 5–15 cm) for locusts to align their direction of travel with one another could account for the sudden, and *spontaneous*, formation of a mobile band (without necessitating individuals to change their behaviour as density changes). Not only did the model provide insight into the type of collective behaviour seen (Fig. 8.1) but it was shown to be quantitatively accurate and revealed that this behaviour does not rely on leadership, or external organizing forces such as attraction to extrinsic features of the environment, or navigation up environmental gradients, or changes in individual behaviour; the sudden direction changes

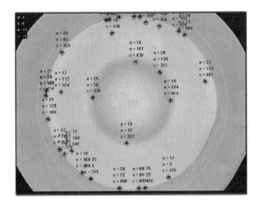

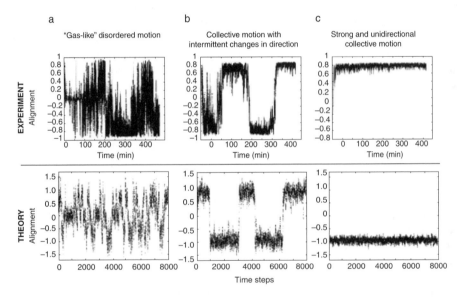

Figure 8.1   Mass migration in the desert locust. The top panel shows an example frame of locusts in the experimental arena (see main text for details) in which the positions have been detected by the computer vision software (drawn as crosshairs over the insects) and the properties used for tracking recoded (here, from top to bottom above each insect, the area and orientation of the animal and how accurately that estimate can be relied on). The lower panels show the collective motion observed (experiment) and predicted (theory) from the self-propelled particle model. At low density (A) both the experiment and model show uncoordinated motion represented by the erratic alignment between the insects. Alignment values close to 1 and −1 represent strong coordinated clockwise and counter-clockwise motion, respectively (see main text for more information). At intermediate densities (B) the group exhibits 'intermittent' motion with sudden transitions between highly-ordered clockwise and counter-clockwise states. Above that density, (C) the group randomly selects a direction and continues in that direction for the entire 8 hour duration of the experiment.

seen in natural hopper bands can result from internally driven stochastic (random) effects.

In a subsequent analysis of these experimental data by Yates *et al.* (2009), it was shown that, although collective behaviour in locusts can be attributed to their tendency to align with near neighbours, individuals changed their intrinsic motion characteristics in response to how well aligned their neighbours were. Modelling of the process was also critical to this study: The authors used experimental data to link the individual-based (Lagrangian) approach with a continuum-based population-level (Eulerian) approach in which they considered motion and interactions between individuals as approximating diffusion and advection processes. Specifically, they assumed that migrating individuals in a swarm have a tendency to exhibit random 'diffusive' motion, and simultaneously a tendency to align with, and thus to move in the same direction as, others—a form of 'advection' or 'drift'. This technique is based on the well-studied Fokker–Planck equation from physics, where it is used to explore the collective properties of gases and liquids.

Using this approach it was revealed that locusts do not have a fixed random (stochastic or noisy) component to their motion. Instead, the degree of randomness depends on the degree of order of the local swarm. The insects appear to increase the degree of random motion as the swarm becomes disordered. Hence, trajectories become more erratic as the swarm loses coherence of motion. Intuitively, one may think that this would result in the swarm breaking apart and the locusts' motion becoming uncoordinated. Remarkably, it has the opposite effect; it allows the system to 'forget' its previous state once order begins to be lost (to a sufficient degree) and thus not go through a disruptive and long transition in which individuals show a conflict in desired directions of travel.

Even though insects align only with near neighbours, swarms can maintain coherent motion over length-scales that are many orders of magnitude greater—such as tens of kilometres. Local alignment tendency is sufficient to explain the motion characteristics of migrating groups, but why do locusts (and other swarming insects) move together in this way? It may, at first, appear like a cooperative collective decision to leave a nutrient poor area and to search more widely for food. However, the underlying mechanism is more selfish. Food shortage, specifically scarcity of protein and salt, results in insects turning to each other as the only source of these essential nutrients. Bazazi *et al.* (2008) demonstrated that locusts are highly cannibalistic and that individuals face a risk of being damaged, and even entirely consumed, by conspecifics in mobile groups. Furthermore, they are specifically sensitive to the approach and contact of others from behind and begin to move in response to these stimuli. The

individuals are, in effect, on a 'forced march'—stop, or move perpendicular to the flow, and you risk being cannibalized. Leave the swarm and you risk starvation and a high predation risk. Individual-based models have improved our understanding of how such local interactions scale to mass migration. Romanczuk *et al.* (2009) simulated the insects' escape and pursuit responses in the face of cannibalism and showed that, even without an explicit alignment term, such behaviours result in coordinated mass migration under a wide range of conditions.

This 'forced march' mechanism is also prevalent in other mass migrating insects such as Mormon crickets, *Anabrus simplex* (Simpson *et al.* 2006). The 'autocatalytic' mutual activation between insects, combined with the risk of being poorly aligned with cannibalistic neighbours, results in migrating bands of both locusts and crickets moving over very long distances in relation to body-length (up to 1 km per day) and these paths tend to be highly linear, interspersed with sudden changes in direction. Such motion characteristics probably play an important role in allowing the insects to search for unpredictably located food patches. In addition, by travelling with conspecifics, locusts in bands can persist (by feeding on others) through very unfavourable environments. Thus multiple selection pressures play a role in shaping migratory strategies in such insects (Bazazi *et al.* 2008).

Swarming insects also benefit from ready access to mates when breeding and lay eggs in close proximity to one another in so-called 'egg fields'. Females die shortly after egg laying and one final adaptation to migratory life is to prepare the offspring for life as either a solitary, asocial insect, or to be ready to be part of a swarm. If gregarized herself, a female secretes a specific chemical into her egg foam, causing the biochemical cascade that results in the onset of gregarization even before birth (Simpson and Miller 2007). This transgenerational transfer of information means migratory swarms can persist across many generations and consequently plagues can persist for months or years.

Individual-based models of animal movement are increasingly used to offer insight into vertebrate mass migration (Conradt *et al.* 2009; Couzin and Krause 2003; Couzin *et al.* 2005; Couzin and Laidre

2009). For example, the tendency for individuals to align their direction of travel with neighbours naturally allows individuals to access information about the environment beyond their own direct interaction range. Just as locusts interacting over the order of centimetres can form coordinated mobile groups that can extend over kilometres, migrating fish, ungulates and birds can create large mobile aggregates. Information about a change in direction can percolate through such groups very rapidly. For example, the detection of a predator by only a small subset of individuals in animal groups can result in an amplifying wave of turning that spreads faster than the maximum speed of the individuals themselves, and this has been observed in insects (Treherne and Foster 1981), fish (Radakov 1973) and birds (Potts 1984). In such circumstances individuals can respond to environmental change (in this example the sudden appearance of a threat) even if they don't detect it themselves.

From the perspective of mass migration, this generation of an 'effective' range of interaction with the environment, which can vastly exceed an individual's own direct perception of the environment, has the potential to be very important. Many migrations depend on individual organisms responding to noisy, long-range and fluctuating gradients, be they thermal fronts, resources, or diffusive gradients, or the Earth's magnetic field. By coordinating motion, organisms effectively form a self-organized 'sensor-array'—each individual sensory unit (individual) is subject to error, but collectively, by integrating their own estimates with the local direction choices of neighbours, aggregates are able to detect and move up gradients that are difficult, or even impossible, to detect from an individual's perspective (Grunbaum 1998a; Torney et al. 2009). This highlights the possibility that this type of information transfer may be ubiquitous in mass migrating species and may play an important role in aiding migration.

8.5.2 A population-level approach: ungulate migration in the Serengeti ecosystem

In addition to individual-based ('Lagrangian' or bottom-up) approaches based on relocation data from individual animals, population-level ('Eulerian'

or top-down) approaches can also be used to make inferences about the drivers of movement and migration. Three examples of this approach can be found in studies of ungulate movement and migration in the Serengeti ecosystem of East Africa (Boone et al. 2006; Fryxell et al. 2004; Holdo et al. 2009b). With the population-level approach, the entire population is subdivided into smaller groups or units (herds, individuals, occupants of a particular spatial location) and these are then redistributed across the landscape according to a set of movement rules. The new spatial distribution of the population can be compared with observed spatial patterns to assess goodness-of-fit, for example using maximum likelihood methods. In two of the Serengeti examples (Fryxell et al. 2004; Holdo et al. 2009b), the movement 'rule' consists of a non-linear emigration function for the probability of departure from a lattice cell. The function evaluates the value of a particular resource (e.g. energy intake) in the currently-occupied cell in relation to its mean value in the surrounding landscape. Groups of animals departing each cell are then redistributed in neighbouring cells in proportion to the abundance of the key resource. Although this movement model does not necessarily result in an ideal free distribution (IFD; for example because movement and perception are assumed to be local rather than global, and therefore less than ideal), it has in common with IFD theory and the related marginal value theorem (MVT) the notion that animal distributions and movement decisions incorporate density-dependence. Foraging animals exert density-dependent feedbacks on their food resources and this effect can be modelled explicitly (Fryxell et al. 2004; Holdo et al. 2009a).

A different population-level approach was employed by Boone et al. (2006) to infer the drivers of the Serengeti wildebeest migration. They used evolutionary programming (EP) to simulate the emergence of alternative migration trajectories, and resource maps were used as selective filters to simulate the emergence of the most well-adapted migration routes. This approach is top-down because the migration pathways are not observed, but rather are used to generate population-level distribution maps that can be compared with the data. Unlike the approach adopted by Fryxell et al. (2004) and Holdo et al. (2009a), density-dependence does not affect migration in this model. In both approaches, however, competing models consist of alternative linear combinations of resources (e.g., water versus forage, or rainfall vs. NDVI) that are used to generate alternative population-level distribution maps, which are then compared with observed distributions to find the model structures and parameters that maximize the fit between models and data.

A potential shortcoming of the population-level approach is that it cannot distinguish specific individuals or groups of animals. Models that adopt this approach may be better described as 'redistribution or flux' models than 'movement' models. With relocation data, whether based on capture–release or radio-tracking data (see Section 8.3.1), true movement trajectories can be inferred for specific individuals. With a population-level approach, movement rules must be inferred from shifts in spatial distributions but, as the example in Fig. 8.2 shows, when the identities of individuals are unknown, alternative movement models can explain any given population-level shift in spatial distribution. In the absence of any external environmental information, models A and B in Fig. 8.2 are equally probable, but imply quite different movement behaviours. Even when environmental data are readily available, some movement parameters may be unidentifiable in the absence of individual-level information (see Section 8.4.2 for a discussion of the identifiability problem).

The individual-based approach has the advantage of allowing inferences to be made on a wider range of processes influencing movement behaviour, including the role of memory versus resource tracking (Dalziel *et al.* 2008; Mueller and Fagan 2008; van Moorter *et al.* 2009) and shifts between discrete movement regimes or modes (Jonsen *et al.* 2006; Morales *et al.* 2004; Patterson *et al.* 2008) that would be difficult or impossible to infer from population-level data. At the same time, the bottom-up approach, because of the nature of the data, must often do without the population-level information that modulates movement behaviour, not only in the context of density-dependence and its effects on forage abundance and predation risk (Hancock and Milner-Gulland 2006), but also because it may convey critically important social cues including

information about resource availability (Couzin *et al.* 2005; Grunbaum 1998b; Hancock and Milner-Gulland 2006).

8.5.3   A hybrid approach

In light of the potential limitations of both individual- and population-based approaches to making inferences about movement behaviour, it is clear that an ideal approach would combine information from both levels. In rare cases, researchers are able to tag very large numbers of individuals and follow their movements in real time. We provided an example in Section 8.5.1, in which video capture and specialized software enabled the researchers to gather data on individual movement behaviour in its social context and to draw striking insights into the links between rule-based movement decisions at the individual level and collective movement behaviour. Most often, these data are lacking, but when population-level census data are available in conjunction with relocation data for individual animals, the potential exists for a hybrid approach to drawing inferences about the mechanics of movement and migration (see also Mueller and Fagan 2008).

Another example of a combined approach is provided in Sibert and Fournier (2001) for application in fisheries. Dart tags provide large amounts of data at the population level, but they do not provide tracking data on the movement behaviour of individual animals; data are often available from 'archival' tags (Sibert 2001). The two data types can be combined into a single likelihood framework by exploiting the known relationship between the advection–diffusion equation (used to model movement at the population level) and the biased random walk (used to model individual tracks). Also, within a likelihood framework, a more general approach might consist of an iterative method that jointly maximizes the likelihood of the modelled distributions given the observed distributions (Fryxell *et al.* 2004) and the likelihood of the modelled displacement kernels given the observed trajectories (Dalziel *et al.* 2008). Both components of the model share environmental data, but they also feed off each other. Such an approach is appealing because it can lead to stronger inference about

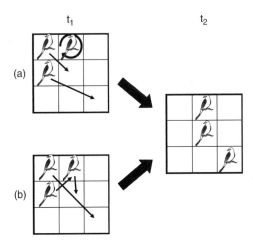

Figure 8.2 Drawing inferences about individual movement behaviour from population-level distributions. The spatial configuration shown at t_1 shifts to a new distribution at t_2. Because individual animals have not been tagged, we do not know the individual movement trajectories that led to this distributional shift. Out of many possibilities, we show two candidates, with mean displacement distances for the three individuals in question of 1.2 (case A) and 1.7 (case B) cell widths.

competing movement models than either the individual- or population-based methods would in isolation.

A rule-based bottom-up model of movement behaviour can result in a hierarchy of predictions at different spatial scales. For a set of candidate models containing assumptions about how individuals respond both to their immediate physical environment and to population-level variables such as conspecific density, population distribution patterns and individual movement tracks could be used to discard candidate models that fit the data poorly at either or both spatial scales (Grimm *et al.* 2005).

8.6 Conclusions

Here we have discussed how models, combined with modern data sources and statistical methods, can be used to test different hypotheses about the causes of migratory behaviour. First we argued that to detect the drivers of migration, it is useful to start by translating contrasting hypotheses into models of movement in order to identify which hypotheses can actually generate migratory patterns. We briefly outlined the controversies regarding the definition of migration and argued that to resolve such disagreements it is necessary to place migration in the wider context of animal movement models (conceptual as well as mathematical). Hence, we discussed the ecological mechanisms that could spontaneously have given rise to migration-like patterns of space use and showed that there are several potential ways in which migratory patterns can emerge from the interaction within and between groups of animals and their environment. On this basis alone, we may argue that observed migratory patterns do not have a single cause and neither will the original cause necessarily remain observable. We highlighted that migration is best seen as lying on a continuum from sedentary to nomadic movement patterns and not as a clearly distinct movement behaviour.

Second, models of migration must be conditioned on data about the movement of animals. We discussed the characteristics of different types of movement data and highlighted the importance of data quality. The processes giving rise to migration are multiple and complex and the methods used to observe migration impose several constraints. Both of these facts may make it difficult to identify unique explanations for observed patterns.

Third, we discussed the use of inferential methods to evaluate the support from the data for different models. On the technical side, model criticism, model understanding and confidence building are fundamental components of modern statistical modelling and take considerably more time to complete than model fitting. More generally, inference is composed of three main aspects: identification of migratory patterns, estimation of migration parameters and selection of migration models. Novel analytical developments may address these three aspects separately but, since they are all non-independent parts of inference, they are best performed as part of the same inferential approach. Furthermore, given the complexity of the migration process, and the difficulties in collecting movement data for migrants, in many cases it will be impossible to identify unique explanations for observed patterns based on location data alone. This problem of non-identifiability cannot be solved by increasing sample size but, rather, by collecting additional independent information (e.g. individual state or behaviour). Thanks to the current revolution in biologging techniques this should be increasingly achievable.

Finally, we illustrated the links between models, data and inference using three case studies. Mass migration in the desert locust provided a clear example of the power of combining individual-based models build on a solid theory (in this case, interacting particle systems developed in thermodynamics) with well-designed experiments on a large number of individuals to identify the causes of the emergence of migratory patterns. Conversely, population-level approaches can also be used to make inferences about the drivers of movement and migration and we discussed three studies using this approach to understand ungulate migration in the Serengeti ecosystem of East Africa. With a population-level approach, competing movement rules are used to generate alternative population-level distribution maps, or alternative migration trajectories, which are then compared with observed distributions to find the most appropriate combinations between models and data. However, as inference is based on shifts in spatial distributions of individuals, in the absence of any external environmental information or of individual-level information, alternative movement models may explain any given population-level shift in spatial distribution. On the other hand, it is more difficult for individual-based models to use important population-level (large scale) information such as population density, forage abundance and predation risk. Thus, a combined approach would be desirable and we discuss two case studies where this has allowed the drawing of striking insights into the links between rule-based movement decisions at the individual level and collective movement behaviour.

In conclusion, the increasing availability of detailed movement data, combined with advances in computing power and statistical software, has led to a renaissance in migration studies. Making the best use of these opportunities to understand the causes of migration involves making choices between a wide variety of different models, data and types of inference.

PART 4

Broader contexts

Pastoral flock on the move. Machin town, Qinghai Province, Tibetan Plateau.
Photograph by Cara Kerven

Migration impacts on communities and ecosystems: empirical evidence and theoretical insights

Ricardo M. Holdo, Robert D. Holt, Anthony R.E. Sinclair, Brendan J. Godley, and Simon Thirgood

Table of contents

9.1 Introduction

How do migratory populations impact the communities and ecosystems that host them? In this chapter, we explore this question through a series of theoretical examples and a brief review of the available empirical evidence. We focus in particular on ungulate migrations, and more specifically on the wildebeest (*Connochaetes taurinus*) migration in the Serengeti (Box 9.1). We first identify the features of

migration that distinguish it from others forms of movement, the ecosystem consequences of which (especially spatial subsidies of nutrients across ecosystem boundaries) have been studied and reviewed at length (e.g., Polis *et al.* 1997; Vanni *et al.* 2004). We then outline the various mechanisms through which migratory animals can impact ecological communities and ecosystem function, and illustrate these effects through a series of theoretical examples

Box 9.1 The Serengeti migration as a case study

The Serengeti ecosystem is an example of a migratory system embedded in a community of resident species. Three ungulate species—wildebeest (*Connochaetes taurinus*), zebra (*Equus burchelli*) and Thomson's gazelles (*Gazella thomsoni*)—undergo an annual migration between the Serengeti plains (grassland) and the woodland savannas of the western corridor and northern Serengeti, over a total area of about 25 000 km^{-2} (Fig. 9.1). The migration is driven by a marked, highly seasonal rainfall gradient, increasing from SE to NW, coupled with strong differences in soil fertility and plant nutritional content between the grassland and savanna habitats (Maddock 1979; Boone *et al.* 2006; Holdo *et al.* 2009b). Other species, including buffalo (*Syncerus caffer*) and topi (*Damaliscus lunatus*), are resident, remaining within relatively circumscribed home ranges on a year-round basis (Sinclair 1977; Murray and Brown 1993).

We illustrate with theoretical examples three facets of the Serengeti wildebeest migration: its effects on the population of a resident competitor (an example of a trophic effect impacting the herbivore community), its effects on fire and tree population dynamics (a downstream trophic effect mediated by a resource), and its effects on net primary productivity and soil fertility (a joint trophic and transport effect). In all cases, we use published models to examine how 'switching off' the migration (i.e., treating migratory

species as residents) might alter community dynamics, ecological processes such as the prevalence of fire, and ecosystem function.

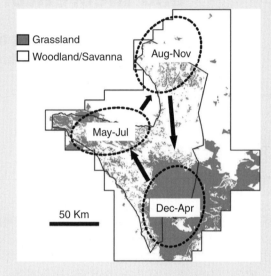

Figure 9.1   The greater Serengeti ecosystem (outer polygon) showing the Serengeti National Park (inner polygon) and a stylized depiction of the migration route followed by wildebeest, zebra and Thomson's gazelles. The two dominant habitat types (grassland, predominantly in the south-eastern plains, and woodland and savanna) are represented.

based on the Serengeti migration. Finally, we insert migratory systems and migration collapse into the broader framework of trophic cascades and explore the implications of migration for community stability, ending with an historic example of migration collapse and its ramifications.

9.1.1   Migration as a special case of movement

A considerable body of work has developed over the last decade on the effects of mobility on food web interactions and ecosystem function (Polis *et al.* 1997; Lundberg and Moberg 2003; Holt 2004; Vanni *et al.* 2004). One intellectual foundation for these studies stems back to the 1970s and the formulation

of the metapopulation concept: discrete populations linked by dispersal, permitting, for instance, regional persistence in ephemeral or disturbance-prone habitats (Levins and Culver 1971; Hanski 1998). Recently, metapopulation biology has been extended to the study of 'metacommunities' i.e., spatial ensembles of communities linked by mobile consumers and seed dispersers at multiple levels (Holyoak *et al.* 2005), and meta-ecosystems, where material and energy flows across space impact local ecosystems (Loreau *et al.* 2005; Varpe *et al.* 2005). These spatial linkages can be expressed by spatial subsidies—the asymmetric transport by organisms or physical transport processes of nutrients and energy across ecosystem boundaries (Polis *et al.*

1997; Anderson and Polis 1999; Stapp and Polis 2003), but also by the cross-ecosystem export of ecological processes; organisms moving from a source ecosystem to a sink ecosystem bring with them changes in levels of predation, competition and mutualism (Holt 2004; Knight *et al.* 2005; Van Bael *et al.* 2008), as well as diseases and new genetic material that change the dynamics of systems compared with what would be expected in closed systems (Lundberg and Moberg 2003).

Despite the fact that much recent research and discussion has been devoted to spatial linkages by organisms between habitats, ecosystems and patches, little work has been devoted specifically to the role of migration as commonly defined. Migration is a special case of movement (Chapters 7 and 8), and we understand migration here not as one-way movement (whether continuous or episodic), but rather as a regular, seasonal pattern of movement that is strongly directional and seasonally reversible (Sinclair 1983; Berger 2004; Mueller and Fagan 2008). Under this definition, we can identify the features of migration, which allow us to disassociate the effects of migratory versus resident animal species on communities and ecosystems (Table 9.1). The aspects of migratory movement that allow us to consider it separately from other forms of movement are:

Spatial scale. Although there is a large amount of variation in migration distance both within and between taxa (Berger 2004), migratory movements often occur over larger distances than other forms of movement within the same species (e.g., natal dispersal or foraging within a specific home range; Mueller and Fagan 2008). Studies of cross-ecosystem effects (including nutrient subsidies) often deal

with movements occurring over small spatial scales (Augustine *et al.* 2003), such as the water–land interface (Knight *et al.* 2005) and thus these movements may have qualitatively different effects in terms of their role as vectors (of disease, seeds, or pollen from genetically-distinct populations, for example) to those of animals moving over large distances.

Timing. Studies of the impact of mobility across ecosystems are often concerned with movement in response to short-term temporal variation, for example diurnal shifts in feeding patterns between habitats (Augustine 2003; Seagle 2003). Migration usually entails phenological differences with respect to other types of movement, in that it is a highly seasonal process, and this timing effect can be of critical importance (Thrush *et al.* 1994; Takimoto *et al.* 2002; Van Bael *et al.* 2008). For example, animals exposed to seasonal changes in the magnitude of interspecific competition or predation pressure are bound to respond differently depending on whether these competitive pressures occur during times of stress and intraspecific density dependence or not (Van Bael *et al.* 2008). Seasonal predation from a migratory species might have strong negative synergistic effects if combined with food scarcity, for example. For African ungulates, animals that share a wet season range (when food is abundant) with migrants are less likely to be stressed by competitive interactions than are species that share a dry season range (when food is limited; Sinclair 1985). In contrast, disease transmission rates in these systems can be higher in the wet than the dry season. The pastoralist Maasai, for example, avoid mingling their cattle with migratory wildebeest in the Serengeti during the rainy season to minimize the risk of transmission of malignant catarrhal fever from wildebeest to their livestock (Cleaveland *et al.* 2008).

In addition to (and as a result of) being a seasonal process, the timing of migration is highly predictable. This predictability represents a forcing function that may be exploited by resident organisms (e.g., predators) at one end of the migratory range. For example, Serengeti lions (*Panthera leo*) time their reproduction to coincide with the presence of migrant wildebeest. Consequently lions in the dry season range of the wildebeest reproduce roughly

Table 9.1 Key aspects of migratory systems that set them apart from systems with non-migratory modes of animal movement, and that have important implications for the effects of migration and migration collapse on communities and ecosystems

Property	Migratory system	Non-migratory system
Spatial scale	Large	Small
Timing	Seasonal/Predictable	Seasonal or aseasonal/ unpredictable
Population size	Larger	Smaller

six months out of phase with those in the wet season range.

Population size. In closed systems, models suggest that seasonal variation in the environment can either depress or increase average population size, depending in a model-specific way on which parameter is fluctuating over time (Holt 2008). But large-scale seasonal variation often tends to depress population size. For instance, if birth rates can be expressed as a saturating function of resource levels then, by Jensen's inequality, temporal variation in resource availability depresses time-averaged birth rates, which in turn tends to depress population size. Migratory species, by avoiding seasons of resource scarcity or heightened mortality risk, may be able to sustain much larger populations than otherwise similar resident species. Indeed, migrants are often far more abundant than their closest resident relatives (Fryxell *et al.* 1988), and their community and ecosystem impacts are therefore bound to be of greater magnitude. Although migration entails costs (e.g., energetic costs, Chapter 5, and a heightened risk of predation or injury), animals that evolve a migratory strategy from a resident one also benefit from more effective exploitation of resources (and therefore escape seasonal limitations in resource availability; Chapter 7) and/or escape from predation and disease (Fryxell and Sinclair 1988b; Bolger *et al.* 2008). Fryxell and Sinclair (1988a), for example, used a mathematical model to show that migratory ungulates in the Serengeti are able to escape top-down regulation by predators, whereas resident ungulates are kept at low population density by predation, a prediction later confirmed by observation (Sinclair *et al.* 2003). A corollary of this is that when ungulate migrations are blocked, this often results in population collapse because the migrant is not adapted to year-round residence in a habitat that is seasonally unsuitable (Bolger *et al.* 2008; Harris *et al.* 2009). The insectivorous parulid warblers that numerically dominate the northern hardwood and boreal forests of North America might face a similar fate if prevented from migrating. These birds can have large impacts upon folivorous insects (Sillett and Holmes 2002), and removing their predation pressure could lead to an upsurge in insect outbreaks, altering forest ecosystem dynamics in a major way. Because species that

undergo mass migrations often become superabundant and play a keystone role in ecosystems, their emergence or disappearance may be of far greater consequence than the emergence or disappearance of similar resident species.

9.2 Impacts of migrants on community dynamics and ecosystem processes

9.2.1 Trophic versus transport effects

The effects of migrants on communities and ecosystems can be broadly divided into two categories; 'trophic' effects and 'transport' or vector effects (Fig. 9.2). Trophic effects are the result of the direct effects of migrants as providers of a pulse of consumers, competitors, and/or prey. In contrast, transport effects are indirect, and are the result of migratory animals acting as vectors for disease, nutrients and energy, and other materials such as seeds across habitat or ecosystem boundaries. Both of these have potential consequences for both local community and ecosystem dynamics. In a recent paper on the role of animal movement in ecosystem function, Lundberg and Moberg (2003) classified animals as resource,

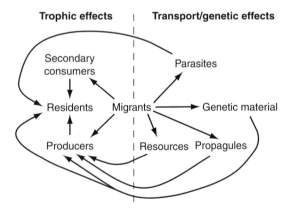

Figure 9.2 Potential effects of a hypothetical migratory species on its resident equivalent. We divide effects into two categories: (i) 'trophic' effects, such as competition for a shared resource or apparent competition through effects of the migrant on the population of a shared predator; (ii) 'transport' or genetic effects, in which long-distance movement of resources, genetic material (of the migrant itself or another organism, e.g., pollen) or propagules (e.g. seeds) can affect the resident producer community and productivity and transport of parasites not locally present can affect the population of resident consumers. In this example we assume that the migrant/resident pair are primary consumers, but comparable effects could be extended to migration at a higher trophic level.

genetic or process linkers. Here we take a somewhat broader view and integrate genetic linkages into the transport category (seeds and pollen moved by animals represent the transport of genetic material).

9.2.2 Effects of migration on communities

9.2.2.1 Competition and facilitation

Migration can impact communities in multi-faceted ways. Even without considering species interactions, many local communities are enriched by migratory species, which persist by utilizing transient pulses of resources, or simply visit en route between their breeding and non-breeding habitats. The outcome of local competitive interactions between species may differ from standard theoretical predictions when one of the species involved is migratory. If a resident species that is a competitive dominant experiences reduced abundance because of seasonal variation, this should free up resources in more benign periods, which could be exploited by a migratory competitive subordinate (Chapter 3). If the migrant is itself competitively superior, its impact on the local community may depend upon the details of its migratory pattern. The competitive pressures imposed by migrants are by definition only present for part of the annual cycle (the seasonality condition above), and the intensity of competition may therefore depend on whether the period of resident/migrant co-occurrence coincides with periods of resource abundance or periods of scarcity (and thus, probable stronger density dependence). In the case of migratory neotropical warblers, for example, some species (black-and-white warblers and American redstarts) are regulated by resources in the wintering range, whereas others (the ovenbird) are regulated by resource availability in the breeding range (Dugger *et al.* 2004). These differences have the potential to affect resident (and other migratory) species in tropical and temperate ranges differentially (competition in wintering areas may have a short-term impact on fitness on adult survival, whereas competition in the breeding range may have a higher long-term impact on fitness by affecting reproductive success).

In the case of the Serengeti, migratory grazers that occupy the south-eastern plains during the wet season move into the central and northern woodlands during the dry season, when food is scarce (Pennycuick 1975; Sinclair 1979; Sinclair *et al.* 1985; Mduma *et al.* 1999). Resident species that occupy the plains portion of the ecosystem year-round interact (and perhaps compete) with the migrating herds only during periods of food abundance; the opposite is true for grazers that reside year-round in the woodland habitat. Competitive displacement thus might be more conspicuous in the latter habitat.

We can examine this competitive interaction with a theoretical example. We first modified an existing model (the Savanna Dynamics, SD, model; Holdo *et al.* 2009a) of grass (both green and dry), fire, and wildebeest dynamics by introducing buffalo as a typical resident herbivore. We estimated the necessary model parameters to model forage consumption and population dynamics for this species from published data (Sinclair 1977). The SD model partitions the greater Serengeti ecosystem into a spatially realistic grid with a spatial resolution of 10 km. Grass growth and decay and herbivore movement and population dynamics are ultimately driven by rainfall, which we model as monthly surfaces generated from rain gauge data. We draw rainfall years at random from the historical record and thus treat rainfall as a stochastic process, embedded in a strong seasonal forcing function. Wildebeest move weekly across the landscape and their movements and local population growth are determined by green forage intake and the protein content of green forage, which varies spatially and is highest in the plains (Holdo *et al.* 2009b). Buffalo, by contrast, do not move between cells and so are residents at the spatial scale of this model; their population growth rates are based on a negative exponential function (with density-independent birth rates) that relates per capita mortality to per capita total forage intake (both green and dry grass). Owing to their larger body mass, buffalo have a higher tolerance to low-quality forage (dry grass in this case) than wildebeest.

For our present purposes, we ran model simulations for 200-year periods under two scenarios: the default, in which wildebeest are fully migratory, and a 'switched off' scenario, in which wildebeest are initially distributed evenly throughout the ecosystem and prevented from moving between cells. We examined the simulated response of both the wildebeest themselves and the buffalo. In the default scenario, our model predicted that after about 50 years, the wildebeest and buffalo populations would asymptote at about 1.5 and 0.18 million animals, respectively (Fig. 9.3(a)). Switching off the migration is expected to affect both species. When prevented from efficiently exploiting the entire landscape, the wildebeest population in the model drops dramatically, to about 0.5 million, or roughly a third of its migratory population size (Fig. 9.3(a)). This occurs because the animals resident in the plains are exposed to an almost complete lack of food during the dry season, and woodland residents fail to benefit from the intake of protein-rich grasses in the plains during the wet season. As a result of this decline in the wildebeest population, the buffalo are predicted to increase to a stable population of about 0.2 million when the migration is switched off, due to decreased competition (Fig. 9.3(a)).

To control for the effect of switching off the migration on the wildebeest, we also repeated the simulations, but fixed the wildebeest at their initial population size of 1.2 million, and distributed uniformly across space. Here the effect on buffalo was reversed; when wildebeest occupy the woodlands year-round, competition for green grass with buffalo is higher during the wet season than would be the case with a migration, and the buffalo are predicted to stabilize at a lower population than would be the case in the presence of a migration (Fig. 9.3(b)). Given that the wildebeest are present, it would appear that migration itself provides a kind of periodic competitive refuge for the buffalo (although one that is not absolutely required for persistence).

Our simulations show theoretically how the seasonal pulse of competition resulting from migration and the effect of a migratory strategy on the size of migrant populations can affect resident competitors. The effect on the buffalo population is not large because dietary overlap (the ratio of green to dry grass, which can alternatively be thought of as low-fibre and high-fibre components, respectively) is incomplete between the two species (Sinclair 1977). We might expect stronger effects on species that are more similar in terms of diet and body size to wildebeest, such as the resident topi (Murray and Illius 2000).

9.2.2.2 Predation

In addition to affecting resource-mediated interactions between resident and migratory consumers, migration has the potential to exert top-down effects on communities of resident species via its effects on predation pressure, and we have not considered this effect in our example. For instance, year-round residence of wildebeest in the woodlands might

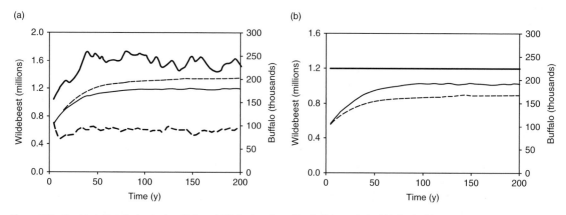

Figure 9.3 Simulated effect of migration by wildebeest (thick lines) on the resident buffalo population (thin lines) of the Serengeti woodlands. We simulated two scenarios: migration switched on (solid lines) and no migration (dashed lines). In (a) we allowed the wildebeest population to respond dynamically to their food resources, and in (b) we assumed a fixed wildebeest population of 1.2 million. The results shown are based on 5-year moving averages (based on means for 20 runs) of 200-year simulations with stochastic rainfall. We assume no hunting in the system.

potentially offset the negative effect of competition for forage on resident species by diluting their risk of predation (Fryxell 1995). On the other hand, because escape from predation and/or more efficient exploitation of food resources enable migrants to become more abundant than their resident equivalents (Fryxell *et al.* 1988), they may subsidise predators in the resident range, allowing them to become more abundant than would otherwise be the case (Packer *et al.* 2005). This subsidy effect could result in increased predation risk for non-migratory species when the migrants are not present, since the former may go from being 'alternative prey' when migrants are present to becoming a preferred food resource when migrants are absent, with important consequences for their population dynamics (Fryxell *et al.* 1988; Fryxell 1995; Sinclair *et al.* 2003). The population explosion experienced by the Serengeti wildebeest following rinderpest eradication in the 1960s may have had this effect. The abundant wildebeest provided a seasonally-predictable subsidy for lions and hyenas in the Serengeti woodlands (Packer *et al.* 2005), and this may have contributed to the near complete disappearance of the very rare and non-migratory roan antelope (*Hippotragus equinus*) since 1980. Cross-boundary subsidies of predators have been documented in other systems, for example across aquatic–terrestrial interfaces (Sabo and Power 2002). These subsidies can also be experienced at great distances. Densities of insectivorous migratory birds in transit can locally spike to high levels, which could inflict substantial mortality on insect populations.

Multi-trophic interactions modulated by migration are also evident in avian communities in savanna ecosystems. In the Serengeti there are 90 species of Palaearctic migrants comprising 70 insectivores and 20 vertebrate feeders (there are no graminivorous migrants from Asia). In contrast, there are 217 resident insectivores and 61 vertebrate feeders. Systematic transect counts of these resident species over the period 1997–2008 during the season when no migrants were present (May–June) recorded 17 748 insectivores and 448 raptors. The same number of transects when migrants were present (Dec–Jan) provided a similar number of resident insectivores (15 079) and raptors (531), and an additional 3697 Palaearctic insectivores and 268 raptors. Thus, there was an increase of some 20% in insectivore numbers and 60% in raptor numbers (A.R.E. Sinclair, unpublished data). Migrants arrive from the north starting in late August and mostly in September. They follow the monsoonal convergence, the Intertropical Convergence Zone (ITCZ), that moves south in August–December bringing rain storms. These storms are followed by migrating insects, and it is these that are used by the migrating insectivores (Sinclair 1978). In summary, the savanna system, already extremely diverse with resident birds, can only support the large influx of migrants when there is a surplus of food during the rains.

9.2.2.3 Disease
A third mechanism through which migratory species might affect communities is by acting as long-distance disease vectors (Morgan *et al.* 2006; Gilbert *et al.* 2008; Koehler *et al.* 2008). Migrants may act as conduits for long-distance transmission of pathogens that may otherwise have remained spatially restricted. In addition, the seasonal influx of migrants and their resultant mixing with either conspecific or heterospecific residents has the potential to exert a forcing seasonal dynamic on rates of infection in local populations, much as seasonal patterns of school attendance affect the dynamics of flu cases in humans. The basic reproductive number of a disease (R_0) is strongly dependent on the pool of susceptible individuals in a population. When a disease is endemic, spikes in infection may occur during periods of migrant influx. The steady-state population size of local populations under these conditions may differ from that expected in the absence of the migratory forcing function. At the same time, the effects of disease (e.g., morbidity or a decline in the ability to mate or disperse) could potentially be dependent on the interaction between dietary stress and the timing of disease. As an example, when rinderpest was enzootic in the Serengeti in the 1950s and 1960s, susceptible calves became exposed to it through contact with livestock during the northern phase of the migration (Talbot and Talbot 1963). This occurred during the dry season, during the time of highest food stress. Talbot and Talbot (1963) speculated that the confluence of dietary stress and infection exerted a synergistic effect, leading to

mortality rates that increased markedly during particularly dry years.

9.2.3   Effects of migration on ecosystem processes

Migratory animals can impact a number of ecosystem processes, such as nutrient cycles and primary productivity, via both direct and indirect pathways. The topic of spatial subsidies, in particular the transport of nutrients and energy across ecosystem boundaries, has received a lot of attention over the last decade (Jonsson and Jonsson 2003; Vanni et al. 2004; Varpe et al. 2005). Significant downstream effects on nitrogen (N) turnover and productivity in sink ecosystems have been demonstrated as a result of nutrient inputs from source ecosystems by a wide taxonomic range of animal vectors, including fish (Helfield and Naiman 2001; Varpe et al. 2005), birds (Post et al. 1998), and mammals (Frank et al. 1994; Schoenecker et al. 2004). These subsidies entail movement, though not always migration. Examples of migratory systems that generate ecosystem-level effects through nutrient transport include anadromous fish (Christie and Reimchen 2005; Varpe et al. 2005), geese (Walker et al. 2003) and elk (Schoenecker et al. 2004). Pacific salmon returning to their natal streams to spawn incorporate large amounts of marine-derived N into riparian habitats, with important consequences for plant primary productivity (Helfield and Naiman 2001).

Less attention has been paid to other effects of animal mobility (especially migration), on ecosystem function. Migrating animals, for example, exert strong effects on their food resources through consumption (Bedard et al. 1986; Sinclair et al. 2007; Van Bael et al. 2008). These direct consumption effects can impact N turnover in ways that differ from the effects of residents (Schoenecker et al. 2004; Holdo et al. 2007). Holdo et al. (2007), for example, combined a two-compartment (plains and woodlands) model of soil N dynamics with functions describing N assimilation, turnover and loss due to herbivory to simulate changes in soil N content and aboveground net primary productivity (ANPP) in the Serengeti woodlands as a function of grazing, migration and fire. The model was based on a series of differential equations describing the dynamics of N

pools (soil organic and inorganic pools, and plant and animal compartments) in the ecosystem (see Holdo et al. (2007) for model equations and details). Herbivores can affect the N dynamics of woodland grasses indirectly by transporting N from the Serengeti plains to the woodlands (a spatial subsidy between plains and woodlands), and directly by consuming vegetation. Grazers affect the N cycle by increasing N turnover; N in dung and urine is more readily mineralizable and made available for plant uptake than N in litter (Seagle et al. 1992; Ruess and Seagle 1994; Holdo 2007). Simulations also suggested that the timing of grazing is important; resident grazers promote N cycling and enhanced productivity at low and intermediate levels of grazing intensity (because N is limiting), but at high levels of grazing, plant standing biomass is kept low and growth is limited by herbivory (Holdo et al. 2007). Excess N is leached out of the system, depressing long-term N availability and ANPP. When the herbivores are migratory, however, they are absent from the woodlands during the growing season, and they therefore do not depress growth during times of maximum productivity (as residents do). This decoupling of the growing and grazing seasons results in a monotonic relationship between herbivore population density and ANPP and soil N in a migratory system, as opposed to a hump-shaped relationship for residents (Holdo et al. 2007).

Here we expand on the analysis of resident versus migrant effects in this model to examine how simulated changes in soil organic N in the Serengeti woodlands vary as a function of grazing intensity (GI, the ratio of consumption to ANPP) with different proportions of resident and migratory herbivores (0, 50% and 100% migratory). Our results (Fig. 9.4) suggest that in the absence of fire, the proportion of migrants strongly influences long-term soil N dynamics. Compared with the case with no herbivory, grazing increases soil N up to an optimum level of GI (about 0.2 in our model). When all the grazers are resident, values of GI greater than about 0.55 result in declines in soil N. Increasing the proportion of migrants appears to have a non-linear effect on this threshold; when all herbivores are migrants, even high levels of GI lead to N increases, but this positive effect begins to decline marginally at high grazing intensity (Fig. 9.4).

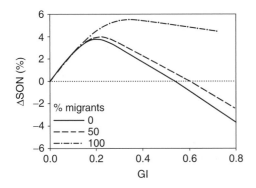

Figure 9.4 Simulated effect of grazing intensity (GI, the proportion of total biomass production consumed) on changes in long-term (100-year) soil organic N (SON) content in the Serengeti (against a baseline of no herbivory), as a function of the proportion of migrants versus residents in the herbivore community. We assume no fire in the system.

A second effect of migrants derives from their inputs of nutrients. When growth (both in terms of body mass increases and population growth) occurs predominantly at one extreme of the migratory range, and mortality and mass loss are higher at the opposite extreme, a net flow of energy and nutrients occurs between the two seasonal habitats. In the Serengeti, wildebeest increase their body mass and calve in the resource-rich plains, and lose mass and have higher mortality in the woodlands. Holdo *et al.* (2007) simulated the effect of grazing with and without spatial subsidies, in the latter case by assuming no seasonal variation in herbivore N budgets. Over the entire woodland habitat, this subsidy effect is insufficient to have an appreciable effect on soil N budgets and ANPP. The distribution of wildebeest across the landscape is highly heterogeneous, however. Based on monthly survey data, we estimate that close to 10% of the total wildebeest population occupies an area of only 200 km² during the dry season, resulting in a local population density (averaged across five months of the dry season) about four times higher than the woodland mean. We modified the simulations in Holdo *et al.* (2007) to compare the average magnitude of the subsidy effect (and the effects of migrants versus residents) with its impact in high-density areas (Fig. 9.5). Our results indicate that, in high-aggregation areas during the dry season, spatial subsidy effects can approximately double the effect of herbivory (versus a baseline of no herbivory) on ANPP and quadruple the effect on soil

organic N. Our estimates show that, at the whole-ecosystem level, the migration results in a net transfer of 0.13 g m⁻² y⁻¹ of N from plains to woodlands, but local influxes can be as high as 0.5 g m⁻² y⁻¹, or about half the combined input of fixation and atmospheric deposition in this ecosystem (Holdo *et al.* 2007). We conclude that the transport of N resulting from migration can therefore be locally important and contribute to enhanced habitat heterogeneity.

9.2.4 Trophic cascades and other downstream effects of migration in ecosystems

In addition to impacting nutrient regimes and regulating their resources, migratory animals can have knock-on effects in ecosystems through cascading effects at multiple trophic levels. Again, the Serengeti migration provides a compelling example of this, as we show that migration collapse in the wildebeest population can lead to coupled changes in grass biomass, fire frequency, and tree cover.

Both empirical (Sinclair *et al.* 2007; Holdo *et al.* 2009c) and theoretical (Holdo *et al.* 2009a) studies have established that wildebeest population size is a key driver of fire frequency in the Serengeti. These effects are mediated by the effect of wildebeest grazing on grass biomass, the main variable limiting the spread of fire across the landscape. Fire, in turn, is the dominant factor driving changes in tree cover (Sinclair *et al.* 2007, Holdo *et al.* 2009a, Holdo *et al.* submitted). We used the SD model to simulate the consequences of migration collapse on ecosystem-wide changes in fire frequency and tree cover. For simplicity, we assumed no elephants and no hunting in the system. As in the earlier example, we conducted 200-year simulations with an initial wildebeest population of 1.2 million animals. The animals were evenly distributed throughout the ecosystem; although it may appear unrealistic not to initially 'confine' the wildebeest to either their wet or dry season ranges, this allows us to isolate the effects of lack of movement from area effects, by effectively providing the population with the same total area in both scenarios. In one scenario, we allowed the wildebeest to move weekly throughout the landscape, and in the other scenario we switched off movement to simulate a collapse of the migration.

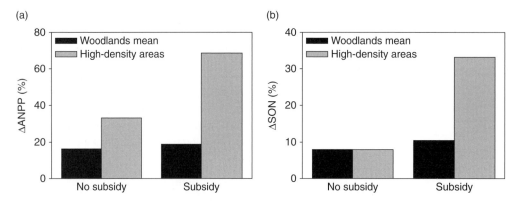

Figure 9.5 Simulated effect (with and without a spatial subsidy component) of migratory ungulates on changes in (a) long-term (100-year) net aboveground primary productivity (ANPP) and (b) long-term soil organic N (SON) in the Serengeti, against a baseline of no herbivory. We assume present-day (1.2 million animals) population sizes, and decompose the effects into total effects (grazing effects plus spatial subsidy) and grazing effects only (no subsidy). We also contrast the ecosystem-wide impact (woodland mean—black bars) with the areas of highest dry season population density, a 200 km² area that hosts ~ 10% of the wildebeest population during the dry season (high-density areas—grey bars). We assume no fire in the system.

As in the competition example (but now without the buffalo), preventing migration from occurring results in a population collapse in the wildebeest (Fig. 9.6(a)). Note that the collapse is more severe than in the example with buffalo present (Fig. 9.3(a)), because wildebeest–buffalo competition in the SD model is asymmetric (Holdo *et al.* 2009a); whereas wildebeest reduce high-quality food availability for buffalo, buffalo consume some low-quality grasses and thus facilitate enhanced intake of green grass by wildebeest (because green grass in otherwise ungrazed areas is enhanced by reduced self-shading from senescing grasses). The wildebeest collapse is predicted to result in a widespread increase in the area burned each year in the ecosystem (Fig. 9.6(b)). Whereas tree cover is predicted to decline initially under default conditions (although it will stabilize after about 50 years), the decline is more severe and longer-lasting when migration is impeded (Fig. 9.6(c)). This suggests that a migration collapse would have implications not only for the population structure of other herbivores in the grazer guild, but more far-reaching implications for the abiotic environment (fire) and for ecosystem structure.

9.2.5 Migration and community stability

There is increasing recognition that spatial processes are fundamental to many ecological processes

(Tilman 1994). Migratory animals, by linking ecosystems, can affect (meta) ecosystem stability (Takimoto *et al.* 2002; Holt 2004) and resilience by acting as sources of 'external ecological memory' (Lundberg and Moberg 2003). This topic has yet to be the focus of sustained theoretical and empirical study, but one can imagine that migratory species could exert strong influences on community stability, both to enhance it and to weaken it, depending on the circumstances. Imposing seasonal variation on to multispecies models that in a constant environment tend towards a stable equilibrium can lead to cycles and even chaotic dynamics (King and Schaffer 1999), with overcompensating density dependence leading to low population densities where extinction may be risked. Migratory species may be able to avoid such excursions, and thus reduce their own risk of extinction. Some species can in turn exploit the regularity of these seasonally regular resource pulses. For instance, Eleonora's Falcon in the Mediterranean has evolved a specialized life history, timing its breeding during the annual cycle to the migratory waves of songbirds each autumn (Del Hoyo *et al.* 1994).

Other resident species may be strongly negatively affected by pulses of consumption, resources, and predation in their local communities, in ways that destabilize communities and ecosystems. A particularly striking example comes from the migratory

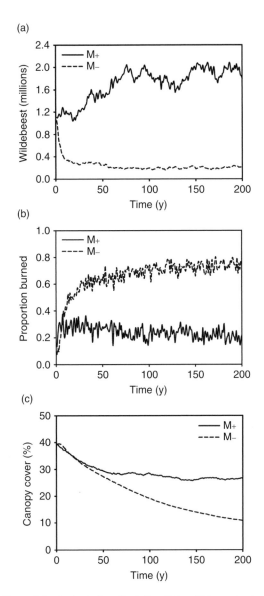

Figure 9.6 Simulated effect of 'switching off' the wildebeest migration on (a) wildebeest, (b) fire, and (c) tree cover (in the Serengeti woodlands). We simulated two scenarios: migration switched on (solid lines) and no migration (dashed lines), with an initial wildebeest population of 1.2 million. The results shown are based on means for 20 runs of 200-year simulations with stochastic rainfall. We assume no hunting in the system.

Snow Goose. A combination of reduced hunting and increased food supply on its wintering grounds in Louisiana and Texas, and en route in the Great Plains, has led to an enormous upsurge in its abundance in the Canadian tundra. This in turn has led

to over-exploitation and even collapse of tundra vegetation, which in places has gone completely, leaving behind only extensive mudflats (Jefferies *et al.* 2006).

Conversely, shifts in resident fauna may alter the importance of spatial subsidies and other influences of migratory species. In the Aleutians, the introduction of the red fox decimated migratory seabird colonies, leading to a reduction in nitrogen inputs and a dramatic shift in the plant community from shrub- to grass-dominated taxa. Many pelagic birds nest on sites very far from where they feed in the non-breeding season, and it is likely that introduced mammals on oceanic islands have sharply altered the strength and pattern of nutrient subsidies on island ecosystems.

9.3 Collapsed migrations and their consequences: empirical evidence

The models described above predict that the collapse of the Serengeti wildebeest migration would have profound impacts on community structure and ecosystem processes. Indeed it has also been suggested that socio-economic feedback loops including reduced revenues from ecotourism resulting from the loss of the migration, and a subsequent decline in resource protection, could lead to increased illegal hunting and habitat loss and the collapse of the whole Serengeti ecosystem from its present state (Harris *et al.* 2009).

Models can be considered as informed speculation about the consequences of future action. But is there empirical support for the hypothesis that ungulate migration collapse would cascade through ecosystems in the ways that the models suggest? Recent reviews synthesizing the available information on the global status of migratory ungulates have demonstrated that, with very few exceptions (e.g., wildebeest and zebra in Serengeti, white-eared kob (*Kobus kob*) and tiang (*Damaliscus lunatus*) in Southern Sudan and some caribou (*Rangifer tarandus*) populations in Canada and Russia), migratory populations of ungulates are in universal decline and a number of populations have been extirpated (Bolger *et al.* 2008; Harris *et al.* 2009). Is there evidence that migration collapse leads to wider impacts at the community or ecosystem level? We focus here on ungulate migrations because of the availability of recent reviews and for comparison with the Serengeti models. We also restrict this focus to those ungulates that migrate in large aggregations on the premise that these species are more likely to have a 'keystone' function in ecosystems and thus their removal may have more obvious consequences.

Harris *et al.* (2009) synthesized global data on 23 species of ungulates that historically migrated in aggregations, attempting to describe migration routes, historical and current population size, ecological drivers of migration and conservation threats in a consistent and rigorous manner. The review was challenged by incomplete and outdated information for the majority of migratory ecosystems, with data in most cases being restricted to historical and current estimates of the size of migratory ungulate populations, and little additional information on the wider consequences of migration collapse. With that caveat, we focus on a particular case study where more detailed ecological research has been conducted.

9.3.1 Wildebeest in Kruger National Park

The wildebeest migration of Kruger National Park, South Africa, presents an illustrative case study of the consequences of migration collapse. We chose this example because of its similarities with the (still healthy) Serengeti wildebeest migration used to illustrate our theoretical examples above. At proclamation in 1926, Kruger contained low numbers of game as a result of excessive hunting and the 1896 rinderpest epidemic. Early management priorities focused on rebuilding game populations through interventions, particularly the provision of permanent water from boreholes (Gaylard *et al.* 2003). Fencing of the park boundaries for disease control purposes and political boundary demarcation commenced with the southern (1959) and western (1961) boundaries and concluded with the eastern (1976) and northern (1980) fences (Bengis *et al.* 2003). The period 1946–1990 has been described as the era of 'management by intervention', when fencing turned Kruger into a heavily managed ecological island. The consequences of fencing, water provision and management culls of both herbivore and carnivore populations were profound (Freitag-Ronaldson and Foxcroft 2003).

Wildebeest historically migrated between the drier lowveld of southern Kruger and the wetter foothills of the Drakensberg Escarpment, a distance of 100 km (Whyte and Joubert 1988). The size of this population prior to the establishment of the park is unknown but historical accounts indicate that it was heavily impacted by hunting in the late nineteenth century. With the completion of the western boundary fence the migration was prevented and reports suggest that the wildebeest population in Kruger declined by 87% (Whyte & Joubert 1988). The ecosystem effects of the collapse of the migration were, however, confounded by a cascading series of management interventions. Concerns over the declines of wildebeest and zebra populations after the completion of the boundary fence, and modelling, which suggested that predation by lions and spotted hyena was linked to the continued decline (Smuts 1978), led to large management culls of lions and hyena in the 1970s (Mills and Funston 2003). These culls terminated in 1980 when it was discovered that they had no detectable influence on lion density or on population trends of wildebeest and zebra. The provision of artificial water points in the dry northern sector of Kruger had more complex impacts. Following a severe drought in 1982/83, wildebeest and zebra moved northwards from their main range in the central region of Kruger, facilitated by the construction of numerous boreholes and dams in previously waterless areas (Owen-Smith and Ogutu 2003). Lion densities in these areas increased either through immigration or increased reproduction, and lion predation was identified as the key factor causing the decline of the rare roan antelope from 452 in 1986 to 42 in 1993 (Harrington *et al.* 1999). The Kruger case study demonstrates that the ecosystem-level consequences of losing migrations are complex and, in this case at least, interact with multiple other drivers of change, not least anthropogenic factors.

9.4 Conclusions and future perspectives

In this chapter, we have attempted to synthesize empirical and theoretical evidence across a range of trophic levels to investigate the broader impacts of migration on ecological communities and ecosystems. Migrations do not occur in isolation; like all ecological processes, they are embedded in a *milieu* of complex biotic and abiotic interactions and drivers. Migratory species can directly compete with, prey upon, or act as food resources for other species, as well as exerting indirect influences on surrounding animal communities as ecosystem engineers through their effects on nutrient cycles, fire regimes and habitat structure. Much work has been conducted on the role of animal movement for the transport of energy, materials, genetic information, and disease in ecosystems, but few empirical studies have specifically explored the impact of migration for a wide range of broader ecological processes. Given that habitat loss and fragmentation have led to disruptions or even total collapse of many migrations, there is a pressing need for further empirical work on the downstream effects of migration collapse in real systems.

CHAPTER 10

Pastoral migration: mobile systems of livestock husbandry

Roy H. Behnke, Maria E. Fernandez-Gimenez, Matthew D. Turner, and Florian Stammler

Human livestock management attempts to mimic the natural migration pattern of large wild herbivores and to take advantage of ephemeral forage resources.

M.A. Little, K. Galvin and P.W. Leslie (1988)

Seasonal migration by white-eared kob is linked to shifting distributions of critical resources. Pastoralists living in the Boma region [of Sudan] also migrate with their live-stock herds between a wet season range west of the Pibor River and the northern swamps during the dry season.... Traditional methods of livestock husbandry presumably designed to maximize secondary production, therefore mimic the evolved behaviour of natural populations of ungulates.

J.M. Fryxell and A.R.E. Sinclair (1988)

Table of contents

10.1   Introduction

Alone in this collection, this chapter investigates the migration of domesticated animals. Our objective is to explain how African and Eurasian pastoralists use migration to exploit the opportunities presented by climate, landscape and animal physiology or, inversely, how they have sought to modify or avoid the hazards posed by these natural factors. Our point of departure is an examination of the extent to which pastoral nomadic movement differs from wild ungulate migration, or mimics it.

In the Arctic the principal domestic and wild migratory ungulate species is identical—*Rangifer tarandus*; the two populations interbreed, individual reindeer shift from a domesticated to a feral state, and wild and domesticated herds occasionally occupy the same areas. In this case it is possible to observe directly how humans have altered migratory cycles by comparing wild and domesticated movement patterns.

In the tropics and the temperate zones, there are some wild analogues to the common domesticated herd species—feral cattle and camels, wild mustangs, undomesticated sheep and goat species—but direct comparisons between pastoral and wild herds are imprecise. Migratory wildebeest are not physiologically equivalent to the cattle of the East Africa plains, nor are saiga antelope exactly comparable to the domesticated sheep of the Central Asia steppes. In the Sahel and in many other areas there remain only remnant populations of wild migratory ungulates that might provide a comparative baseline. In these circumstances direct comparison is insufficient and we must instead identify the factors that drive pastoral movements and consider how these factors depart from or replicate those that direct wild ungulate migrations.

Both wild ungulate and livestock migrations are influenced by at least three different sets of considerations:

- by the distribution, quality and quantity of food supplies,
- by competition from other livestock/ungulates for feed resources and
- by a wide variety of factors that may impede or facilitate access to resources.

The potential effect of these factors on pastoral movement is summarized below in a series of propositions. The middle sections of this chapter examine these propositions in light of field reports on pastoral migration in tropical Africa, temperate Asia, and the Arctic. Our objective is to identify the distinctive aspects of domestic herd migration with respect to the utilization of feed supplies, the regulation of access to natural resources, and the influence of non-forage constraints and incentives to movement. The chapter concludes with a discussion of the role of human agency in the determination of pastoral livestock migrations.

10.2 Propositions

10.2.1 Forage abundance and quality

Most migratory livestock are ruminants, including reindeer, cattle, sheep, goats, yaks and camels. Ruminant physiology makes grazers sensitive to variations in feed quality. With cattle, a modest increase in forage digestibility from 50 to 55%, produces an estimated 32% increase in the amount of energy digested per day. This increases the amount of energy available to the animal above maintenance by almost 200% and produces a 100% increase in weight gain (Malechek 1984, based on van Dyne et al. 1980 and Blaxter et al. 1961). This process of amplification, which has been termed the 'multiplier effect,' has also been documented among subarctic reindeer (White 1983). For ruminants (and for their owners should these exist) there are strong productive and reproductive reasons to select an optimal diet.

For the migratory ruminant, the abundance and quality of natural forage varies both in time and space. Temporal changes in forage can be summarized in terms of the forage-maturation hypothesis: 'The protein content and digestibility of forages are often related negatively to maturation state, so immature plants found in areas with low vegetative biomass may be nutritionally superior to mature, high-biomass vegetation' (Fryxell 1991, p. 479). The declining feed value of older plant material is caused by the accumulation of plant cell wall components, which dilute the concentration of nutrients and obstructs their digestive absorption (Owen-Smith and Novellie 1982; Georgiadis and McNaughton 1990). As a result, the nutritional value of grass tends to decline as leaves age, and old grass swards generally offer higher biomass but lower nutritional quality than younger swards (Langvatn and Hanley 1993). The inverse relationship between forage quality and quantity presents graziers with a dilemma—to maximize forage intake or diet quality, or to compromise between these objectives (Chapter 6).

Spatial gradients in forage quality and quantity are caused by regional variations in soils and climate, the main determinants of the growth and species composition of pastures. In the Arctic and temperate zones, regions with cool, dry summers often produce the best quality forage (Langvatn and Albon 1986). Limited water supply or lower temperatures and light conditions at higher latitudes (van Soest 1982; Deinum 1984; Hay and Heide 1984; Wedin et al. 1984) or altitudes (Wilmshurst et al. 1995) reduce the proportion of plant cell walls relative to carbohydrates and protein, producing more digestible forage. But these higher latitudes or altitudes often experience severe winter weather that makes it difficult for large ungulate populations to survive in them year-round. Similarly, in the semi-arid tropics, regions of low rainfall tend to produce small amounts of high quality plant biomass, while wetter areas generally produce abundant forage of increasingly inferior quality (Breman and de Wit 1983; Holdo et al. 2009b). But the low-rainfall areas where animals can maximize their growth in the wet season are routinely places where water and feed for livestock are severely restricted in the dry season. While a small number of individuals can survive year-round in refuge habitats, what migrant animals seek in their seasonal ranges is a nutritional boost that permits breeding (Sinclair 1983). They therefore target high quality, low volume forage sources in seasons of biomass abundance, in spring in the northern latitudes and during the wet season in the semi-arid tropics. Grazing preferences become less selective and animals shift to high-volume forage sources in an attempt to meet maintenance nutritional requirements in seasons of forage scarcity, in the winter in cold climates and in the dry season in the semi-arid tropics:

> In…African ecosystems there is a strong rainfall gradient with an accompanying gradient in grass primary productivity and maximum standing crop…. Wildebeest, white-eared kob, and domesticated livestock migrate down the productivity gradient during the wet season, when surface water is plentiful, but return up the productivity gradient during the dry season, when surface water is scarce. This could reflect an innate preference by…tropical species for grass swards at an intermediate stage of maturation that would parallel the elevational migration of many montane herbivores (Wilmshurst et al. 1999, p.1230).

Seasonal ungulate migrations occur when animals survive periods of plant senescence in places that are distant from where the animals go to gain weight and reproduce. At the very least, migration links together two kinds of habitats—one that promotes animal growth and reproduction in seasons of forage abundance and favourable weather, and another that facilitates survival in seasons of dearth and severe weather. Subsequent sections of this chapter will argue that pastoral migrations achieve much more than keeping animals alternately well fed and out of harm's way, but this is the irreducible minimum that must be achieved.

10.2.2 Density dependent distributions

Even in a simplified, hypothetical migratory system consisting of three zones and three seasons, individual herds could follow eight different migratory routes and still—in the aggregate— produce zonal stocking rates consistent with resource availability in each zone in all three seasons (Behnke and Scoones 1993; Chapter 3). The free choice of diverse migratory routes by individuals may therefore produce an apparently uniform adaptive response to environmental conditions. This uniformity is achieved at the population level by the summation of various movements driven by different combinations of motives, 'a predictable pattern on a gross scale but entirely unpredictable on a fine scale' (Sinclair 1983, p. 251) Sinclair was referring to the movements of wildebeest and topi antelope; the ability of humans to explain their actions makes the movement patterns of domestic herds more readily intelligible at the individual level, as numerous anthropological accounts of pastoral migration have demonstrated (Stenning 1957; Cunnison 1966; McCabe 1994; Kerven et al. 2008.). The challenge for pastoral studies has rather been to move from illustrative individual cases to adequate accounts of the movements of entire livestock and human populations. In this respect, the broad arrows often used to depict pastoral movement patterns are cartographic crutches that aid communication but project a spurious uniformity (McCabe 1994).

Theories of the ideal free distribution or density dependent habitat selection (IFD/DDHS) systematically link individual choices to overall population distributions (Fretwell and Lucas 1970; Farnsworth and Beecham 1997; Sutherland 1983; Veeranagoudar et al 2004; Ward et al. 2000). The fundamental idea behind these theories is that resource consumers respond simultaneously both to resource distributions and to the shifting distributions of other consumers. Consistent with this perspective, the role of animal density in sustaining mixed sedentary/migratory wildlife distributions has been expressed as follows:

> 'Why, then, does an individual decide to be a resident or a migrant? The decision depends on the sizes of the two populations relative to their stable sizes. If all animals were migrants, then any individual who stayed behind in an area capable of supporting him year round, would have an initial advantage of superabundant food and hence no competition.... This advantage, of course, is only temporary, lasting until numbers have increased to the stable level; but for the first individuals it is a real advantage. Similarly, if all animals were resident, there would be an initial advantage to becoming migrant. Hence, the advantage depends only on what the other members of the populations are doing' (Sinclair 1983, p. 256).

The final sentence of this quotation expresses the essential idea behind IFD/DDHS theory. Compare this statement with an anthropological description of pastoral decision-making, and the parallels are immediately obvious:

> Assessments also have to be made about the likely movements of other herds, for a potentially good area may become over-popular and over-crowded, reducing the availability of resources and perhaps creating competitive conflict between herdsmen, while a less favoured area may be under less pressure and therefore become preferable (Gulliver 1975, p. 372).

Pastoral societies possess a variety of institutions—land tenure rules, systems of territorial control, social networks and organized violence—to manage resource appropriation and use. The interplay between these institutions, movement patterns, and the biological tendency towards free distribution has yet to be adequately explained.

10.2.3 Constrained resource matching

Herds may be chasing protein and calories across savannas, steppes or tundra, but this is not all they are doing. From a population-wide perspective, regional forage availability is the 'common denominator driver' of pastoral migration, a recurrent concern that is supplemented or supplanted by a variety of other biophysical or socio-economic considerations.

In the semi-arid tropics, movement systems can be sorted into two basic groups according to water availability: those in areas with abundant drinking water and unimpeded access to regional forage supplies, and those that are restricted to using that subset of the total forage supply that lies in the vicinity of scarce water supplies (for these water-constrained systems see Coppolillo 2000, 2001; Hary et al. 1996; Western 1975); systems can shift seasonally between these two poles, water-limited in dry seasons but not in wet seasons. In the better-watered parts of the temperate zone and in the Arctic, the importance of regional forage availability persists, but water constraints are replaced by considerations regarding the prevalence of predators and pests (wolves and flies for reindeer) and exposure to cold and snow (transhumant altitudinal systems). Especially at northern latitudes, the cost of humans moving may become an important consideration because substantial portable dwellings are needed to protect people from the cold, and these are expensive to move around. Irrespective of climatic zone, pastoral movements are adjusted to permit access to markets, services and employment opportunities, or to reflect land tenure restrictions or administrative boundaries. The location of arable land relative to pastures is a recurrent concern in agro-pastoral systems.

Pastoral migrations, like agriculture in general, exploit biophysical regularities to achieve a variety of different human purposes, and are modified to reflect these purposes. The relevance of socio-economic variables to the organization of domestic herd movement introduces a level of complexity that obscures the expression of the biological processes that clearly structure undisturbed, wild migrations.

The following discussion combines ethnographic description and biological information to construct illustrative case studies of pastoral migration in the semi-arid tropics of sub-Saharan Africa, in temperate Asia, and in the Arctic. This review focuses on 'horizontal' migratory systems common in desert, savanna, steppe, tundra and plains environments. 'Vertical' movement systems that exploit extreme elevation gradients are common in temperate Asia, particularly along the arch of mountainous terrain that stretches from Iran through the Pamir, Tien Shan, and Himalaya mountains to the Tibetan Plateau and the western borders of China. These altitudinal movement systems are under-represented in the following discussion.

10.3 Case studies

10.3.1 Sudano-Sahelian West Africa

The Sudano-Sahelian region of West Africa, stretching eastward from the Atlantic Ocean to Chad, is the bioclimatic region lying south of the Sahara Desert and north of the Guinean savanna zone. Vegetative structure in the region grades from Sudanian open savanna to the south (maximum long-term average annual rain of 1000 mm falling from June to September) to steppe vegetation to the north (minimum long-term average annual rain of 200 mm falling from July to August). Herbaceous vegetation is dominated by annual grasses in the Sahel with increasing presence of perennial grasses in the southern Sudanian zone.

Domesticated and wild grazers move across these rangelands in order to maintain access to both water and quality forage. During the rainy season, animals will also move to gain access to minerals (salt licks). Characteristics of the region's climate and vegetation patterns strongly influence these movements. These include the following.

1. Vegetative production is patchily distributed as a result of the high spatial variability of rainfall. The degree of patchiness generally declines as one moves south into the Sudanian zone where vegetative productivity is limited less by moisture.

2. The forage quality of herbaceous patches is ephemeral. The nutrient concentration of herbaceous vegetation deteriorates rapidly at the end of the rainy season after seed set (Penning de Vries and Djitèye 1982). Vegetation stays greener longer to the south due to a longer rainy season and to the increased prevalence of perennial grasses that will resprout during the dry season if there is sufficient residual soil moisture.

3. Soils are ubiquitously poor in the region, while rainfall is variable and declines from south to north. The growth of herbaceous vegetation in the north of the Sudano-Sahelian zone is limited by water availability and tends to be sparse but of relatively high nutritive quality. In the south, plant growth is more abundant due to higher rainfall, but of lower quality (Penning de Vries and Djitèye 1982).

4. Animal presence has historically been limited by the increased challenge of trypanosomiasis as one moves south from the southern Sudanian to Guinean zone (average precipitation >1000 mm/year). Increased clearing of lignaceous cover for farming has however reduced tsetse fly habitat and therefore exposure to the disease (Bassett 1986; Bassett and Turner 2007).

5. The availability of surface water significantly declines with the end of the rainy season. Generally speaking, the period of surface water availability shortens as one moves north in the region. Supplying water to animals through alternatives (wells) often requires significant investments of labour by human caretakers.

6. Persistent heavy grazing pressure during the rainy season will have effects, particularly on the species composition of vegetation (Grouzis 1988; Hiernaux 1998; Penning de Vries and Djitèye 1982; Turner 1999c; Valenza 1981), increasing the prevalence of the short-cycle species (e.g. *Zornia glochidiata, Tribulus terrestris*) and those of lower palatability (e.g. *Sida cordifolia, Cassia tora*).

These characteristics place broad constraints on animal nutrition and the strategies for livestock husbandry. Seasonally, forage quality drops significantly with drying and continues to decline as more nutritious plant parts (younger leaves) become rare in the vegetation due to a combination of grazing, termite activity and shattering. As a result, for animals unsupplemented by imported feed, most of the annual nutrition occurs during and immediately after the short rainy season (3–6 months), with the importance of selective grazing to acquire any nutrition increasing throughout the dry season. It is not uncommon in the zone for vegetative material remaining on heavily-grazed rainfed pastures to have minimal even negative nutritional value during the last couple of months of the dry season (April–May). These nutritional constraints are shown dramatically by studies tracking weight changes in livestock, which find weight gains during the rainy season followed by a couple of months of weight maintenance, followed by a longer period of weight loss sometimes exceeding the weight gains during the previous rainy season (Ayantunde 1998; Ayantunde *et al.* 2001).

The period of weight loss is predictable and, therefore, livestock husbandry is largely focused on maximizing gains during the short rainy season and extending the period of weight maintenance during the beginning of the dry season. This typically involves moving livestock to areas distant from areas of higher human population densities, most particularly to the northern Sahel where pastures are more extensive and vegetation of higher nutritive quality (Schlecht, Hiernaux, and Turner 2001; Penning de Vries and Djitèye 1982). Due to changes in the relative quality of forage and water sources at particular locations in the north and south, appropriate management will often require movements between these locations as the rainy and dry seasons progress (Breman and de Wit 1983; Le Houérou 1989).

10.3.1.1 *General patterns of livestock migration*

Historically the greatest concentrations of human population in the region have existed in the southern Sahelian and Sudanian zones (600–900 mm/year)—an area allowing both farming and livestock husbandry. The characteristics listed above help shape patterns of livestock movement from this band at different spatiotemporal scales. At the broadest scale (≈100–500 km), livestock have historically moved to the northern pastures during the rainy season to take advantage of the brief, but highly nutritious, growth of annual grasses (these long-range seasonal movements are commonly referred to as 'transhumance'). After the end of the rainy season, animals move south from the northern pastures to take advantage of the greater availability of surface water. In some cases this movement will extend into the Guinean zone at the end of the dry season (April–May) to take advantage of the earlier rains there.

The strong rationale for movements along a N–S axis is illustrated by the fact that this pattern of seasonal movement is common no matter where a herding family lives along the axis (Benoit 1979; Gallais 1975; Bassett and Turner 2007; de Bruijn and van Djik 1995; Bonfiglioli 1988). Pastoralists in the north will move southwards from their home area at the end of the rainy season to pastures in the southern Sahel and Sudanian zone, sometimes being the guest of those that they have hosted during the rainy season. Those located in middle latitudes along the N–S axis may show two periods of transhumance away from home base: to the north during the rainy season and to the south during the late dry season. The corridors followed to make these movements are best seen as braided networks of paths linking different water/encampment points. The route followed within these braided networks depends in part on the forage/water availability at potential way stations. In areas of heavy cultivation pressure, those paths that remain are narrower, less weblike and more linear than those observed in the pastoral zone (Cissé 1981; Turner 1999b; Marty 1993; Garin *et al.* 1990). The actual choice of the paths to follow reflects not only fluctuating forage/water conditions but the geography of the social networks of those managing herds (Bassett and Turner 2007). Livestock are a mobile store of wealth that is vulnerable to theft, and the risk of livestock loss increases in areas where herd managers have few social contacts.

At an intermediate scale (≈15–100 km), livestock movements during the dry season have been oriented toward floodplain pastures—utilizing these pastures for periods of 0.5–6 months. On floodplain

pastures, livestock take advantage of the increased availability of surface water and greener vegetation while avoiding deep water and diseases prevalent during the wet season (Cissé 1981; Beauvilain 1977; Schmitz 1986). Reliance on floodplain pastures has declined over time as these pastures have increasingly been converted to crops. Away from floodplain areas, the extension of cropped fields in some areas of the Sudanian zone necessitates intermediate-scale movements to areas where natural pastures are less interrupted by cropped fields (Fig. 10.1).

Observers of remnant wildlife populations in the region have identified analogous N–S movements and, at more intermediate scales, movements to and from floodplains (Barnes 1999; Green 1988; Le Pendu and Ciofolo 1999; Poche 1974). However, finer-scale daily movements (≈3–15 km) of domesticated livestock diverge from those of wildlife in their point-centred nature. Cattle, sheep, and goats are managed not only for meat but for milk production. Under the open range conditions of the region, this goal necessitates separating dams from their offspring in order to capture for humans a portion of the milk produced. As a result, finer-scale grazing movements generally are loops departing and returning to fixed points (village, hamlet, pastoral encampment, water point) where the mothers are milked before being reunited with their young (Turner and Hiernaux 2008).

While access to forage and water is a major determinant of the movements of livestock in the region, it is important to recognize that actual patterns of livestock mobility may diverge from the patterns just described. Livestock herders are interested not only in locating their herds in areas with water and high densities of quality fodder, but also in being able to convert livestock production to grain and cash in order to support their families. Therefore, movements will not only be affected by the conversion of pastures to croplands but also by the variable access to milk markets and levels of security offered by different destinations. Moreover, livestock movements, particularly those distant from the home base, require the allocation of an adequate number of herders, which may or may not be feasible given the demands of the herd manager's other productive activities (e.g. farming

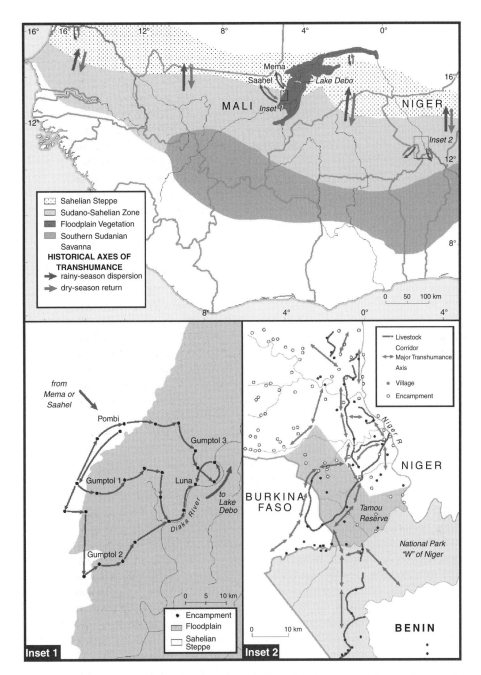

Figure 10.1 Sahelian migratory patterns. Seasonal movements (transhumance) of livestock herds in Sudano-Sahelian West Africa typically lead to northerly movements during the rainy season with a return at the beginning of the dry season (top panel). Herds follow corridors that join encampment points. Major deviations from this pattern occur in some places along the southern portion of the zone where herds move southward during the mid-to-late dry season to catch the earlier rains and return northward during the mid-to-late rainy season. An example of this pattern is followed by the FulBe of the Say region of Niger (Inset 2; bottom right panel) who move their herds southward into Burkina Faso or Benin. These movements are increasingly blocked not only by cropland expansion but by stricter enforcement of protected area boundaries. Historically, floodplains have represented major dry-season pasture resources. Pastoralists' access to these pastures has been eroded over time. One area where this movement pattern persists is the Inland Niger Delta of Mali. Along the western edge of the floodplain (Inset 1; bottom left panel), herds move to either the Mema or Saahel during the rainy season; return to the floodplain in the early dry season; and converge toward the deeper part of the floodplain (Lake Debo) as the dry season proceeds. Over the last 40 years, growing land-use competition and reduced flooding have necessitated shifts in the floodplain paths (*gumptol*) followed by the herds (Inset 1 presents shifts in gumptol location (1 to 2 to 3) due to such pressures).

etc.) and his effective access to labour (Turner and Hiernaux 2008). Socioeconomic trends in the region have contributed to; a growing fraction of the regional herd being owned by those who are not herding specialists, and allocations of labour away from herding to other productive pursuits (Habou and Danguioua 1991; Turner 2009). As a result, longer-range livestock mobility has declined, with a growing fraction of livestock managed year-round in the Sudanian zone (Turner and Hiernaux 2008).

10.3.1.2   Nutritional trade-offs of livestock movements
Despite the fact that the region's biogeography requires livestock movements, decisions by herd managers whether to move to particular destinations are often not clear cut. Not only are there risks to livestock movements from human-caused insecurity, but the nutritional calculus is not always clear (Schlecht, Hiernaux, and Turner 2001; Colin de Verdière 1995; Andriansen 2003). This is because the energy expended by animals as well as the stress put on them in moving longer distances with less opportunity to eat and drink (e.g. at the end of rainy season) may outweigh the benefits of the greater access to forage/water at the destination. Despite the romanticism attached to the pastoral lifestyle by western observers, this calculus very much shapes the movement decisions made by pastoralists themselves. Moving with livestock is not driven by cultural attachment but by conscious and sometimes agonizing decisions. The nutritional trade-offs surrounding livestock movements become increasingly unforgiving at the end of the dry season and beginning of the rainy season. At the end of the dry season, livestock movements generally decline despite increasing pasture scarcity reflecting not only the general poor quality of natural pastures but the weakness of livestock. Early rains lead to patches of early grass growth. Decisions on whether or not to move to these areas are some of the most difficult that a herd manager faces throughout the year. Miscalculations can result in the loss of a significant portion of a herd. Under better conditions, decisions to move, while beneficial to most of the herd, may result in the death of the weakest animals along the way.

10.3.1.3   Livestock mobility and the environment
Vegetation and soil structure in the region is most sensitive to livestock grazing during the short rainy season (Hiernaux and Turner 1996; Penning de Vries and Djitèye 1982; Turner 1999c). The effects during the long dry season are much less pronounced given that grasses have seeded and dry soils are less vulnerable to compaction. Therefore, dispersal of grazing animals over a wider area during the rainy season is less likely to result in ecologically-damaging grazing pressure than if livestock are congregated near human population centres. With all else equal, this supports the historic pattern of south-to-north transhumance movements during the rainy season (Breman and de Wit 1983). Consistent with general arguments made for rotational grazing and holistic rangeland management, short bouts of heavy, non-selective grazing pressure on Sahelian annual grasslands have less effect on species composition, grassland productivity and soil structure than more persistent grazing pressure during the rainy season (Turner 1999c). Such temporal patterns of grazing pressure are more likely to be produced by more mobile pastoral systems of livestock management where livestock often remain for not more than three days at any one location. One important caveat to this conclusion is that when distinct waves of herd visitations occur throughout the rainy season, the grazing pattern experienced locally may not diverge significantly from more sedentary systems.

10.3.1.4   The role of human management and livestock mobility
One important question is the role of human management in the efficiency of adjustments of livestock grazing to the changing spatial distributions of available forage. While variation exists at the species and individual level, domestic livestock show abilities (similar to those of wildlife) to move to areas with higher forage potential outside of their immediate sensory range. For example, Sahelian herders learn to look for lost cattle at the beginning of the rainy season along transhumance corridors to the north and likewise along the same corridors to the south at the end of the rainy season. At more local scales, goats are particularly aggressive seekers of palatable forage across village landscapes, compared to cattle and

sheep. Sheep, when left on their own, may choose to remain near villages during the dry season rather than venturing forth in search of forage.

Herders play an important role during the rainy season by keeping livestock out of cropped fields. But is this their only role? How does herding affect the ability of livestock to access forage on open pastures? In the abstract, one could argue that herders will play a positive role in this regard. First, herders arguably can access a wider range of geographic information about forage/water availability through their social networks than can the animals under their care. Second, herders aim to offer animals in the herd a range of forage types to meet the needs of all members of the herd. For example, during the transition periods from dry-to-rainy and rainy-to-dry seasons, the rumens of different animals adjust at different rates to forage of different qualities and moisture contents. In their daily grazing orbits, herders will intentionally lead animals along routes that provide a mix of dry and green vegetation from which animals can choose. They are well aware of the types of vegetation preferred by different individuals in the herd during such transition periods. An important empirical question is whether wild ungulate herds would access similar mixes of fodder types from orbits led by lead animals. While it is difficult to find cases of domestic livestock going on transhumance without herders, studies of local grazing patterns of herded and free pastured animals in the region have found that herding does have statistically significant effects, not only increasing the dispersal of animals around villages and water points but increasing the time spent in areas of high densities of palatable fodder (Turner and Hiernaux 2008; Faugère *et al.* 1990a, 1990b).

10.3.2 Temperate Asia: Steppes of Mongolia

Pastoral environments in Mongolia and northern China are arid to semi-arid steppes with maximum elevations of 4000 m. The distribution of precipitation is unimodal with most rainfall falling in the summer months, and the dry season coinciding with winter. Spring snowstorms are not uncommon and severe winter weather is a regular feature (Fernandez-Gimenez 1997). Winter minimum temperatures reach below −40 °C and summer temperatures can exceed 32 °C (Fernandez-Gimenez 1997; Erdenebaatar 2003; Wang 2003; Bedunah and Harris 2005). Most pastoralists in this region herd a diversity of livestock species, including sheep, goats, cattle, horses and camels. At higher elevations and more northern latitudes, yaks replace cattle and camels are scarce or absent, while cattle are often scarce in the more arid desert steppe.

Contemporary migrations are primarily elevational (Fernandez-Gimenez 1997; Erdenebaatar 2003; Wang 2003; Bedunah and Harris 2005) although, in historical times, Mongolian pastoralists in some regions made long-distance latitudinal migrations (Fernandez-Gimenez 1999; Humphrey and Sneath 1999). Both historical and current movement patterns vary greatly with local geography, and distances moved are often highly variable within and between locales, and within and among years in a given location (Fernandez-Gimenez 1999; Humphrey and Sneath 1999; Erdenebaatar 2003; Fernandez-Gimenez and Batbuyan 2004; Fernandez-Gimenez *et al.* 2007).

Key environmental drivers of herd migrations include the following.

1. Plant production and forage quality and quantity vary over space at multiple scales, as well as seasonally and inter-annually. At the broadest spatial scale, Mongolia's ecological zones are arrayed along latitudinal and elevational gradients with increasing precipitation at higher latitudes and elevations, corresponding to higher production and greater species richness, but not necessarily higher nutritional quality. Within ecological zones, soils, topography and grazing pressure drive species composition and production in the mountain steppe and steppe zones, with spatial and temporal variability in rainfall playing a greater role in the desert steppe (Fernandez-Gimenez and Allen-Diaz 2001). Within years, forage quality is highest in late spring and declines as plants senesce. At longer time scales, prolonged droughts, which affect forage abundance and quality, drive long-distance moves and result in temporary or permanent relocation of herds and households.

2. The quality and quantity of water for livestock and domestic consumption is a key driver of herd movements. Water is most abundant during the summer rains. However, water quality in stationary sources such as desert springs declines in the summer, leading to increased gastrointestinal problems for livestock and people, and influencing movements towards free-flowing water (rivers) and deep wells. In the winter, herders in many areas rely on snow for domestic water, and this may limit their use of some pasture if snow is insufficient. The abundance of natural surface water sources has declined significantly over the last decade potentially due to climate change (Batima 2006). In addition, many wells constructed during the socialist collective era have fallen into disrepair since privatization in 1992 and are no longer functioning. The declining availability of surface and well water is a serious constraint on mobility and pasture use.

3. Mongolia has an extreme continental climate with warm summers and cold, dry winters. High winds are common, especially in the autumn and spring. These climatic conditions motivate herders to seek sheltered campsites during the cold and windy seasons, especially when animals are giving birth in the early spring.

4. Severe winter storms known as *dzud* have historically occurred on a 5–8 year cycle. These storms often render forage inaccessible if snows are deep, and act as a density-independent limitation on livestock populations. When dzud occurs, herders who are able to, move their animals to avoid livestock mortality, sometimes long distances.

The environmental and social drivers of herd migrations on the Mongolian steppe are illustrated in the following case study of Jinst Sum (district) located in Bayankhongor Aimag (province) in the Gobi desert-steppe region of western central Mongolia. Pastoral herd movement patterns in this area have changed over time with shifting political regimes and administrative boundaries (Fernandez-Gimenez 1999; Humphrey and Sneath 1999). However, basic elements of the movement patterns have remained consistent, even as the average distance moved and the range of habitats used has become more limited, or fluctuated with environmental and economic conditions.

Jinst Sum is located on a broad plain between the Khangai and Gobi Altai Mountain ranges, and averages 1380 m in elevation. A low range of foothills called Narin Khar Ridge bisects the district from the west. The Tuin River flows through Jinst from its headwaters in the Khangai mountains to the North towards the large inland lake, Orog Nuur, located just south of the district's southern boundary. Orog Lake, like several smaller springs in the district, is surrounded by a lush desert marshland. Natural water sources in the district include the Tuin River, and two natural marshes, Khar Us (Black Water) and Khuis Us (Navel Water). In addition, herds water at a number of hand-dug and mechanical wells scattered across the landscape.

Pastoralists in Jinst typically move at least four times annually between 3 or 4 distinct seasonal pasture areas (Table 10.1, Fig. 10.2). Here, we describe the broad patterns of movement of households and herds dominated by small stock. It is important to note, however, that within seasons different livestock species are herded to different plant communities within a seasonal pasture area, as appropriate to each species' dietary needs. Spring and early autumn pastures are located in desert marsh riparian areas with good spring water supplemented by numerous wells at their peripheries. The vegetation in these areas is dominated by the tall, coarse-textured grass *Achnatherum splendens*, which is high in structural carbohydrates and relatively low in protein content. Production in these areas is high, and forage quality is low during most of the year. *A. splendens*, called *ders* in Mongolian, is most palatable to a wide range of livestock in early spring, when tender new growth is sprouting. Its tall structure and the large amount of standing dried biomass in *ders* areas affords shelter from spring winds for small stock and young animals, and high-volume forage for mature large stock (primarily horses and camels). Often these riparian oases are located adjacent to surrounding salt-shrub communities, which are low in grass production, but contain a number of halophytic species with moderate to high nutritional value (e.g. *Salsola passerina, Anabasis brevifolia*), which are preferred by camels and goats. The salt content of plants and soil in these communities is thought by herders to be important in helping animals consolidate their fat in the autumn, and the mineral salt is also collected for human use. Standing biomass in these salt shrub communities is relatively low, 16.5 g/m^2, but still higher than the

Table 10.1 Summary of production objectives, environmental and social pasture/campsite selection, and movement criteria for the Jinst Soum, Mongolia case study (Source: Fernandez-Gimenez (1997)).

	Spring	Summer	Autumn	Winter
Production objectives	Early: conserve fat; late: recuperate from winter; early and late: good birth and survival rate	Put on fat; milk production	Consolidate fat; gain fitness/resistance for winter	Conserve fat
Environmental criteria	Warm (south-facing slope); sheltered (lee-side); deep, dry bedding ground; water nearby; early snowmelt; early grass growth	Cold, clear water; few insects; forage quality and quantity	Open steppe Salt and mineral licks; highly nutritious plants (*Allium*, *Artemisia*); good cold water; cool, hard-surfaced bedding grounds	Warm; sheltered; deep, dry bedding ground; standing forage reserve; shallow snow or water (wells)
Social/economic factors	Labour; transportation; campsite possession or access rights; access to markets and services	Number of neighbours; access to markets and services	Number of neighbours; labour; transportation; access to markets and services	Labour; transportation; campsite possession or access rights; access to markets and services
Habitat	Desert marsh	River banks and lake shores	Open steppe	Sheltered mountain valleys, canyons, or foothills; usually far from natural fresh water; cold temperature; snow for domestic use
Movement criteria	Reproductive cycle (before lambing/kidding becomes intense) water availability; Greening of grass	Grass growth; weather; water quality	Weather; phenological changes in plants; animal behaviour	

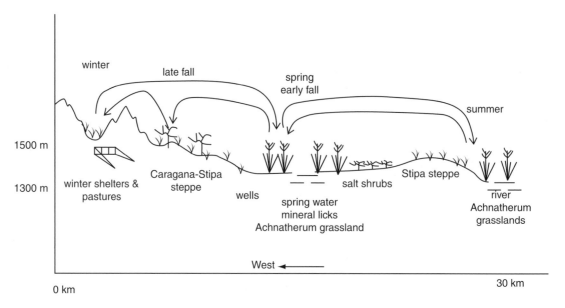

Figure 10.2 Schematic cross-section of seasonal movements in Jinst Soum, Mongolia.

herbaceous production of the upland steppe (Fernandez-Gimenez and Allen-Diaz 2001). Rights to spring camps are usually governed by customary use and informal tenure arrangements, although increasingly households hold formal possession licences over spring campsites.

In the summer, herds move away from still-water riparian oases as the water grows brackish and foetid, making people and animals sick, and the insects become intolerable. During the hot summer months in the Gobi, Jinst herders camp on the banks of the Tuin River, where their herds have access to riparian grazing lawns (typically *Elymus chinensis* and various *Carex* species), as well as the upland desert-steppe vegetation (perennial bunchgrasses, primarily *Stipa gobica* and *S. Glareosa*, interspersed with forbs and scattered leguminous shrubs), water is cool and plentiful, and the breezes keep the insects away. Summer pastures are essentially open access, with no exclusive rights to graze permitted. Thus, herders distribute themselves where they choose, but take into account both the benefits and drawbacks of proximity to and abundance of neighbours. Herders' main objectives in the summer are to maintain high levels of milk production in lactating animals, and to enable animals to recover and fatten after winter.

In the autumn, especially late autumn when temperatures cool, herders move out into the expansive desert-steppe uplands, obtaining water from wells, or trekking their herds to the springs every 3 days. Here, they seek highly nutritious plants to fatten their animals, including wild onions, *Artemisia* subshrubs, and salt shrub species. Typical herbaceous standing biomass in these desert-steppe communities is 10.75–13.6 g/m^2 (this estimate does not include the shrub component; Fernandez-Gimenez and Allen-Diaz 2001).

Migration from autumn to winter camps is triggered by colder temperatures and the arrival of snow. Winter camps are the most permanent bases in this mobile society and, by the end of the 20th century, most households or herding camps held long-term possession licences giving them formal exclusive rights to their winter campsites. In Jinst, most winter campsites are located in sheltered mountain valleys and canyons, usually on south-facing slopes and protected from the prevailing winds. Water is obtained from snow, wells, or from ice blocks transported from the frozen river. Winter forage is standing dry grass, usually the upland desert-steppe bunchgrasses, *S. gobica* and *S. glareosa*. In the desert steppe, little fodder is cut and stored, although *ders* is occasionally stockpiled as hay, and some herders prepare small amounts of home-made concentrate feed from a combination of grasses and high protein forbs. In winter, the shallow but extensive snow cover (for water), finite forage supply, and requirements for a deep, dry bedding ground for livestock, lead herders to disperse their camps, composed of only a few households, over a wide area, in contrast to the aggregated settlement pattern along the river during the summer.

In addition to this regular pattern of seasonal transhumance, herders of Jinst and other parts of Mongolia sometimes take their herds on treks that are much longer in distance and duration, usually in response to prolonged drought or severe winter conditions. Changing economic opportunities and constraints following the transition to a market economy in the early 1990s also influenced herd mobility, often in ways that are suboptimal from a forage matching standpoint. That is, herders remained longer in areas that were over-crowded and where the quality or quantity of forage was insufficient, because they lacked access to transportation, or desired proximity to markets, schools, or other social services. Patterns of mobility in Jinst are dynamic in response to both climate-driven forage availability and economic factors (Fernandez-Gimenez *et al.* 2007). In the absence of formal individual or collective tenure over pasture areas, droughts and forage shortages have led some herders to migrate far outside of their customary transhumance cycles, encroaching on the customary territories of other groups of herders in neighbouring districts or provinces, consuming winter grazing reserves and creating conflicts over pasture (Fernandez-Gimenez *et al.* 2008).

There are few wild ungulates in Jinst today, however important populations persist in other regions of Mongolia, including some with similar ecological characteristics. Recent research on the migration patterns and diets of Mongolian gazelle (*Procapra gutturosa* Pallas) permits a tentative comparison of influences on livestock and wildlife movements on the Mongolian steppes, and particularly the similarities and differences in diet due to seasonal foraging patterns and the influence of herding on domestic stock. Mongolian gazelles today primarily inhabit Mongolia's eastern steppes, but their range extends into the eastern part of the Gobi (Yoshihara

et al. 2008). The diet of the Mongolian gazelle varies seasonally, in ways similar to the seasonal foraging patterns observed for herded small stock in Jinst. In a study of Mongolian gazelles in China, protein consumption was highest in spring, wild onions dominated the diet in summer, and grasses were the primary food source in winter (Jiang *et al.* 2002). A study of food resources used by Mongolian gazelles and livestock at three sites in Mongolia found that in the easternmost steppe site there was a high degree of diet overlap between sheep/goats and gazelles. However, diets of domestic stock and gazelles diverged significantly at the desert-steppe site, leading the authors to conclude that herding of sheep and goats strongly influenced their diets at this site (Yoshihara *et al.* 2008). Other studies have found that production alone (estimated by NDVI values) did not fully explain gazelle migration patterns (Ito *et al.* 2006), or that gazelles appeared to prefer areas of intermediate productivity (Mueller *et al.* 2007), indicating support for a trade-off between quality and quantity of the food available. NDVI values best predicted habitat use in the winter months (Leimgruber *et al.* 2001; Ito *et al.* 2006).

In summary, Jinst herders balance a variety of biophysical and social factors in making movement decisions. Forage availability and quality govern some decisions, especially at smaller spatial scales, whereby pastoralists herd different livestock species to different plant communities and microhabitats based on their dietary preferences and needs. Although specific data on seasonal diet composition are lacking, we infer, based on our knowledge of herd movements and habitat use, that quality is highest in the late spring and early summer, when herds use salt-shrub communities, and new growth of *Achnatherum splendens* and *Stipa* species is available. In early autumn, quality is also likely to be higher, giving herds a nutritional boost just prior to and during the breeding season for camels and small stock, as animals consume more protein-rich forbs (*Allium* species) and shrubs (e.g. *Artemisia xerophytica*), which are believed by herders to help animals consolidate fat. Both diet quality and forage quantity are lowest during late winter and early spring, when animals rely on standing dry biomass for maintenance. These patterns show overall similarity to one diet study of Mongolian gazelles, and both wildlife and domestic stock make use of pastures with high amounts of relatively poor-quality standing biomass in the winter and early spring, the annual forage bottleneck. However, livestock production is water-limited, and herders in the semi-arid Gobi region clearly balance considerations about water quality and quantity with those of forage quality and quantity, as well as economic factors. As changing climate and land use affect the availability of surface water, water has become an increasingly important determinant of, or constraint to, movement. Herders consider the density of other households when deciding where to camp in summer, when abundant fresh water is a limiting factor, and distribute themselves broadly across the landscape in winter, when forage is more limiting than water. However, social and economic factors, such as the need to be close to settlements for access to schools, markets or services, increasingly override biophysical considerations, resulting in increasing degradation around settlements (Okayasu et al. 2007). These mounting social and economic constraints may account, in part, for observed divergence of Mongolian gazelle diets and the diets of small stock in the Gobi region (Yoshihara et al. 2007).

10.3.3   Arctic case study: Nenets reindeer nomads East and West of the Polar Urals

Reindeer pastoralism is practised in Eurasia by approximately 20 indigenous groups in the Arctic and subarctic involving roughly 100 000 people and 2 million reindeer (ACIA 2004). Traditionally, reindeer pastoralism was highly mobile but few groups continue to practise long-distance nomadism.

In the Arctic and subarctic, unlike in many other pastoral regions, diversification of the species composition of herds is not an option due to difficult climatic conditions. The exception is eastern Siberia, where an ethnically mixed community of Eveny and Sakha herd reindeer, horses and genetically unique breeds of northern cattle under extreme conditions (Maj 2009; Takakura 2002; Granberg et al 2009). In all other regions of the North, herders who want to spread risk or dampen the volatility of natural disasters engage in additional non-pastoral activities such as fishing, hunting, gathering and wage labour.

The Arctic as a pastoral habitat is characterized by:

1. low plant productivity limited by a climate that leaves the pastures free of snow roughly between May and September;
2. a low number of plant functional types with value for reindeer grazing, mainly tundra grassland (e.g. *Carex* spp., *Eriophorum* spp., *Poaceae* spp), shrub tundra (e.g. *Salix* spp, *Betula nana*) and lichen ranges (*Lichenophyta, e.g. Cladonia* spp. *Cetraria* spp);
3. mostly frozen soils throughout the year. In many Asian areas of the Arctic in summer the active layer above the permafrost level is less than 1.5 m, though there are very few permafrost areas under reindeer pastures in the European Arctic;
4. precipitation between 370 and 500 mm in the high Arctic, slightly more in the subarctic. The yearly distribution of that precipitation is changing with a changing climate (Frey and Smith 2003);
5. generally low temperatures but much surface water from countless rivers and lakes, so that water supply is never a problem for livestock. The combination of low precipitation and abundant surface water makes the arctic a 'dry wetland.'

Migratory reindeer herds generally follow plant growth, advancing north with the greening of spring pastures and retreating south as plants senesce in autumn. The most common (but not the only) migration pattern is for reindeer to move between lichen-rich winter pastures in a forest zone and herbaceous summer pastures at windy locations on the coast, where insect harassment is reduced, or at higher altitude. The diet of reindeer can be broadly characterized as a gradient between these two extremes, shifting from a dependence on lichen consumption in winter to maximizing the consumption of green plants in summer, supplemented in autumn by berries and mushrooms. Summer is the season for gaining weight, often on salty coastal pastures that, according to Yamal herders, 'increase the appetite of the reindeer and make them grow faster and fatter'. Lichen pastures are useful for surviving the winter by maintaining the body mass built up in summer. Late winter/spring pastures are often the seasonal 'bottleneck' determining the survival rate of the herd over the year.

Among both domestic and wild reindeer we find highly mobile populations as well as localized populations, with the reindeer in Taimyr covering the largest distances for wild reindeer—up to 1500 km annually (Geller and Vorzhonov 1975; Klokov 1997)—and the Siberian Nentsy and European Komi reindeer covering the longest distances for domestic reindeer accompanied by herders—1200 km annually (Stammler 2005; Habeck 2005; Dwyer and Istomin 2008). In many cases pastures serve as habitats for both 'migrant' and 'resident' (more localized and less migratory) reindeer, such as on the coast of the Barents Sea (European Russia), the Northern Yamal and Taz Peninsulas of West Siberia, and the Taimyr Peninsula (wild reindeer). Where the two patterns of movement coexist, 'resident' animals are fatter in good years because 'migrants' lose body weight on their long-distance movements. On the other hand, in bad years 'migrant' animals benefit from greater flexibility in obtaining forage during extreme weather events such as the icing-over of pastures.

Less migratory herds do not change ecological zones, but instead over-winter in treeless areas in the high North, or stay year-round in the forest in the more temperate zones (Stammler 2005 for northern Yamal Nentsy; Yoshida 1997 for Gydan Nentsy; Dwyer and Istomin 2008 for Taz Nentsy). Migration between these localized herds is influenced by the same factors as long-distance migratory systems, but they play out on a different geographical scale and herders have more scope to determine their own hierarchy of criteria for pasture quality at particular times of the year. For example, some coastal pastures suitable for summer use are also lichen-rich and suitable for winter occupation. As lichens are more valued than salinity by most herders, coastal locations that are open year-round are often preserved for winter grazing and summer is spent elsewhere.

Vegetation is only one—and not always the most important—consideration governing movement (Kitti *et al.* 2006). Depending on the relative importance of reindeer herding and fishing for household income, migration routes for the former are adjusted to match the seasonality of the latter. On the West Siberian Yamal Peninsula, many private reindeer herders choose their migration routes to match

good fishing lakes in the summer. Even though there is significantly more mosquito harassment around these lakes, the value of the fishing in summer outweighs the disadvantages of insect harassment for the herd. A better option for small herd owners is to place their animals with bigger herds for efficient summer fattening on wind-exposed pastures, while they fish in rivers and lakes. Such herders resume reindeer herding in autumn, when they move back south to pastures that again are chosen for their proximity to good locations for ice-fishing. In other cases, herders whose reindeer follow localized grazing patterns can detach camp movements from herd migration, in which case the camp follows a nomadic migration route that combines hunting/fishing considerations with the occasional need to be close to the herd (see the second case study, below). In general, as herd mobility is reduced, hunting and fishing become more important in the household economy.

The following case studies describe a typical long-distance migration corridor used by mobile camps with large herds in the Yamal Peninsula, and a second more localized migration pattern exemplified by a Nentsy community on the shore of the Barents Sea in the European North of Russia.

10.3.3.1 Case 1: Long-distance migration by large 'migrant' herds
Movements of people and animals among the Nentsy are influenced by a complex set of non-botanical factors, most important of which are pollution by industry, different land rights regimes and post-Soviet reorganisation, distance from human harassment/poachers, proximity to markets, soil humidity/dryness, pasture elevation/windiness, mosquito harassment, coastal salinity, noise harassment (pipelines, roads, railroads, wind parks), freezing over of pastures and snow depth (Fig. 10.3). Table 10.2 summarizes the interplay between vegetation condition and these additional factors in establishing a system of long-distance migration.

10.3.3.2 Case 2: Short-distance migration of a smaller 'resident' herd
Table 10.3 summarizes the interplay of socioeconomic and biophysical factors in establishing a system of short-distance migration. In this second

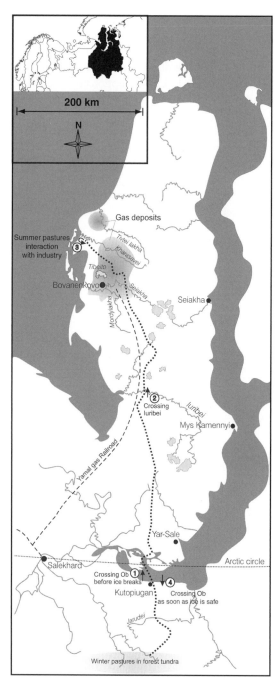

Figure 10.3 The seasonal migration cycle of a Nentsy reindeer nomad camp, Yamal Peninsula. Modified from Stammler (2005, p. 104).

example, the close observation of the whole environment, including the pastures, permits herders to make optimal use of a confined territory.

This herd moves in an area that is small enough to detach the camp location from herd location where necessary. Herders make use of the territory for fishing, hunting and gathering, which influences camp location. In localized migratory systems, herders commonly use large rivers and lakes as 'natural fences', a practice that is effective from May until October, when these bodies of water freeze over and no longer form a barrier to animal movements. In the example described here (Table 10.3 and Fig. 10.4), when the herd is on a peninsula during the rut, water forms a barrier from three sides, while the camp location on the only open side 'closes the bag' and traps the herd until it calms down. Only then is the herd released and guided towards lichen pastures on the coast. These herds also graze on pastures that are seasonally flooded by salt water, which minimizes the need for herders to supply additional salt for their animals.

10.3.3.3   Decisionmaking in migration

The primary bottlenecks that impede both localized and long-distance movements are constricted passageways in early spring and late autumn. Humans push their herds through these pastures as quickly as possible. In the first case, this happens according to a sophisticated timetable, as these passageways are used by successive waves of camps and herds. Delayed exit from these passages by early waves of animals can lead to mixing with incoming herds. Separating mixed herds is laborious and herders try to avoid it if at all possible.

Particularly important for structuring migration are decisions about whether to cross rivers. In the previous example of long-distance migration, the Yuribei River must be crossed before the ice breaks in spring, otherwise camps and herds are delayed for months waiting until the water level falls. Further north, the Se-Yakha River must be crossed before the main mosquito eruptions, which requires the herds to navigate through difficult industrial gas installations along the migration route. Arriving at their northern destination on time promotes the optimal weight gain of the animals. Lingering on northern pastures before moving back south is also

Table 10.2 The seasonal migration cycle of a Nentsy reindeer nomad camp, Yamal Peninsula. The herd is of approximately 5000 reindeer, managed by 50 people, travelling an overall distance of 1100 km. Southern winter pastures at 65.30 NL, northern pastures at 70.55 NL (Source: Stammler (2005, pp. 91–118); see summary table pp. 101–103)

Season and approximate months	Movement	Pasture conditions	Other considerations
Early spring: March–April	Start migration northwards from forest zone with frequent movements and new camps every 2–3 days.	Lichen-rich pastures in the forest zone; with movement north changing from forest zone to treeless tundra; migratory corridors narrow and are grazed intensively by several herds one after another.	The 'Day of Reindeer Herder' festival, the year's biggest party and main yearly stop in village to get supplies.
Late spring: April–May	Rush to the north before calving as long as watercourses are frozen and easy to cross. If rivers break-up before crossing, then passage to summer pastures is blocked for months.	Transitory pastures, used by many camps; a mix of lichen and first snow-free green patches. By many herders seen as the main bottleneck of the year.	Calving often happens on the move; herders must balance the need for frozen snow to ease movement at night, versus the need for times of rest on snow-free patches for calving.
Summer: June–July	Continue north to the coast, reaching it with the start of mosquito harassment. Careful balance of feeding, drinking, resting time required for optimal weight gains. Staying at northernmost coastal area one week in August.	Selective grazing of the first leaves of dwarf bushes (*Betula*, *Salix*), and fresh grass. Salty coastal grass and pastures exposed to wind are important for reindeer appetite.	Movement is more and more restricted by the presence of extractive industry (e.g. crossing Bovanenkovo gas deposit).
Early autumn: August–September	Turn back south, crossing major rivers before water levels rise in autumn.	Feeding on mature green plants, being distracted by mushrooms and berries leading to increased herd restlessness.	Navigating through industrial installations and trading with their workers.
Late autumn: October–November	Continue south, crossing rivers as soon as they freeze over.	Proportion of lichen in pastures increases as snow falls. Narrow transitory corridor pastures, heavily used by competing camps. Danger of pastures icing-over with sudden temperature and precipitation changes.	Change to winter gear and set up the stove. Bulls are restless because of rut. Prepare for the annual reindeer count and select corral for slaughter; figure out property relations in animals and socialize with neighbours.
Winter: December–March	Cross into the forest zone; the pace of movement slows, herds are not watched every day, and camps are not relocated for 2 weeks or more.	Lichen-rich pastures to sustain animals during winter cold down to -50°C. Snow-depth possibly restricting pasture access.	Counting and slaughter animals and fetch from town yearly income and welfare payments.

Table 10.3 The seasonal migration cycle of a Nentsy reindeer nomad camp, Kolokolkova Bay, Barents Sea. The herd is of approximately 500 reindeer, managed by 12 people, of which four are active herders. Overall distance: approximately 100 km for camp, 150 for herd. Pasture area between 68.30 and 69.00 NL (Source: Stammler fieldwork 2004 summer, Stammler and Vitebsky fieldwork 2005 spring)

Season (and approximate months)	Herd movement	Camp movement	Pasture conditions	Other considerations
Early spring: March–April	Move to the shore.	Moving from campsite encircled by lakes to a site 5 km east of the herd.	First green plants appear on salty pastures that flood later during the spring thaw.	Bulls move ahead of the rest of the herd.
Late spring: April–May	Females move inland to calving ground.	Move 5 km south from a hill to the shore of a small river, to prevent the herd returning to its previous calving ground.	Slightly elevated terrain with the most available snow-free patches.	Humans observe but do not interfere in calving except to adopt calves that are abandoned by their mothers.
Summer: June–July	Undivided herd moves to highest elevation in the area.	The camp and herd are united on Khekurseda hill. From this site people make fishing trips to a large nearby lake.	A windy site that minimizes mosquito harassment, but is also littered with stones that injure the animals' hooves.	Could move to the coast, but use of these lichen pastures is deferred until winter.
Early autumn: August–September	Move down slope along small rivers flowing north.	Southernmost camp site near the areas biggest river, for fishing; from here the camp is moved northwards to a large lake for autumn fishing.	Herd feeds on shrubs along rivers—last snow-free green food.	Mushroom season; the herd is restless but controlled by open water that forms a natural barrier.
Late autumn: October–November	Move to a small coastal peninsula for the rut.	Move west to a lakeshore closer to the rutting grounds. Herders from this camp move 15 km to 'lock' the herd on the 'rut-peninsula'.	Poor, lichen-free forage located on sandy lowland.	The peninsula is convenient for keeping the restless herd together. Avoid lichen areas to prevent trampling during rut.
Winter: December–March	Move to northern coastal areas in late winter, turning inland along the valleys of small rivers.	Stay close to the northern coast to guard the deer, or alternatively guide the herd northeast where more extensive pastures are available to accommodate herd growth.	Lichen-rich, salty pastures—ideal forage. The last green feed of the year is available, preserved under the snow along rivers.	There are alternative winter pastures from which to choose, depending on when large lakes with connections to the sea finally freeze over.

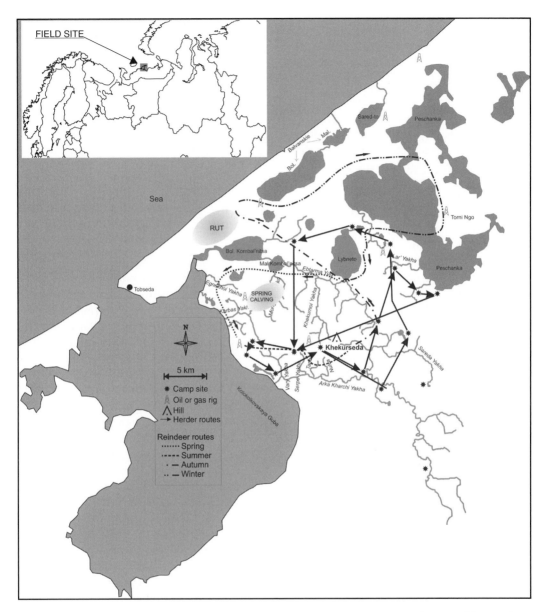

Figure 10.4   Schematic migration map of a Nentsy reindeer herding camp in a confined territory in European Russia (Source: Hand-drawn map by Petr Taleev during fieldwork of Stammler and Vitebsky, May 2005. Digitized by Arto Majoinen).

attractive, but must be weighed against the 450 km trek that must be completed in autumn to bring the herds close to the slaughterhouse.

In Case 2, the migration route described in the table was established only in 2003 after a private cooperative separated their herd and camps from a former state farm. The movements of this group

demonstrate how people 'design' migration routes in a relatively confined area. Unlike the long linear corridor migrations, localized herders make use of their limited area by moving in loops and figure eight patterns. The two fixed points in these annual cycles are the calving and the rutting sites. Both are identified by the herders and introduced to the herd in a process of mutual learning and forgetting. In this case humans establish new habits for the reindeer herd, which in the early years tends to wander off to previous calving and rutting places that have become inaccessible for political reasons. By observing herd behaviour at the new locations, humans learn the favourable characteristics of the new pastures while the herd forgets the old pastures, an intimate partnership in migratory decision-making involving both humans and animals. Far from being unstructured 'without any established migration route' (Dwyer and Istomin 2008), these localized movements are established in a process of trial and error over several years (see Beach and Stammler 2006).

Similar processes take place whenever political or administrative decisions render pastures inaccessible for herders and herds, following the reorganization of land rights in Russia (Stammler 2005), the fencing of herding territories in Finland, or the closing of national borders in the Nordic countries (Kumpula 2006). As natural resource and infrastructural development advances across the Arctic, these processes are likely to recur in many more cases (Tuisku 2002; Stammler and Peskov 2008).

10.3.3.4 'Push and Pull' factors in migration

To structure the rhythm of their migration, herders may alternately push the herd from behind, pull it from the front, or stop it from advancing too quickly. Placing a camp or a herding team in front of the herd to slow its progress is one way to decrease the extent of mobility among domestic reindeer. In autumn during rutting, the herders may place themselves in front of the herd in order to 'brake' the pace of bulls advancing for lichen-rich pastures. Alternatively, reindeer among the Nentsy may prefer to linger in spring on lichen-rich pastures until herders push them to begin the spring migration. Having to cross a river by a certain time causes herders to push their animals, and pushing is particularly important during calving, when weak calves may be tied to sledges to prevent them from lagging behind. Herders may also decide to move on when they find that it becomes increasingly difficult to keep their herd together on a particular pasture, as restless reindeer spread out. In this situation herders may find it convenient to keep their herd

reindeer happy and move with them to different pastures, before unhappy reindeer abandon their human keepers. In other situations the instinct of females to return to their former calving places may facilitate herd movement.

Lichen-rich pastures are preferably not used during the rut due to the trampling pressure of restless bulls (see above Case 2), which destroys lichen grounds. Lichen-preservation is a broadly accepted strategy by the people leading herd movements. Camps should not be placed on lichen grounds, and reindeer should not graze lichen in summer when the lichen are dry and of less value (Stammler 2005).

10.3.3.5 Day-to-day 'symbiotic domesticity'

Finding a good pasture on a day-to-day basis is a symbiotic activity between herder and herd, herder and lead animals. Driving a reindeer in harness exemplifies this relationship. At the macro level, an experienced reindeer driven in harness will know the general direction of an established migration route. At the meso level, the sledge driver will use the harness to determine the path taken, which is then followed by other sledge caravans and the herd, if camp and herd migrate together. This meso-level decision-making incorporates human knowledge of the land, for example avoiding contaminated sites, large concentrations of predators or humans, iced-over pastures, busy roads, or taking into account human needs such as passing through a trading post. On the micro level, however, it is again the reindeer that help determine the detailed direction and pace of movement—for example, using their judgment to avoid dense shrubs or stop at a particularly good feeding spot.

During the worst periods of harassment by mosquitoes, herders and herd together determine times of feeding, drinking and rest. Reindeer are guided back to the barren ground site in front of the camp several times a day, where they lie down and rest. Annoyed by the insects, they would otherwise run far against the wind, and 'forget' resting and feeding and lose weight. After feeding, herders may direct the herd in circular movements to a lake or river for drinking, before they go back to the barren ground site. In this way humans act in partnership with their animals to protect them.

10.3.3.6   Domesticated versus wild reindeer

Reindeer are exceptional because significant numbers of both domesticated and wild reindeer share the same habitat. Reindeer domestication began at least four centuries ago but is thought to be in an early phase and incomplete; sub-species identification is still under discussion (Roed *et al.* 2008). Comparative studies of wild and domestic variations of *Rangifer tarandus tarandus* emphasize differences in foraging behaviour, morphology, and intra-species diversification between different varieties of domestic reindeer and between wild and domestic animals (Zabrodin *et al.* 1979; Krupnik 1993; Geist 1999; Stammler 2005).

Some authors credit the differences between domestic reindeer populations to the selection process of indigenous herders (Baskin 1991). Geist argues that 'humans could give the reindeer habits that are useful in its control and exploitation' (1999, p. 317). Among these habits is a less mobile foraging behaviour, leading domestic reindeer to use up to 40% of available forage per unit of pasture, whereas wild reindeer use only 1–4% (Yuzhakov and Mukhachev 2001). Socio-culturally, there is little doubt that reindeer herders and hunters perceive wild and domestic reindeer as different animals. In most native languages, the terms for wild and domestic animals are not etymologically related (Vitebsky 2005; Beach and Stammler 2006). Herders say that wild reindeer are stronger and bigger than their domestic counterparts. Domesticated reindeer that are not closely managed become feral but can be re-domesticated, which is impossible with wild reindeer (Beach and Stammler 2006).

In the late 20th century, domestic and wild reindeer herds have mixed in many places in the Arctic. In these cases the domestic animals lose out: during the rut, wild bulls are stronger than domestic ones, leading to a 'half-wild' generation of calves that is not tractable for herding. In places that experience large increases in wild reindeer populations, many herders have lost their entire domestic herds to bigger wild herds, causing financial damage as well as cultural change (Klokov 1997 for Taimyr; Gray 2006 for Chukotka; Finstad *et al.* 2006 for Alaska). In Northwest Yakutia, however, coexistence may be possible and even beneficial for herders, who hunt wild reindeer for meat and also maintain their domestic herds as a more predictable resource base (Ventsel 2006).

Most domestic reindeer in the Russian half of the Arctic are managed in what is called 'close herding', while most reindeer in Fennoscandia and Alaska are managed in 'loose' herding (Baskin 1991) or ranching (Ingold 1980; Beach 1981), while some herders in Russia employ a combination of both systems (Takakura 2004). In close herding, reindeer are supervised and their movement monitored round-the-clock for most of the year except in late winter/early spring (Takakura 2004; Stammler 2005; Dwyer and Istomin 2008), which means that herders and reindeer negotiate on a day-to-day basis (and in summer even an hourly basis) the pasture to be used by the herd. In 'loose herding' and ranching, on the other hand, the herd is gathered only periodically to perform basic veterinary activities, cull animals for market and brand young animals. In close herding one of the main objectives of human influence is to prevent herd dispersion and mixing with other herds, a practice that has become unnecessary with the introduction of fencing and ranching in Fennoscandia. In these extensive systems the relationship between owners and their herds resembles the interaction between reindeer hunters and the wild animals that they ambush in autumn for slaughter. Property rights are the main distinguishing feature—pastoral animals are recognized as property when they are alive, while hunted animals are owned only after they are killed (Ingold 1980).

Building fences in Fennoscandia has reduced human mobility, with nomadic reindeer herders settling down in the period after WWII up to the 1960s. In the polar Ural Mountains of Russia, reindeer herders have also reduced their migration due to partial sedentarization in winter time, which led to the abandonment of winter pastures and the shortening of migration routes by around 150 km. There are still cases of fully nomadic migration of people and animals over thousands of kilometres (Stammler 2005; Habeck 2005; Dwyer and Istomin 2008). Long-distance migration has tended to survive when herders are able to minimize the costs of movement by using reindeer for transport. Conversely, the use of purchased inputs for transport, such as snowmobiles, generally contributes to a reduction in mobility and eventually to sedentarization (Pelto 1987; Stammler 2009).

Over the centuries reindeer pastoralism has moved through various phases. It apparently began with humans accompanying wild herds along routes determined by natural environmental factors and was gradually modified with the domestication, pacification and control of behaviourally modified domesticated reindeer. In the last century migration cycles have been shortened in response to technical innovations such as fencing, cost considerations, and administrative restrictions, a process that has frequently culminated in the settlement of human populations and the confinement of reindeer behind fences.

10.4 Conclusions: similarities and differences between wild ungulate and livestock migrations

This book synthesizes current knowledge about the migratory habits of wild animals. At an impressionistic level, pastoral and wild ungulate migrations are remarkably similar, and it is not immediately clear what sets domesticated migratory systems apart from those of other migratory species. At least three sets of variables drive both wild ungulate and livestock migrations: the distribution of resources, competition for these resources, and constraints on access to them. Both the distinctive features of pastoralism and its common heritage with other migratory systems are revealed by the way in which pastoral movements address these issues of resource availability.

Old World pastoralism spans the tropics to the Arctic, making productive use of biomes as varied as deserts and savannas, marshes, mountains, high meadows, plains, taiga and tundra. The cases presented in this chapter provide a small but geographically dispersed sample of this diversity, and support the following conclusions.

10.4.1 The management of forage supplies

Forage quality is often inversely related to abundance, both over time as plants mature and along fertility and productivity gradients in heterogeneous landscapes. There is a broad similarity between the way in which pastoral livestock and wild ungulates exploit this variability. For both, optimizing reproduction is a function of maximizing nutritive intake by targeting areas of high quality forage when food is abundant and shifting to areas that produce peak vegetative biomass when food is scarce. Long-distance meridional pastoral movements in Eurasia and Africa conform to the predicted oscillation between refuge areas during periods of plant dormancy (the dry season in the semi-arid tropics or winter in the temperate or Arctic zones) versus breeding grounds in periods of plant growth (the rainy season in the tropics and summer in the northern latitudes).

Like remnant wildlife populations in the region, cattle populations shift northward into the Sahel during the rains to exploit the brief but nutritious florescence of annual vegetation, retreating south to floodplain pastures as the dry season intensifies. Both wild reindeer/caribou and domestic reindeer herds move to herbaceous pastures for calving as 'snow melt triggers green-up, the burst of plant growth which has the highest nutrient value' (Gunn 1998, p. 320). Before the collapse of the Soviet Union, the same nutritional considerations drove long-distance sheep and saiga migrations on the Kazakh steppes, which tracked the advancing wave of green-up northward in spring and retreated to the south in autumn with advancing snow cover (Singh *et al.* 2010; Bekenov *et al.* 1998; Robinson and Milner-Gulland 2003; Fedorovich 1973).

Localized migratory movements—exemplified in this discussion by migrations on the Mongolian steppe and among Nentsy herders along the Barents Sea—are influenced by the same forage considerations as long-distance migratory systems, but play out on a restricted geographical scale (Coughenour *et al.* 1990; Coppock *et al.* 1986; Behnke 1999; Alimaev 2003; Behnke *et al.* 2008; Wacher *et al.* 1993). Extremely localized systems may not look migratory at all. In a normal rainfall year village cattle in Zimbabwe move no more than a few kilometres up or down a catena, exploiting the distinct vegetation resources on offer in different microenvironments at different times (Scoones 1995).

The seasonal trade-off between forage quality and quantity is also reflected in the way contemporary migratory pastoralists provide supplementary feed for their animals. In Fennoscandia, where herd movement is limited by fences, additional feeding is used to mitigate extreme weather events. Natural hay and home-made concentrates are a component of the winter diets of Mongolian herds; crop residues and other forms of supplementation are important additions to the dry-season rations of Sahelian livestock. What sets contemporary migratory husbandry systems apart from more intensive forms of livestock farming is not the absence of supplementation, but how it is used. In intensive systems, feed supplementation and cultivated forage are used to maintain high, steady levels of growth and output. Under these conditions, prolonged periods of weight loss or low performance simply waste feed and money and are indicative of poor management.

The seasonal oscillation between scarcity and plenty in migratory systems sets a different husbandry challenge. The fluctuating weights of Sahelian cattle illustrate the problem. In the dry season cattle in the Sahel routinely start to starve, and in drought years they continue starving until they die (see Ellis and Swift 1988 for the Turkana of East Africa). In the northern latitudes, the comparable crisis period occurs in late winter or early spring, when emaciated animals struggle to survive until warmer weather brings fresh grazing. Provided that they make it through the crisis, migratory breeds quickly regain condition and weight when favourable conditions return. The trick is to keep as many animals as possible alive until that time.

If they use preserved feed supplies at all, pastoralists typically use them to maintain animal numbers during periods of forage deficits, in the expectation that growth and reproduction will return when natural forage is plentiful, provided enough animals survive. Animals are fed not to make them fat, but simply to keep them alive. By buffering pastoral livestock against fluctuations in forage supplies, feed supplementation can maintain larger livestock populations than those attained by natural ungulate populations that are subject to the full rigours of climatic variation (Kerven 2004; Bayer and Waters-Bayer 1994). The occasional use of stored feed or grazing reserves (Fernandez-Gimenez and Le Febre 2006) epitomizes the transitional nature of migratory pastoralism—an agricultural

production system, but one that attempts to exploit, and occasionally supplement but not replace, naturally occurring livestock feed sources.

The burning of tropical pastures provides another example of the gradual pastoral transition from natural to managed grazing. Because they are generally located in areas of higher rainfall where stock water is more readily available, tall-grass savannas provide dry-season refuges for Sahelian herds. In the dry season these pastures are mature and high in bulk but low in quality. Pastoralists frequently burn these pastures early in the dry season to stimulate protein-rich re-growth based on residual soil moisture, or to render fresh growth accessible to grazing by removing unpalatable stubble (Laris 2002; Hiernaux and Diarra 1984). In order to restart natural plant maturation processes artificially, herders set intentional fires that are subtly different from the natural wildfires that would otherwise occur in these regions.

10.4.2 Patterns of movement and distribution

Rarely is there a single right way to move.

> . . . If a move is to be made, then there must be a choice of timing, of direction, of distance, and of new location. This choice is not in practice simply a matter of reaching a decision through the assembly and assessment of information on resources, for usually there is no obviously single best choice but a variety of possibilities. Information is seldom complete in any case, and the techniques of its assessment are not wholly efficient. Opinion is involved, and this can vary from herdsman to herdsman. But in any case seldom is there but one time and direction of movement clearly indicated. There is a range of opportunities of roughly equal pastoral advantage (Gulliver 1975, p. 371–2).

The preceding quotation refers to the Turkana pastoralists of East Africa, but is applicable to a wide variety of pastoral systems. As previously noted, in the Sahel the early rains are a period of particular uncertainty. In the Arctic unpredictable icing-over and re-freezing can make forage suddenly unavailable (Forbes and Stammler 2009; Rees *et al.* 2008). In inherently risky environments there often are no obviously superior, safe options; opinions differ, herds move in different directions, and chance rewards some and punishes others. This occurred in the winter of 2006/07 when a spell of warm weather and heavy rain were followed twice by a rapid fall in temperatures in the West Siberian Yamal Peninsula. During the 2006 icing event, herders visited or called ahead to villages along their migratory routes to enquire about pasture conditions. Having received encouraging news about pastures further to the south, some herd owners decided to continue their migration in that direction, while others cancelled half of their normal yearly migration and remained north of the iced pastures. On their return trip north the following spring, those who had chosen the southern option were hit by a further unexpected icing event that depressed calf survival rates to less than fifty percent. Herders who had decided that year to cancel their migration and remain north of the iced pastures had a normal calf survival rate of over 80%.

Apart from differing estimates of risk and imperfect knowledge, herders' movement decisions are also influenced by their variable resource endowments and tenure rights. Even within a reasonably homogeneous pastoral community, this diversity can foster complex and apparently contradictory movement patterns. A recurrent source of intra-community variability in agro-pastoral systems is the location of arable fields and homesteads. Based in different locations, herders who are also farmers tend to view the landscape from the perspective of their home location, and adjust their movements in response to their individual situation (de Boer and Prins 1989). Intra-community variability in movement patterns is also caused by variations in herd wealth, with large flocks or herds tending to be more mobile than small ones (Robinson and Milner-Gulland 2003; Kerven *et al.* 2008). Stock owners' distinctive social networks and land tenure rights and the variable species composition or age and sex structure of their livestock holdings also sends herds moving in different directions. In sum, 'human, livestock, environmental, and political factors, [result] in a pattern of movement that varies

from herd owner to herd owner and from year to year' (McCabe 1983, p. 117).

When there is one obviously superior movement pattern, coordinated use often becomes necessary and queuing emerges along busy trek routes or at migratory bottlenecks, as illustrated in the Arctic case study in this chapter, or around congested water points in semi-arid areas. When it is clearly advantageous for them to do so, pastoralists have the capacity to coordinate movement and resource use. With respect to mobility and residence, however, decision-making in pastoral societies tends to devolve to independent social units—individual households or small groups of households. Something approximating free distributions would appear to emerge from the multiple, serial negotiations between these geographically dispersed, differentially endowed and relatively free agents. Rigorous testing of this hypothesis is difficult since it requires movement records on entire pastoral populations relative to seasonally shifting natural resource configurations (for attempts in this direction see Behnke *et al.* 2008; Finke 2000).

The heterogeneous, temporally unstable and unpredictable distribution of resources favours reciprocity in harsh environments where producers can ill afford to under-exploit transient resource concentrations. This openness is often culturally embedded in indigenous resource tenure and use systems. Fernandez-Gimenez has described the 'vagueness, permeability, and overlap of boundaries around pastoral resources and user groups' in Mongolia (Fernandez-Gimenez 2002, p. 49). Referring to pastoral Africa as a whole, Behnke characterized pastoral tenure systems as having 'fuzzy or indeterminate social and territorial boundaries' (Behnke 1994, p. 15; Goodhue and McCarthy 1999); and for West Africa Turner describes 'porous systems of pastoral usufruct' that are 'socially malleable' and 'typically display an unbounded, point-centred spatial pattern' (Turner 1999a, pp. 103, 108). In economic terms, what pastoralists are trading among themselves are not products or services, but access rights to the ephemeral natural resource concentrations that are needed to sustain production. As in more conventional trading systems, when debts accumulate, culturally prescribed notions of reciprocity encourage an eventual settling of accounts, from which all parties stand to benefit. These are not the bounded, resource-conserving systems commonly described in the literature on common property resource management. Indeed, these are in neoclassical economic terms a theoretical oxymoron—systems of tenure rights that regulate open access—a uniquely migratory perspective on the notion of property.

10.4.3 Non-forage constraints

Wild ungulate populations do not perfectly match their forage resources and neither do livestock—again a fundamental similarity (Senft *et al.* 1987 Bailey *et al.* 1996). Water is a recurrent constraint, but restricts movement in different ways in different climatic zones. In the semi-arid tropics, water dependency is a physiological barrier to the free movement of any species—domestic or wild—that must drink in the dry season. In the Arctic and subarctic where surface water is common and drinking is not an issue, bogs, rivers and lakes create physical barriers to passage, separating both domestic and wild herds from potential grazing areas.

Insect pests and parasites affect livestock in all three case study regions covered in this chapter, but the extent to which pests inflict sufficient damage to alter movement patterns is variable. In Turkmenistan, for example, sheep flocks that normally reside in the desert are forced into irrigated agricultural areas in drought years. Here the flocks find abundant feed from roadsides, ditches and crop residues, but are infected by liver and intestinal parasites from polluted drinking water. According to shepherds, mortality rates are roughly proportional to the length of time flocks spend in the area, so shepherds try to leave as soon as possible, and most do not use infected areas in years of normal rainfall (Behnke 2008). In some instances moving away from the problem is neither advantageous nor necessary. In The Gambia of West Africa, cattle preserve high levels of green grazing in their diets by moving seasonally into areas with the highest rates of tsetse infestation when tsetse populations are at a maximum. The in-coming cattle pay very little 'penalty'

in terms of increased trypanosomiasis infection, apparently because large numbers of migrant animals share the fly burden and dilute the individual risk of infection (Wacher *et al.* 1993).

There are also some distinctively human constraints that impinge on the movement of domestic herds. As all the preceding case studies emphasized, optimizing pastoral production may be only one among several objectives for pastoral households, such as access to wage employment or alternative income sources, markets, education and social services, or the amenities afforded by a settled way of life. Matching animal numbers to feed resources may be compromised in an attempt to satisfy these multiple (often non-pastoral) objectives. Mongolian herders cluster around overgrazed settlements that provide services and potential employment; Arctic herders go fishing.

On the other hand, human knowledge, foresight and technical ingenuity can also serve to remove natural barriers to pasture exploitation. In both the Sahel and Arctic, herding has a quantifiable positive impact on foraging by domesticated herds. The deferred use of critical pasture reserves, the scheduling of complex movements through topographical or industrial bottlenecks, and infrastructure (boreholes in the tropics or animal shelters in cold climates) minimize the impact of natural constraints. Interventions can be as sophisticated as modern veterinary medicine, or as simple as the smoke from a cattle byre that wards off biting insects. These activities improve the degree of matching between feed resources and animal populations or overcome species-specific physiological limitations, e.g., with respect to dependence on drinking water, exposure to heat or cold, susceptibility to disease:

> While wild ungulates cannot usually fully utilize landscapes, production-oriented systems attempt to sustainably extract as much from an ecosystem as possible. Wildlife distributions are often limited by localized deficiencies of water, minerals, and navigable terrain. Production livestock systems attempt to remove these landscape constraints by manipulating resources or animal behaviour (Coughenour 1991, p. 539).

10.4.4   The human difference

The similarity between wild ungulate and pastoral movements is close enough that one might be tempted to conclude that there is no difference, and that pastoral livestock distributions are a mechanistic reaction to environmental conditions. This would be a mistake. Pastoral migrations may appear to be a natural response to the exigencies of the landscape, but they are sustained by social institutions that encourage discipline within and between communities of herders (deferred pasture use, inter-tribal raiding, land tenure restrictions, etc.). Pastoralists also do not simply match livestock to existing forage resources; each year and over the long term they shape their feed supply, removing 'landscape constraints by manipulating resources or animal behaviour.…[to]…significantly alter the plant–herbivore balance' (Coughenour 1991, p. 539).

Appearances aside, these are culturally regulated agricultural systems operating in human-modified landscapes. The illusion of naturalness rests on the relatively light impact of pastoralists on the landscapes they use, in comparison with other forms of land use, and on the unobtrusive style of pastoral livestock management. Given clear-cut biological incentives, skilful herding routinely facilitates the movement of animals in directions that they would anyway be inclined to go, rather than restricting or coercing their movement, or delivering inputs to the animals, as occurs in more intensive livestock production systems. Human vigilance and sustained intervention are intermittently needed to maintain advantageous patterns of herd movement, for example when crops are standing in fields close to pastures. Herd tracking studies have demonstrated that herded animals are better distributed with respect to available forage within a grazing radius than free pastured animals (Turner *et al.* 2005). But when predators are not a risk, pastoral livestock (and especially the larger-bodied herd species such as cattle, yaks, horses and camels) may spend from hours to months on their own without these movement systems breaking down into chaos at the landscape scale.

Mechanistic biological models explain aspects of these migratory systems. This chapter has shown that ungulate migrations can be usefully analysed in terms of spatial and temporal fluctuations in forage availability, irrespective of whether the animals are domesticated or wild, long or short-distance movers. But there are limitations to this approach. The capacity of density dependent habitat selection theory to explain the spatial distributions of domestic herds remains untested, but it will probably provide only a partial explanation for land use systems that must balance the pressures for free and optimal livestock distributions against the restrictions of property rights and administrative boundaries. More generally, the diversity of socio-economic variables that impinge upon pastoral migration complicates any attempt to generalize about the organization of these movement systems. Movement is a livestock husbandry practice used by pastoralists to achieve wide ranging cultural, social and economic goals, and it is subordinate to these larger concerns.

10.4.5   A closing note on the environmental consequences of large-scale migratory pastoralism

The advantages for rangeland conservation of intermittent use and resting versus continuous use are documented across a wide spectrum of grazing environments (Galvin *et al.* 2008). 'It is no accident that the world's large populations are of migratory species' writes Sinclair with obvious approval (1983, p. 250). Misgivings by some scientists and policy-makers about the sustainability of migratory livestock systems tend, perhaps unsurprisingly, to focus on the issue of livestock numbers. It is recognized that large wild migratory populations can have equally large environmental impacts (Chapter 9). Writing about the resident wildebeest populations of the western Serengeti, Sinclair observed:

> The residents have the double disadvantage of increased predation and increased food competition when the migrants arrive and literally remove all the food. When this happens, the migrants move on (an example of a paradoxical strategy where the intruder has an advantage over a resident) (1983, p. 256).

Sinclair was referring to wildlife; the tone is often less dispassionate when similar observations are made about the actual or modelled environmental impacts of migrant livestock. What is at issue here is not science but aesthetics or economic self-interest. Environmental impacts that are acceptable and 'natural' when perpetrated by wildlife are often characterized as unacceptable degradation when potentially caused by migratory livestock (Illius and O'Connor 1999; Murray and Illius 1996; Prins 1989), or by someone else's migratory livestock (Adams 1982).

Alternatively, it is possible to view both migratory wildlife and mobile livestock as fence-wrecking despoilers of modern, sedentary, commercial ranching:

> Recently there has been [in tropical African savannas] some collectivization of nomads into ranches and discussion of the problems of overgrazing, but there is little emphasis on management in the sense of rational utilization of the grazing resource. The problem is complicated by substantial economic returns from tourists, who come to see the wild animals. But the needs of wild animals in large herds conflict with sedentary pastoralism, particularly if the pastoralists fence their land (Fisher et al. 1996, p. 396).

Unfashionable in scientific circles, the biases expressed in this statement probably represent the majority opinion among policy-makers and administrators in developing Asian and African countries.

And herein lies a final parallel between wild ungulate and pastoral migratory systems—both are under threat. In part, this threat comes from habitat loss and resource fragmentation caused by expanding human populations and economic development (Behnke 2008), processes that research on pastoral migration probably can do little to mitigate. From Ethiopia to China, however, central government authorities also mistrust and denigrate that minority of their citizens who are migratory pastoralists. The Ethiopian government would like to settle these people, if only it had the power to do so. The Chinese government does have that power and is doing so, on an unprecedented scale with state-sponsored fencing programmes, the clearance of 'degraded' rangelands and the forced resettlement of entire indigenous communities (Yeh 2005; Yan et al. 2005; Brown et al. 2008). Official justifications for these policies recite the shibboleths in the preceding quotation—the dangers of overgrazing and irrational resource utilization versus the technical and economic advantages of fencing. Provided government-sponsored policy is based on evidence and susceptible to reform, a clearer understanding of the causes and consequences of pastoral movement will serve both the interests of science and better policy.

CHAPTER 11

Conservation and management of migratory species

Jennifer L. Shuter, Annette C. Broderick, David J. Agnew, Niclas Jonzén, Brendan J. Godley, E.J. Milner-Gulland, and Simon Thirgood.

Table of contents

11.1   Introduction

Migration is a phenomenon that has evolved independently amongst a wide variety of taxa throughout the animal kingdom (Baker 1978; Dingle 1996; Alerstam *et al*. 2003; Chapter 2). The widespread nature of migratory behaviour is generally believed to be linked to the advantages it confers on organisms that live in seasonal environments characterized by cyclic variation in the quality and/or quantity of resources (Chapter 7). In such environments, migration is an adaptation that enables individuals to exploit seasonal resource peaks at different locations, while avoiding the seasonal resource depressions that occur at any one of them (Southwood 1962; Solbreck 1978; Dingle and Drake 2007).

The last two centuries have been marked by declines in migratory populations and losses of migratory behaviour (Harris *et al*. 2009; Wilcove and Wikelski 2008; Brower and Malcolm 1991; Bolger *et al*. 2008) that appear to transcend the 'evolutionary flexibility' in expression that has characterized the natural history of the phenomenon (Alerstam *et al*. 2003). For example, Berger (2004), Berger *et al*. (2006) and Bolger *et al*. (2008) emphasize the frequency of migration loss amongst large terrestrial mammals such as the Great Plains bison (*Bison bison*), the African springbok (*Antidorcas marsupialis*) and the Asian elephant (*Elephas maximus*). Additionally, the spawning migrations of aquatic

organisms such as Pacific salmon (*Oncorhynchus spp.*; Hanrahan *et al*. 2004), Lake Sturgeon (*Acipenser fulvescens*; Auer 1996; Ferguson and Duckworth 1997) and the American eel (*Anguilla rostrata*; Haro *et al*. 2000) have also suffered truncation and loss.

The detailed causes of migration loss are case-specific (Wilcove and Wikelski 2008); however one common element is that most threats to the persistence of migratory populations and migratory behaviour are anthropogenic in nature. Anthropogenic threats include: habitat loss or degradation; over-exploitation; climate change and the interruption of migratory movements (Wilcove and Wikelski 2008; Bolger *et al*. 2008; Kirby *et al*. 2008). In a natural world that is increasingly impacted by human development and exploitation, the advantages conferred by migration seem increasingly outweighed by the mounting costs associated with the behaviour.

Traditionally, the field of conservation biology has focused on biodiversity loss in the form of species extinctions or extirpations (Soulé 1985; Soulé and Kohm 1989). While species loss is undeniably worthy of conservation concern, the disappearance of biological phenomena such as migration can also have significant ecological impacts at the population, community and ecosystem levels (Chapter 9). For example, grazing activity by migratory wildebeest (*Connochaetes taurinus*) in the Serengeti Plains leads to a substantial reduction in the biomass of green plants on the grasslands, but it also prevents senescence and stimulates net primary productivity (McNaughton 1976). The influence of the wildebeest migration on the Serengeti ecosystem is so strong that Harris *et al.* (2009) predict that migration loss would lead, first, to a decline in herbivore populations, second, to a decrease in the populations of their carnivorous predators and third, to the collapse of the Serengeti ecosystem as a whole, as wildlife tourism-based incentives for environmental protection evaporate. The potential severity of such effects indicates that threatened migrations are themselves worthy of conservation effort (Brower and Malcolm 1991; Dingle 1996; Wilcove and Wikelski 2008), regardless of whether migration loss directly compromises the long-term viability of the affected population.

While the ecological consequences of migration loss are numerous, the disappearance of migratory behaviour can also have substantial economic consequences. From a human perspective, animal migrations are spectacles that have both aesthetic and utilitarian value (Dingle 1996). The spectacle created by animals engaged in mass migration can generate a sense of awe in those who witness it. Many people relish the opportunity to view animals (aggregated or not) that are only transient visitors to an area while they migrate to far-flung destinations. Examples of migrations that support valuable wildlife tourism industries include the north and southward migrations of passerines at Point Pelee National Park in Ontario, Canada (Hvenegaard *et al.* 1989), the annual migration of blue wildebeest between the Masai Mara Reserve in Kenya and the southern portion of Serengeti National Park in Tanzania (Norton-Griffiths 2007) and the annual arrival of migratory humpback whales (*Megaptera novaeangliae*) to Hervey Bay in Queensland, Australia (Wilson and Tisdell 2003). The utilitarian value of migration is not limited to non-consumptive wildlife viewing. The relatively high population densities of migrants (vs. resident species; Fryxell *et al.* 1988) and the predictable timing and location of their migratory movements (Bolger *et al.* 2008) can render them very valuable for harvesting. Additionally, the mobility of migratory populations often results in exploitation by multiple human populations in different locations. Migration loss from an exploited population can have a substantial

impact on the human populations that rely on hunting migrants for subsistence or commercial purposes, as the population size of the harvested organisms decreases and/or they cease migrating to locations where they were once hunted (Harris *et al.* 2009). Therefore, the consequences of lost migrations are likely to extend well beyond affected migrants, to include effects on the ecosystems that they are part of and the economies that depend on them. Given the economic importance of migration and its aesthetic and intrinsic values, recent population declines and migration losses amongst migratory species are worthy of concern and focused conservation actions.

Efforts to manage migratory organisms are often complicated by challenges associated with the migration process. For example, managers of long-distance migrants face a set of difficulties that are different from those faced by managers of short-distance migrants. Other characteristics, such as partial migration, solitary or highly aggregative behaviour during migration and site fidelity, can also have an impact on the relative effectiveness of different population management strategies. Therefore, a detailed understanding of migratory behaviour is essential for successful management of populations and their migrations.

Continued threats to the persistence of migratory populations and the migratory process itself have led to several recent taxon-specific publications on the conservation and sustainable management of migratory organisms (Bolger *et al.* 2008; Kirby *et al.* 2008; Mehlman *et al.* 2005; Berger 2004; Shillinger *et al.* 2008; but see Wilcove 2007). In this chapter, we attempt to employ a broader (i.e., multi-taxa and multi-environment) perspective to explore how the medium that organisms migrate through (i.e., water, land and air) along with key attributes of migratory behaviour, affect the vulnerability of migratory populations to threatening processes. We provide an overview of some of the mechanisms available for conserving and managing migratory species, with emphasis on different applied approaches for conserving existing migrations and restoring lost migrations. We also offer an assessment of whether migratory organisms are particularly dependent on conservation and protection measures, and whether migrants that are commercially exploited or considered 'pest species' are more difficult to manage than their sedentary counterparts. We finish by discussing a crucial mechanism for successful conservation of long-distance migrants—inter-jurisdictional legislation, policies and agreements that facilitate the development and implementation of coordinated protection and/or exploitation strategies across a migrant's entire range.

11.2 Migration loss: vulnerability to key threats

There have been notable declines in both the frequency of migrations and the abundance and distribution of the species that exhibit them. Although recent trends suggest that both migration and migratory species face a heightened risk of decline (Brower and Malcolm 1991; Wilcove and Wikelski 2008; Bolger *et al.* 2008; Kirby *et al.* 2008), the extent to which migrations or migrants are particularly vulnerable to threatening processes is not clear. In some instances, migratory behaviour can serve as an advantage to organisms faced with anthropogenic threats. For example, migration can enable temporary escape from habitat loss or overexploitation and the relative vagility of migratory organisms might make it easier for them to relocate in response to climate change-induced shifts in habitat (Malcolm *et al.* 2002). In other cases, migration can render organisms more vulnerable to threatening processes. By definition, migratory behaviour involves travel between disjunct ranges (Fryxell and Sinclair 1988; Berger 2004), across distances that vary from considerable to vast, relative both to the body size of the migrant and its range size during its non-migratory seasons (Baker 1978). Consequently, in comparison with non-migrants, migratory animals are likely to be exposed to a more diverse array of threats associated with each of the multiple environments they inhabit and travel through (Brower and Malcolm 1991). In all cases, an assessment of the relative vulnerability of specific migratory populations requires careful consideration of both extrinsic factors (the nature of the threat, the habitats traversed during migration) and intrinsic factors (the nature of the migratory behaviour and its intent, the life-history strategy of the animal).

11.2.1 Extrinsic factors

11.2.1.1 Habitat loss/degradation

The loss, degradation and fragmentation of habitat due to human development or activities are some of the primary threats facing all species (Pimm and Raven 2000; Seabloom *et al.* 2002; Brooks *et al.* 2002) including those that migrate (Robinson *et al.* 2008; Osborn and Parker 2003). For example, of the 20 threatened populations of ungulate mass migrants identified by Harris *et al.* (2009), 17 are negatively impacted by habitat loss. Additionally, an assessment of the conservation status of 60 migratory raptor species in the African–Eurasian region revealed that the main threat facing the 32 species with 'unfavourable conservation status' was the loss or degradation of suitable habitat (Tucker and Goriup 2005).

There are several direct causes of habitat loss/degradation, including agricultural land conversion, residential and commercial development, pollution and the construction of transportation corridors (Kirby *et al.* 2008; Bolger *et al.* 2008). Less direct causes of habitat loss include the construction of barriers (e.g., fences or dams) that prevent travel to previously used areas (Section 11.2.1.4; cf. Chapter 10) and climate change-driven alterations in habitat suitability (Robinson *et al.* 2008). Most migrants are dependent on a variety of different habitat types during the course of their migratory journeys (e.g., breeding, foraging or wintering grounds, stopover sites, migration corridors). The loss or serious degradation of any of these critical habitats could produce declines in migratory behaviour and/or population abundance, however the severity of such impacts varies depending on which habitat type is affected.

Research on migratory birds indicates that the population impacts of habitat loss are greatest when the losses are from the habitat type for which density-dependent constraints on population size are strongest (Sutherland 1996). While residents are expected to exhibit population declines that are proportional to the amount of habitat lost, migratory populations that are limited on their breeding range may exhibit little or no response to a notable loss of winter habitat. Conversely, the destruction of a comparable amount of breeding habitat might lead to population declines amongst migrants that are more severe than expected, relative to the proportion of the migrant's total range that is affected (Newton 2008; Sutherland 1996). Therefore, effective habitat-focused conservation efforts will vary between migratory species and populations, and should be directed towards the habitat type where population bottlenecks occur. However, clear identification of such habitat can be challenging, due to: the difficulties associated with tracking individuals over their entire migration cycle; the influence that conditions in one habitat can have on mortality rates and/or reproductive success in another habitat (i.e., carry-over effects; Norris and Taylor 2006); and inter-annual variation in the habitat type that is limiting (Newton 2008).

In addition to the diversity of habitat types that are essential to their survival, many migratory animals have other characteristics that make them particularly vulnerable to habitat loss/degradation. These include both the large size of individual home ranges and the inflexibility that often restricts choice of range locations. For example, the broad spatial extent of a migrant's annual range makes it difficult to implement traditional conservation approaches. While it may be feasible to protect the habitat of some non-migratory populations by establishing protected areas, effective reserve design is more complicated for migratory populations, due to the sheer size of the area required to protect their ranges (Thirgood et al. 2004). In addition to range size, the lack of flexibility that some migratory organisms show in their selection of some or all of the different habitats that they use (Andersen 1991; Sutherland 1998) can render them more susceptible to habitat loss or degradation. If sites to which migrants show fidelity are lost or degraded, their inability to seek replacement habitat could lead to migration loss (Berger 2004; Harris et al. 2009) or population declines due to reduced reproduction or survival in affected areas (Watkinson and Sutherland 1995; Sutherland 1998).

Thus, while the anthropogenic loss and degradation of suitable habitat is a major threat to all non-human organisms, migratory animals are likely to be amongst the most vulnerable. However, the extent of that vulnerability depends on the characteristics of a particular migration. While traditional habitat protection strategies are an option for some migratory populations (albeit often more complicated than those for residents), the root cause of continued habitat loss and degradation (growth in the global human population and per capita consumption; Seabloom et al. 2002) shows no sign of abating.

11.2.1.2 Over-exploitation

For centuries, humans have exploited wildlife for subsistence or commercial harvest, frequently at unsustainable levels. Exploitation is considered one of the major drivers of biodiversity loss (Worm et al. 2006; Chapin et al. 2000). It can have a direct impact via the extinction of harvested species and an indirect impact through the extinctions of unharvested populations that are reliant on over-exploited species. Additionally, even moderate levels of exploitation can lead to declines and extinctions amongst populations that have low resilience and have already been reduced to critical levels by other threats. Of marine species, 55% of documented population extinctions are thought to have been as a result of over-exploitation (Dulvy et al. 2003). Species with a slow life history are particularly vulnerable to becoming over-exploited (Mace and Reynolds 2001), but do those that are also migratory face an even greater risk of extinction? While migration can provide a temporary respite from exploitation pressure, specific attributes of migratory behaviour might make a migratory population more vulnerable to unsustainable harvesting (Jarenmo 2008).

Migratory species may be at risk from exploitation during migration, stopovers, and/or at the end point of migration. For some species, the predictability of migration routes and timing can contribute to over-exploitation, while for others, large aggregations formed during or after migration might render migrants more easy to detect and thus more attractive and vulnerable to hunters (Bolger et al. 2008; Sadovy and Domeier 2005; Jarenmo 2008). For others still, the phenomenon of partial migration (where only a portion of a given population is migratory; Dingle 1996) can make populations particularly vulnerable to over-exploitation in circumstances where migratory aggregations are composed solely of reproductive adults (Box 11.1).

Box 11.1 Conservation of aquatic migrants: marine turtles

1. General

Biology

There are seven recognized species of marine turtle, and whilst all migrate to breeding sites, at sea their life cycles vary. Leatherback turtles (*Dermochelys coriacea*) are perhaps the most difficult to manage or protect given their nomadic existence. From hatching, they remain in the pelagic zone, where they forage on jellyfish and other gelatinous organisms. Diving to depths of more than 1000 m, they occupy all dimensions of the oceans, nesting in the tropics and foraging at latitudes of 50 °N (Ferraroli *et al.*, 2004; Hays *et al.*, 2004). Other species, in particular the well-studied loggerhead turtle (*Caretta caretta*), appear to use different foraging strategies depending on size. Larger individuals satellite tracked from the Cape Verde Islands travelled to coastal regions of West Africa where they remained, foraging benthically, whilst smaller individuals foraged widely in the pelagic waters of the region (Hawkes *et al.* 2006). Loggerhead turtles have also been recorded undertaking different migratory strategies, some moving to the colder waters off the eastern seaboard of the USA to forage in the more productive waters, returning to warmer waters for the winter months, whilst others migrate to one location where they remain (Hawkes *et al.* 2007). Both green (*Chelonia mydas*) and loggerhead turtles in the Mediterranean have been tracked to the same foraging locations after subsequent breeding events, suggesting that they remain faithful to these sites throughout the inter-breeding years (Broderick *et al.* 2007).

Threats and vulnerabilities

Marine turtles face numerous threats, including direct harvest of turtles and their eggs at breeding sites, nesting habitat loss from coastal development, the potential impacts of global climate change, coastal squeeze, skewed sex ratios (the sex of offspring is determined by temperature, with females produced at hotter temperatures) and changes in oceanographic conditions. Indirect harvest through fisheries bycatch (i.e., accidental over-exploitation) is, however, considered one of, if not the, greatest threat to marine turtles worldwide. Indeed, Lewison *et al.* (2004) estimated that globally 200 000 loggerhead (*Caretta caretta*) and 50 000 leatherback turtles were caught in the pelagic long-line fisheries in 2000. The leatherback turtle has declined dramatically in

the Pacific Ocean over recent decades, its demise thought to be a result of the intensive long-line fisheries in that region (Spotila *et al.*, 2000).

Conservation and management solutions

The foraging locations used by green and loggerhead turtles in the Mediterranean are typically shallower coastal regions that fall within geopolitical boundaries. Their clear jurisdictional status makes it relatively easy to develop and implement protection measures. The diverse foraging behaviour exhibited by marine turtles necessitates very different conservation and management strategies. For those species or populations that forage in the pelagic, reduction of the bycatch rate from long-line and gill nets is essential (Southwood *et al.* 2008). For those that forage in the neritic, interactions with bottom trawlers may prove fatal. The development of the Turtle Excluder Device (TED) in the USA (effectively a trap door within the trawl net that allows turtle to escape) is thought to have reduced turtle strandings as a result of shrimp trawl bycatch in the Gulf of Mexico (Cox *et al.* 2007). However, bycatch from trawls and gillnets is still a significant threat to marine turtles globally (Lewison and Crowder 2007).

2. Green turtles, Ascension Island, UK

Biology

Green turtles, like other marine turtles, take up to 40 years to reach reproductive maturity, and once they are ready to breed they exhibit natal philopatry, returning to the same beaches from which they hatched to reproduce. Males and females migrate to mate off the nesting beaches, females then excavate their nests and deposit their eggs. Green turtles are also faithful to their foraging grounds and repeatedly return to the same seagrass beds after nesting migrations (Broderick *et al.* 2007). Fidelity to breeding and foraging locations and the aggregative behaviour at both locations makes them vulnerable to targeted harvesting, but has also enabled the implementation of effective conservation strategies.

Threats and vulnerabilities

Historically, marine turtles, in particular the green turtle, have been an important source of meat, and in many

continues

Box 11.1 (*continued*)

regions adults are still directly harvested for meat and, in some cases, eggs are removed from nests for consumption. Natal philopatry and aggregations formed at mating and egg-laying sites make this species vulnerable to exploitation, more so as females emerge on to the beach to lay their eggs, where they are slow moving and easily captured. At Ascension Island (a remote oceanic island in the South Atlantic Ocean—UK 7°57′S 14°22′W) female green turtles were harvested for their meat from the 16th century until the 1940s, when the harvest was no longer commercially viable owing to the low number of females coming ashore (Broderick *et al.* 2006). Turtles were taken in large numbers, up to 1500 in peak years, with the majority traded to passing ships, which would use the fresh meat (animals were kept alive until needed) to feed their crew or transport it back to the UK to meet demands for turtle soup, which was considered a delicacy. Records of the number of animals harvested were kept by the British Royal Marines stationed on Ascension Island since 1815 (Fig. 11.1). Broderick *et al.* (2006) collated archival harvest data and, using a deterministic age–class structured model, predicted that the initial number of breeding females present in the population prior to 1822 had to have been at least 19 000–22 000 individuals in order for the population to have survived the level of harvest recorded. Whilst the records indicate that nesting numbers during the 1930s were too low for a commercially viable harvest, data gathered in 1998–2004 suggest that the current breeding population of green turtles at this site contains an estimated 11 000–15 000 females.

Conservation and management solutions

While the Ascension Island population, along with other green turtle populations, was previously threatened by over-exploitation, more recent population trends are consistent with recovery and exponential rates of population growth (Broderick *et al.* 2006; Chaloupka *et al.* 2008). The results of Broderick *et al.* (2006) illustrate the dramatic recovery of green turtles at Ascension Island. This population is still increasing exponentially and shows no evidence of slowing, suggesting it has not yet reached 50%

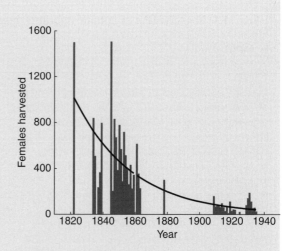

Figure 11.1 The number of female green turtles (*Chelonia mydas*) harvested at Ascension Island, 1822–1935, for years when data were available.

of its carrying capacity. Similar recoveries have been documented for green turtle populations in Australia, Costa Rica, Hawaii, Japan and the USA (Chaloupka *et al.* 2008). The reasons for these recoveries may be a result of both the species' life cycle and a reduction of threats. The relatively localized inshore foraging grounds of the herbivorous green turtle may enable this species to be more effectively protected in the coastal habitat in which it forages. Although interactions with coastal fishers may occur, large-scale industrial bycatch from pelagic long-lines is less likely to cause a major impact upon green turtles while they are foraging in seagrass beds, unlike the carnivorous leatherback and loggerhead turtles (Lewison *et al.*, 2004). Satellite telemetry has revealed that the main foraging grounds of the Ascension Island green turtles are located in the coastal waters of Brazil (Luschi *et al.*, 1998), where small-scale traditional fisheries for marine turtles once existed and a pioneering conservation programme has dramatically reduced bycatch (Marcovaldi *et al.*, 1998). In the case of the green turtle then, it appears that conservation management at both the nesting and foraging grounds of the adult life stages can lead to rapid population recovery.

An additional factor that can exacerbate the impacts of exploitation on migratory populations is as much attributable to human political institutions as it is to the migratory attributes of a given population. The migratory trajectories of long-distance migrants often cross boundaries between different political jurisdictions, each of which may have a different approach to regulating exploitation of the species (Jarnemo 2008). In such cases, developing an inter-jurisdictional approach to ensure the sustainability of harvesting throughout an organism's entire range would be optimal, but this can be challenging to achieve. Therefore, while human exploitation threatens the long-term persistence of migrants and non-migrants alike, migratory behaviour can affect the extent to which a population is susceptible to this threat.

11.2.1.3 Global climate change

Migration can be seen as an adaptation enabling animals to survive and benefit from spatiotemporally fluctuating resources (Chapter 7). Climatic changes expose organisms to novel environmental conditions and affect the magnitude and frequency of the environmental events that shape the life history of individuals and population demography. In migratory organisms the timing, distance and direction of migration, as well as the propensity to migrate in the first place, all have the potential to be affected (Chapter 6).

From a conservation perspective, a general concern is that climate change may cause temporal shifts in the timing of life-history events, which can uncouple trophic interactions and have a strong negative effect on population viability. Long-distance migrants are thought to be especially vulnerable. For example, they may not be able to predict changes in resources and/or conditions on the breeding ground while still on the wintering ground, and therefore, they may fail to respond accordingly (Both *et al.* 2006).

The extent to which we can expect species to adapt to the new conditions via phenotypic plasticity and micro-evolution is largely unknown, as is the expected change in the timing of migration. Despite the complexity and the high degree of uncertainty associated with the problem, the potential threat of climate change-induced disruption of phenologies should not be ignored. For instance, European populations of migratory bird species that have not shown a phenological response to climate change are, in fact, declining (Møller *et al.* 2008).

Whereas disrupted phenologies across trophic levels are a potential threat to migratory species, climate change may also affect the phenomenon of migration per se. As winter temperatures increase in parts of the Northern Hemisphere, the migra-tory fraction of partially migratory bird populations has been predicted to decline due to increased survival among residents, even if data are ambiguous (e.g., Nilsson *et al.* 2006). Historically, however, habitat loss, anthropogenic barriers in the landscape, and hunting—rather than climate—have contributed to the loss or truncation of migrations, at least in ungulates (Bolger *et al.* 2008). The most important threat may be the interaction between climate change and habitat destruction (Kirby *et al.* 2008; Sanderson *et al.* 2006), where the former may select for migration but the latter could limit the possibility. The relative importance of climate change in shaping the future of animal migration remains to be seen.

11.2.1.4 Interruption of migration

While migratory populations are exposed to a mounting number of threats, the phenomenon of migration is itself increasingly at risk, with numerous documented losses in historic times (Berger 2004). Migrations are vulnerable to loss for several reasons. The relative linearity and long-distance nature of many migrations renders them vulnerable to physical barriers and obstructions (Berger 2004; Berger et al. 2006; Wilcove and Wikelski 2008), particularly in the case of terrestrial and riverine migrants. Barriers can include blockages of migratory routes or degradation of traversable areas to a point where they are avoided because they are associated with a high degree of risk (e.g., of predation, mortality, starvation). Barriers threaten the persistence of migration by physically obstructing movement, causing migration route abandonment, and/or causing increases in mortality during migration (Serneels and Lambin 2001; Berger 2006; Nellemann *et al.* 2009).

The ability of barriers to alter migratory behaviour has not gone unnoticed by wildlife managers who are interested in redirecting or even eliminating the migrations of targeted species. Constructing obstructions to influence the expression of migratory movements has been effective at achieving management goals in some cases (Andreassen *et al.* 2005; Mbaiwa and Mbaiwa 2006). However, such alterations can come at a great expense to the status and condition of migratory populations (Mbaiwa and Darkoh 2005; Mbaiwa and Mbaiwa 2006).

In addition to physical barriers to movement, changes in the underlying mechanisms that drive migration might also result in the alteration or loss of migratory behaviour. For example, for migrations that evolved as a strategy to optimize the quantity or quality of available food resources (Fryxell and Sinclair 1988), an increase in food quality and/or quantity during periods when food was historically scarce could result in the loss of migratory behaviour. This hypothesis is central to efforts to manage the migratory behaviour of moose in Norway (Gunderson *et al.* 2004; Andreassen *et al.* 2005; Section 11.4.1.4). In theory, for species for which migration serves as a strategy to avoid predation risk (Fryxell and Sinclair 1988), a decrease in predation pressure (e.g., due to a decrease in predator population size) might produce comparable results. However, the primary cause of most current and historic examples of migration loss appears to be the creation of obstructions to migratory travel between disjunct ranges, via human development and environmental management practices (Wilcove and Wikelski 2008; Berger *et al.* 2006; Berger 2004).

11.2.1.5 *General variation between media*

The specific source and nature of threats such as habitat loss, over-exploitation and climate change might vary between organisms that migrate through different media (i.e., land, water and air). However, there appears to be no consistent variation between terrestrial, aerial and aquatic migrants regarding their relative vulnerability to such threats. In fact, the case studies interspersed throughout this chapter have been selected to illustrate the wide diversity of issues that face both the migratory populations living in each of the major media and the people trying to conserve them. However, there are some general distinctions that can be made based on the two- or three-dimensional nature of the migration route. For example, migrations that occur in more two-dimensional environments, such as terrestrial migrations and some aquatic migrations (i.e., those that occur in river systems), are more vulnerable to physical disruption than migrations that occur in environments that are more three-dimensional in nature (e.g., aerial and marine migrations). The ability of terrestrial and riverine migrants to cir-

cumnavigate physical barriers, such as fences (Andreassen *et al.* 2005), railroads (Ito *et al.* 2005; Berger 2006), roads (McLellan and Shackleton 1988), urban development (Berger 2004) and hydroelectric dams (Coutant and Whitney 1999; Mahoney and Schaefer 2002), is minimal. In comparison, organisms that migrate through aerial and marine media are less likely to have their migratory journeys obstructed by physical obstacles, because the creation of effective barriers in such environments is much more difficult. However, oceanic and aerial migrants are unlikely to be any less susceptible than terrestrial or riverine migrants are to other threatening processes. In addition, specific characteristics (e.g., site fidelity, which is discussed in more detail below) might diminish the apparent advantage that migrants in three-dimensional environments have when circumnavigating barriers.

11.2.2 Intrinsic factors

Migration is a phenomenon that has evolved amongst a wide variety of taxa and, as a consequence, the form it takes can differ substantially between migratory animals (Dingle 1996; Baker 1978). Explicit evaluations of whether a particular characteristic renders a migrant more or less vulnerable to the key threats outlined in Section 11.2.1, are rare (but see Matthiopoulos *et al.* 2005; Dolman and Sutherland 1995). However, the existence of numerous studies documenting the diversity of migratory behaviours has enabled us to develop hypotheses regarding the potential susceptibility of specific characteristics to different extrinsic threats. The findings of this assessment, both documented and speculative, are summarized in Table 11.1.

11.2.2.1 *Site fidelity*

Site fidelity is generally defined as the tendency of individuals to remain in, or return to, a particular area (Schmidt 2004; Schaefer *et al.* 2000). For animals that migrate, fidelity could be applied to routes followed during migratory journeys and/or the seasonal ranges that they travel between. While a genetic basis for fidelity to specific ranges has been demonstrated for some migratory species (Pomilla and Rosenbaum 2005; Berthold 1991), the fidelity

Table 11.1 The relative vulnerability associated with different migratory characteristics in relation to the major threatening processes that migratory organisms face. An 'x' in the 'threats' column signifies enhanced vulnerability to the corresponding threat, relative to other potential traits. An 'x?' indicates that associated vulnerability is speculative (no evidence in the peer-reviewed literature). The symbols associated with each specific threat type (†, β, ψ, *) are used to identify relevant studies in the 'sources' column.

General characteristics	Specific characteristics	Threats				Sources
		Habitat loss/ degradation†	Over- exploitation^β	Climate change^π	Interrupted migration*	
site fidelity (to ranges, migration routes)	high	×	×	×?	×	Matthiopolous et al. (2005)†; Andersen (1991)^ψ; Nelson and Mech (1981)^ψ; Hjeljord (2001)†; Sutherland (1998)†; Berger et al. (2004)†^ψ; Berger et al. (2006)†^ψ; White et al. (2007b)†^ψ; Bolger et al. (2008)^β; Raine et al. (2007)^β; Stanley and Doyle (2003)*; Kiffney et al. (2009)*
	low/flexible					
grouping behaviour	aggregative	×	×	×?	×	Hamilton (1971)^β; Mitcheson et al. (2008)^β; Jaeger et al. (2005)^β; Pegg et al. (1996)^β; Nocera et al. (2008)^β; Sala et al. (2001)^β; Sadovy and Domeier (2005)^β; Hasegawa and DeGange (1982)^β; Bolger et al. (2008)^β; Williams et al. (2009)†^ψ; Matthiopolous et al. (2005)†
	solitary		×			
migration	full		×		×	Craig and Herman (1997)^β; Broderick et al. (2006)^β; Bell et al. (2007)^β
	partial		×			
behavioural plasticity	flexible					
	rigid	×	×?	×?	×?	Dolman and Sutherland (1995)†; Sutherland (1998)†
cues and timing of migration	fixed		×	× if out of sync with phenology of resource		Robinson et al. (2005)^ψ; Kirby et al. (2008)^ψ; Walther et al. (2002)^ψ; Jenni and Kery (2003)^ψ; Ahola et al. (2004)^ψ; Both et al. (2006)^ψ; Inouye et al. (2000)^ψ; Bolger et al. (2008)^β
	variable			× if phenology of resource availability changes		
migration distance	long	×	×	×	×	Bolger et al. (2008)†; Kirby et al. (2008)†^ψ; Jarnemo (2008)^β; Jenni and Kery (2003)^ψ; Both and Visser (2001)^ψ; Jonzén et al. (2006)^ψ; Jonzén et al. (2007)^ψ; Berger (2004)*
	short	×?	×?	×		
responsiveness to resources and conditions during migration	high					Bolger et al. (2008)†
	low	×	×?			

that other species show to particular ranges or migration routes appears to be a learned behaviour (Hjeljord 2001).

The repeated use of specific locations has been hypothesized to confer several benefits to individuals, including familiarity with resources and conditions and the avoidance of predators (Greenwood 1980; Switzer 1997). However, in addition to the advantages it can offer, site fidelity could also make individuals more vulnerable to threatening processes than migrants with more variable space use patterns might be. Migrants that exhibit fidelity to at least some portion of their migratory trajectories are probably more likely to suffer negative consequences due to habitat loss or degradation than migrants whose movements are targeted towards specific locations. Matthiopolous *et al.* (2005) simulated the impact of fidelity to breeding grounds on the metapopulation dynamics of far-ranging animals. Their results indicate that site fidelity slows the colonization process and inhibits population growth because individuals must surmount their tendency to select familiar areas to establish new colonies. Although this effect is temporary, it can lead to metapopulation collapse (Matthiopolous *et al.* 2005). Therefore, site fidelity is likely to impede the ability of animals to respond to habitat degradation or loss by occupying or travelling through more suitable, but novel, locations.

Empirical evidence that is consistent with such an effect has been documented amongst numerous migratory cervids (e.g., Nelson and Mech 1981; Hjeljord 2001). For example, Andersen (1991) describes a Norwegian moose (*Alces alces*) population that has followed the same migratory route and used the same wintering area for approximately 7000 years, despite a progressive decline in the quality of their winter range. This tendency towards the repeated use of particular ranges and corridors was believed to have prevented or at least delayed a spatial shift to higher quality habitat and consequently had negative impacts on population dynamics. Similar patterns have also been documented amongst some avian populations (Sutherland 1998). In some cases, the option of using alternative ranges or migration routes has been all but eliminated by anthropogenic habitat alteration (Berger 2004; Berger *et al.* 2006). In such

situations, where site fidelity is at least partially attributable to the lack of alternative possibilities, migrants could be particularly vulnerable to migration blockage or the loss of remaining habitat (Berger 2004; White *et al.* 2007).

In addition to migration barriers and the direct loss and degradation of habitat, site fidelity can also make migrants more vulnerable to other threatening processes. Some of the predicted impacts of climate change are latitudinal or altitudinal shifts in the ranges of tree species and entire biomes (Malcolm *et al.* 2002; Davis and Shaw 2001; Kirilenko and Solomon 1998). Such changes are likely to result in the loss and degradation of habitat for both migratory and non-migratory animals. Although the relatively high vagility of migratory animals might improve their ability to cope with such changes, for some migrants this advantage might be counter-balanced by the constraints on movement that site fidelity can produce. Finally, fidelity to specific corridors or ranges can increase the predictability of animal locations, thereby making them more accessible to hunters. This is especially true for migratory populations, where it can be difficult to coordinate sustainable management strategies across multiple jurisdictions (Shillinger *et al.* 2008; Klaassen *et al.* 2008).

11.2.2.2 *Aggregative vs. solitary behaviour*
For many migratory vertebrates and invertebrates (e.g., sharks (*Selachimorpha* spp.) in the Gulf of Mexico, cranes (*Gruidae* spp.) in the northern hemisphere, desert locusts (*Schistocerca gregaria*); Wilson *et al.* 2001; Buhl *et al.* 2006), the migratory phase is characterized by aggregative behaviour, even if solitary or territorial behaviour is common during the rest of the year (Ramenofsky and Wingfield 2007; Buhl *et al.* 2006; Simons 2004). For others, migrations might be relatively solitary, while grouping behaviour occurs on the breeding or over-wintering grounds (Ramenofsky and Wingfield 2007; Cumming and Beange 1987). There are several advantages associated with group formation, some of which apply to all phases of the migratory cycle and others of which are phase-specific. For example, aggregative behaviour exhibited during migratory and/or sedentary phases can lead to a decrease in natural predation and perhaps, even anthropogenic exploi-

tation risk. Group formation can spread risk across multiple individuals (Hamilton 1971), enhance predator avoidance (Nocera *et al.* 2008) and enable defensive strategies such as the mobbing of predators (Curio *et al.* 1978). While dilution of predation risk is likely to be one of the main advantages of aggregation during more sedentary phases, aggregation during migration can confer another major advantage. When groups form, the transfer of information between individuals can lead to a rapid transition from disordered individual movement to highly aligned collective movement (Buhl *et al.* 2006) and an improvement in navigational accuracy (Simons 2004). Thus, aggregation can increase the efficiency and precision of the migratory journey.

Although there are several benefits associated with aggregation, there are also disadvantages that could render migratory individuals more vulnerable to extrinsic threats. For example, aggregated populations can be much more vulnerable to mortality from accidental events (e.g., falls due to unstable terrain; Bleich and Pierce 2001). Additionally, group formation during migratory and/or sedentary phases (and the resulting increase in density) could make migratory populations easily detectable and readily targeted and thus more vulnerable to exploitation (Mitcheson *et al.* 2008). Over-exploitation of spawning aggregations has been implicated as a causal factor in the population declines observed in several migratory fishes, including the sauger (*Sander Canadensis*; Jaeger *et al.* 2005, Pegg *et al.* 1996), Nassau grouper (*Epinephelus striatus*; Sala *et al.* 2001) and several other reef fishes (Sadovy and Domeier 2005). Population declines observed amongst other migratory taxa, including the short-tailed albatross (*Diomedea albatrus*; Hasegawa and DeGange 1982) and several terrestrial ungulate species (Bolger *et al.* 2008) have also been at least partially attributed to heavy exploitation during aggregative periods.

The consequences of aggregative behaviour are more ambiguous in terms of whether they make migrants more or less vulnerable to habitat loss and are probably partially dependent on the degree of fidelity that individuals exhibit to the habitats they use (Matthiopolous *et al.* 2005). If a specific area that is traditionally used as a migration route or breeding or over-wintering range by a migratory group is lost, degraded or blocked, many individuals will be rendered vulnerable to the consequences described in Section 11.2.2.1, as opposed to the few that might be affected if individuals were more solitary in nature (Williams *et al.* 2009). Additionally, if climate change produces an increase in extreme weather events (Rosenzweig *et al.* 2001), high density concentrations of individuals in discrete locations could leave entire populations vulnerable to catastrophic weather events (Williams *et al.* 2009). However, if migratory individuals are more solitary (and thus occur at low densities on the landscape) more extensive areas of suitable habitat might be required for persistence than are needed to support a similar-sized aggregated population concentrated in a comparatively small area. Consequently, more solitary behaviour might render a migratory population more susceptible to habitat loss and degradation than an aggregative population might be.

11.2.2.3   *Partial migration*

A population that includes both migratory and non-migratory members is referred to as being partially migratory (Kaitala *et al.* 1993). Evidence from some partially migratory populations indicates that an individual's migratory or non-migratory status is relatively inflexible because it is either genetically controlled (Biebach 1983; Berthold 1984; Kaitala *et al.* 1993) or bound by learned tradition (Andersen 1991). However, there is also evidence indicating that, for many partially migratory populations, migratory and resident behaviours are conditional and responsive to a variety of factors, such as asymmetry in social position or resource-holding power, social cues or assessments of environmental conditions (White *et al.* 2007). Threatening processes that have a negative impact on habitat quality (i.e., climate change, direct anthropogenic loss or degradation of habitat), also have the potential to either induce partial migration or increase the proportion of resident individuals in populations that are already partially migratory (Kokko and Lundberg 2001).

Regardless of the mechanism that underlies partial migration and the degree of individual flexibility associated with it, the phenomenon can render a population more vulnerable to over-exploitation. For example, Craig and Herman (1997) point out that many earlier demographic studies of humpback whales (*Megaptera novaeangliae*) in the North Pacific Ocean assumed that all animals migrate between summer and wintering grounds. However, they discovered that adult females were partially migratory and that earlier reproductive and population growth rate estimates derived from observations made on the wintering grounds were likely to be inflated (Craig and Herman 1997). If care is not taken to ensure that multiple migration strategies are accounted for when estimating vital rates, over-optimistic assessments of population growth and sustainable harvesting levels might result, which in turn can lead to over-exploitation. The risk of over-exploitation is also heightened in situations where the migratory portion of a population comprises reproductive adults (Box 11.1). Harvesting that targets these adults or their offspring could have a devastating effect on population growth, leading to low or no recruitment following exploitation. For example, whilst marine turtles are often harvested at sea at their forage sites, in situations where harvest has been directed at adult females and/or their eggs at the nesting beach, populations have been dramatically reduced or extirpated (Fig. 11.1; Broderick *et al.* 2006; Bell *et al.* 2007; McClenachan *et al.* 2006).

While partial migration has the potential to increase susceptibility to over-exploitation, threat-induced increases in the frequency of the phenomenon or in the proportion of resident individuals amongst partially migratory populations (e.g., due to habitat loss, climate change or a change in the underlying mechanisms that drive migration) could have a negative impact on the ecosystems that migrants travel through and to. Specifically, a reduction in the strength of the linkage between the ecosystems included in ranges of migratory individuals could have a variety of impacts, depending on the role of migrants in these ecosystems. Possible effects include a decrease in nutrient inputs or a decline in seed dispersal (Chapter 9).

11.2.2.4 *Behavioural plasticity*

The extent to which individuals are capable of altering their movement strategies can affect how vulnerable they are to threatening processes. Two migratory characteristics that vary in this respect include site fidelity and migrant/resident status amongst partially migratory populations. For some

organisms, fidelity to migration corridors or ranges used during more sedentary phases has a genetic basis (Pomilla and Rosenbaum 2005; Berthold 1991). For example, Lohmann *et al.* (2008) argue that salmon (*Salmonidae*) and sea turtles (*Chelonioidea*) imprint on the specific magnetic field associated with their natal range and use this, in combination with local cues, to navigate up to thousands of kilometres back to these specific areas to spawn. In contrast, amongst other migratory organisms, individuals appear to learn the location of specific habitats via information gleaned from interacting with conspecifics (Box 11.2). For example, juvenile moose and white-tailed deer (*Odocoileus virginianus*) in North America are believed to learn the locations of traditional winter ranges and migratory routes by accompanying their mothers on their migratory journeys during their first year of life (Andersen 1991). As discussed above (Section 11.2.2.3), the basis for migrant or resident status amongst individuals that belong to partially migratory populations appears to be as variable as site fidelity. For example, a genetic basis for the two strategies has been documented amongst some bird species (Berthold 1991) and in some fish species the strategies appear to be partly developmental and partly genetic (Jonsson and Jonsson 1993). Yet amongst several partially migratory species, including birds (Pulido *et al.* 1996; Sutherland 1998) and ungulates (Nelson 1998; White *et al.* 2007), migrant and resident status appears to be conditional in nature.

A few studies have tried to determine the costs and benefits of site fidelity (Pärt 1995) or assess the relative fitness of migrants vs. residents in partially migratory populations (e.g., Hebblewhite and Merrill 2007). However, explicit assessments of the extent to which behavioural flexibility (in site fidelity or migratory or resident behaviour) renders populations or migratory behaviour more or less vulnerable to extrinsic threats are not common. An exception is Dolman and Sutherland (1995), whose modelling study indicated that the ability of migratory birds to alter their migration routes in response to environmental change might be essential for their long-term viability. Empirical support for this finding comes from a later publication by Sutherland (1998), who found that bird populations with a strong genetic basis for fidelity to migration routes and winter

Box 11.2 Conservation of an aerial migrant: trumpeter swans in North America

Biology

The trumpeter swan (*Cygnus buccinator*) (Fig. 11.2) is the largest native waterfowl species in North America (Mitchell 1994). Adult trumpeter swans can live for up to 30 years (Baskin 1993), feed almost entirely on aquatic plants and breed in large shallow ponds and wide slow rivers in the north-western and central regions of the continent (Mitchell 1994). When European settlement of the continent first began, the species is believed to have been widely distributed across the USA and Canada (Baskin 1993; Lumsden 1984). However, the market for their meat, feathers and tanned skins led to a pattern of over-exploitation that resulted in substantial population declines and range retraction by the turn of the 20th century (Baskin 1993; Sladen and Olsen 2005).

There are three populations of trumpeter swans that are currently recognized in North America: the Pacific coast population; the Rocky mountain population; and the Interior population (CWS undated). The Pacific and Rocky mountain populations are both composed of descendents of remnant flocks that survived the historic period of heavy exploitation (CWS undated; Sladen and Isen 2005). Most

members of these two populations continue to migrate along traditional routes, some of which span distances of over 1500 km (Sladen and Olsen 2005). The flocks that comprise the Interior population were established as part of over 30 years of re-introduction efforts and, unlike the populations that inhabited this region prior to the historic population decline (Lumsden 1984), they are generally non-migratory (Sladen and Olsen 2005; Mitchell 1994).

The relative rarity of long-distance migrants amongst re-introduced birds has been attributed to the learned nature of migratory behaviour amongst trumpeter swans (Sladen and Olsen 2005). In migratory populations, parents lead their offspring both to and from wintering grounds (Sladen and Olsen 2005; Lumsden and Drever 2002; Sladen *et al*. 2002) and transmit information about over-wintering locations and the traditional migratory routes. In standard population restoration programmes, young swans generally have no migratory parents from which they can learn migration and traditional migratory routes (Lumsden and Drever 2002) and, thus, re-established populations often remain resident year-round in the vicinity of introduction locations.

Figure 11.2   A trumpeter swan on the nest. If nesting efforts are successful, the young cygnets will depend on their parents to teach them migratory routes and the location of over-wintering areas. The learned nature of trumpeter swan migration has been a challenge for re-introduction programmes aimed at restoring populations and migratory behaviour. However, restoration efforts that attempt to teach young trumpeter swans to migrate have had some limited success (Photo credit: US National Parks Service).

Threats and vulnerabilities

Over-exploitation is believed to have been the primary contributing factor associated with the historic population decline of trumpeter swans (Shea *et al*. 2002). After the 1918 *Migratory Bird Treaty Act* ended the unregulated exploitation of the swans in the USA and Canada (Shea *et al*. 2002), there were continued concerns over small population sizes and the loss and degradation of winter habitat (Shea *et al*. 2002; Baskin 1993). Beginning in the early 1900s, these concerns have motivated diverse conservation and restoration efforts, including the establishment of winter feeding programmes and species–specific protected areas, as well as re-introduction programmes (Shea *et al*. 2002; Baskin 1993; Banko 1960). Although the trumpeter swan still faces threats such as habitat loss, lead poisoning (Shea *et al*. 2002) and competition with introduced species (McEneaney 2001), these efforts are believed to have contributed to the range expansion and population increases observed in recent decades (Mitchell 1994; Sladen and Olsen 2005).

continues

Box 11.2 *(continued)*

Traditional re-introduction programmes in particular, have been credited with re-establishing viable populations in parts of the trumpeter swan's historic range (e.g., in the North American interior; Shea *et al.* 2002; Mitchell 1994). However, the inability of such programmes to restore migratory behaviour and historic migration routes has been a major shortcoming (Lumsden and Drever 2002). Migration loss is not entirely restricted to re-introduced populations. For some native populations (e.g., the flock that inhabits the Red Rock Lakes National Wildlife Refuge in Montana), the historic exploitation of migratory adults, in combination with supplementary feeding of the remaining individuals during the winter season, are believed to have contributed to the loss of migratory behaviour and unsustainable over-winter densities (Baskin 1993). In addition to the potential ecological impacts that the loss of trumpeter swan migration has had in the locations where migratory birds used to over-winter (which to date, have received little attention), migration loss has also had negative ecological impacts at inhabited locations. These include: the over-grazing of aquatic plants and a series of negative cascading effects on fish species that inhabit resident areas (Baskin 1993); population growth limitation; and even threats to the long-term persistence of trumpeter swans in locations where winter food availability is low or dependent on winter severity (Shea *et al.* 2002; Baskin 1993). Therefore, while historic conservation measures have eliminated some of the major factors that once threatened trumpeter swans, the manner in which they have been applied has contributed to a loss of migratory behaviour, which itself serves as a barrier to population expansion amongst resident flocks, as well as a potential threat to their long-term persistence.

Conservation and management solutions

Decades of conservation efforts have achieved considerable success in reducing the impact of the major factors that once threatened the long-term persistence of trumpeter swans (Shea *et al.* 2002; Mitchell 1994). However, while conservation efforts have been relatively effective at arresting and reversing population declines, the restoration of trumpeter swan migrations has proven to be much more difficult. One potential approach for dealing with this issue focuses on training resident birds to leave their breeding grounds in autumn and migrate to suitable wintering locations (Ellis *et al.* 2003; Lumsden and Drever 2002). In 1990, efforts to conserve the resident Red Rock Lakes NWR flock were shifted to focus on teaching birds to migrate to more suitable, ice-free over-wintering locations, by capturing, crating and trucking them to several different wintering areas (Baskin 1993). The majority of the translocated birds did not continue to migrate to these new locations after the initial 'forced migration', however approximately 15–20% made the return journey the following year (Baskin 1993). While conservation approaches that focus on teaching individuals to follow historic or new migration routes have had some success with trumpeter swans and other large birds for which migration appears to be a learned behaviour—i.e., Canada geese (*Branta canadensis*) and sandhill cranes (*Grus canadensis*)—there are also considerable drawbacks to such approaches, not the least of which is the major investment of time and resources that they require. Additionally, for trumpeter swans, the gap between resources invested and pay-off achieved is likely to be particularly wide, because they are relatively difficult to rear, train and lead in comparison with several of the other avian migrants to which such approaches have been applied (Ellis *et al.* 2003) and they are reluctant to fly in the calm wind conditions necessary for ultralight flight (Ellis *et al.* 2003). However, the modest successes that have been achieved in teaching trumpeter swans to migrate suggest that further efforts might be worthwhile, especially if methods that are less resource-intensive are developed (Sladen and Olsen 2005; Ellis *et al.* 2003).

ranges were likely to remain faithful to suboptimal habitat. In contrast, birds that learned migration routes from other members of the population were less likely to show fidelity to degraded habitats and to experience the negative consequences associated with such conditions (Sutherland 1998).

In some circumstances, the impacts of threats like habitat loss or degradation can unfold in a gradual, long-term manner while, in others, they can occur and escalate quite quickly. For example, the latitudinal shifts in habitat and extreme weather events that are predicted to accompany climate change are predicted to occur quite rapidly (Malcolm *et al.* 2002; Rosenzweig *et al.* 2001). Some studies indicate that movement strategies that are learned are more amenable to rapid change in response to changes in environmental conditions than those that have a genetic or developmental basis (Sutherland 1998, but see also Andersen 1991, and Nelson and Mech 1981). Partially migratory populations with a strong genetic basis for migrant or resident status can exhibit relatively rapid changes in the proportions of these two space use strategies (e.g., 25 years amongst partially migratory bird populations; Berthold 1991). However, changes might not occur at a fast enough pace to allow populations to avoid or substantially reduce the impacts of sudden and or extreme anthropogenic threats. In contrast, if individuals are capable of altering their own movement strategies in response to changes in environmental conditions (i.e., conditional migration), they might be more resilient and better equipped to respond to extrinsic threats.

11.2.2.5 *Migration timing and proximate cues*
Migratory animals use a variety of different cues to initiate migratory behaviour (Chapter 6). Some appear to respond to climatic variables such as temperature and precipitation (Jenni and Kery 2003; McCormick *et al*. 1998; Boone *et al*. 2006) that, although they exhibit broad seasonal trends, can be relatively unpredictable in timing and intensity at finer scales. Amongst other organisms, the initiation of migration appears to be associated with endogenous biological rhythms or temporally predictable conditions like photoperiod (Gwinner 1996).

The cues that a population relies upon to initiate migration can affect its vulnerability to different types of threats. Migratory species for which the initiation of migration appears to be driven by endogenous biological rhythms or photoperiod (e.g., long-distance migratory birds; Jenni and Kery 2003), could be more vulnerable to the impacts of climate change (Both and Visser 2001). For example, if climate change affects the phenology of plant growth or the timing of insect emergence, a mismatch could develop between the timing of migratory movements and the availability of food resources (Robinson *et al*. 2005; Kirby *et al*. 2008). If their migratory behaviour is triggered by more variable environmental conditions, animals might be more capable of adjusting the timing of their migrations in response to climate-induced environmental

change (Walther *et al*. 2002). However, if animals respond to environmental cues that reflect deterioration of resources and conditions on the range from which they are departing, and these no longer correspond to an increase in the quality of resources and conditions at their destinations, then their reliance on variable environmental conditions as proximate cues might result in a mismatch between resources available at the destination range and the timing of their arrival (Jenni and Kery 2003; Ahola *et al*. 2004; Both *et al*. 2006; Inouye *et al*. 2000). Thus, in some circumstances, the responsiveness of migration to variable environmental conditions might not reduce an organism's vulnerability to the impacts of climate change.

The type of proximate cue that triggers migration can also affect the vulnerability of migratory organisms to over-exploitation. For some migratory species, exploitation pressure is strongest during the migratory period, when the location of individuals is easier to predict than it is during other behavioural phases (e.g., Enloe and David 1995). If the timing of migration is highly consistent and predictable between years and/or seasons (as it would be if photoperiod or biological rhythms were serving as proximate cues), then exploited populations could be more vulnerable to over-harvesting, as exploitation efforts would be likely to have a greater probability of success (Bolger *et al*. 2008).

11.2.2.6 *Migration distance*
While some organisms migrate thousands of kilometres between different continents and biomes, others migrate between ranges that are separated by relatively short distances (Kirby *et al*. 2008; Bolger *et al*. 2008). Organisms that migrate long distances are likely to incur greater costs associated with the longer journeys that they undertake. These include greater energy expenditure (Alerstam *et al*. 2003), longer exposure to predators (Lank *et al*. 2003) and parasites (Alerstam *et al*. 2003), and the risks associated with exposure to extreme weather events (Morrison *et al*. 2007). To some extent, such costs must be outweighed by fitness-related benefits in order for the behaviour to persist. A recent study of Arctic shorebirds indicates that birds that migrate further distances to northerly locations experience lower nest predation risk than those that migrate

shorter distances (McKinnon et al. 2010). However, negative population trends amongst birds that exhibit inter-continental migrations suggest that they might also be more vulnerable to extrinsic anthropogenic threats in comparison with short-distance migrant or resident populations (Sanderson et al. 2006).

Assuming that the probabilities of habitat loss/degradation or the occurrence of barriers are equal along each segment of an animal's migratory journey, longer migration routes would have a greater chance of exposure to such threats. Affected areas could exacerbate the mortality costs associated with migration or serve as blockages to migratory movement. Ungulates that migrate long distances generally spend more time migrating than short-distance migrants and thus they are more dependent on the availability of suitable resources along the journey to replenish the energy they expend (Bolger et al. 2008). Additionally, in landscapes increasingly altered by land conversion and development, the number and width of the migration routes available to long-distance ungulate migrants has decreased, leaving them ever more vulnerable to obstruction and interruption (Berger 2004). Birds that undertake long-distance migrations are dependent on the availability of suitable resources at stopover sites along the way, but the quality and quantity of resources at the ranges they migrate between can be even more important. This is because food resources at breeding and winter grounds determine the amount of fat accumulated prior to migration, as well as the extent to which depleted fat reserves are replenished and nutritional requirements are attained after the migratory journey (Kirby et al. 2008). Therefore, long-distance migrants might be more vulnerable than short-distance migrants to habitat loss and degradation.

Long-distance migrants that are exploited for commercial, recreational or subsistence purposes might be more vulnerable to over-exploitation if they are harvested in more than one political jurisdiction (Box 11.3). To ensure that harvesting remains sustainable, coordinated efforts to develop and implement cooperative management plans, that involve all of the states or countries that a migratory population travels through, might be necessary (Jarnemo 2008). However, the logistical and politi-cal difficulties associated with such an approach might make this difficult to achieve. Short-distance migrants that remain within the same political jurisdiction are not exposed to these threats.

The rate, direction and variability of forecasted climate change varies considerably between regions (IPCC 2001). As a result, long-distance migrants might be moving between environments that are affected very differently by climate change. Consequently, they might be more likely to experience a climate change-induced divergence between the timing of migration and resource availability at their destination range, as compared with organisms that migrate within a single region (Jenni and Kery 2003; Kirby et al. 2008). Additionally, some of the changes in resources and conditions that are predicted to accompany climate change (i.e., prolonged breeding seasons and increases in breeding habitat suitability) might increase the fitness of some short-distance migrants (Jenni and Kery 2003). Comparable benefits for long-distance migrants might be limited by the lower flexibility that they often exhibit in terms of the timing of reproduction and migration (Jenni and Kery 2003; Both and Visser 2001; Both et al. 2006). The findings of these studies notwithstanding, it should also be noted that empirical evidence regarding the relative advantages of short-distance migrants (vs. long-distance migrants) with respect to their abilities to adjust to the effects of climate change is not unequivocal. European passerines that engage in long-distance migrations have exhibited a rapid advance in both the timing of spring migration and their arrival dates in northern Europe (Jonzén et al. 2006). It has been suggested that these changes may represent an evolutionary response to climate change, rather than a phenotypic response to changes in foraging conditions (Jonzén et al. 2006). While this evidence indicates that the capacity of long-distance migrants to adapt to climate change may be stronger than expected, it still remains to be seen whether observed shifts will be sufficient to avoid population declines (Jonzén et al. 2007). Additionally, migratory organisms with longer maturation times (in comparison with passerines, which reach reproductive maturity at one year of age; Jonzén et al. 2006) are unlikely to exhibit such a rapid evolutionary response to environmental change.

Box 11.3 Conservation of a terrestrial migrant: saiga antelopes in Central Asia

Biology

The saiga antelope, *Saiga tatarica*, inhabits the steppes, semi-deserts and deserts of Central Asia. It has two subspecies—*S.t. tatarica* is predominately found in Kazakhstan and Russia, in four populations, one of which migrates from Kazakhstan to Uzbekistan in the winter, and occasionally to Turkmenistan in particularly bad winters. The much less numerous *S.t. mongolica* is found only in a small area of western Mongolia. In Kazakhstan, the species undertakes seasonal migrations of up to 1000 km, from southern desert zones where it over-winters to the northern steppe zones, where it spends the summer making nomadic movements in response to precipitation (Bekenov *et al.* 1998). These migrations are driven by precipitation and vegetation productivity, with the animals taking advantage of lush summer grasses in the north, and the relative lack of winter precipitation in the south (Singh *et al.*, 2010). In Russia, the population still moves but its distribution is poorly known and migrations seem to be constrained by substantial human-induced reductions in the extent and quality of its range (Lushchekina and Struchkov 2001; Leon 2009). The species gathers in very large aggregations to give birth for around 10 days in May, during the spring migration, such that births are extremely tightly synchronized, both spatially and temporally. Spatial aggregation is thought to be a predator-swamping mechanism, and temporal aggregation is likely to be the result of a trade-off between maximizing access to peak forage productivity and avoiding cold weather early in the season (Milner-Gulland 2001).

Threats and vulnerabilities

The saiga suffered a catastrophic population decline in the 1990s, caused by unregulated hunting following the break-up of the Soviet Union (Milner-Gulland *et al.* 2001). Due to a 95% reduction in numbers over a 10 year period, it was uplisted in 2002 from Near Threatened to Critically Endangered in the IUCN Red List (www.redlist.org). Because hunters preferentially targeted males for their horns, used in traditional Chinese medicine, the sex ratio became extremely female-biased, leading to reproductive failure in some years of the early 2000s (Milner-Gulland *et al.* 2003). Aggregations are currently much smaller than historically recorded, and were not observed at all for several years in the hardest-hit populations (Fry 2004). This

(a)

(b)

Figure 11.3   (a) A herd of saigas during their spring migration; picture zoomed in from a photograph taken during an aerial survey. Saigas are very hard to count by eye from planes, particularly in fragmented groups, which has led to biased estimates of population recovery (McConville *et al.* 2008). Photo credit: A.J. McConville. (b) Up until the late 1990s, and the break-up of the Soviet Union, saigas migrated in herds of many thousands, leaving paths in the steppe, and producing a dramatic spectacle and a major source of meat and income. This picture is from the 1970s. Photo credit: Institute of Zoology, Kazakhstan.

continues

Box 11.3 (*continued*)

is thought potentially to have led to increased mortality among juveniles (Milner-Gulland 2001). The reduction in both population densities and group sizes has also made monitoring difficult, by increasing sampling error and biasing counts downwards through reduced herd detectability (Fig. 11.3; McConville *et al*. 2009). Poaching is still ongoing, and remains a major concern range-wide (Kuhl *et al*. 2009, Saiga News 2009). Despite the population crash, the migration itself appears still to be intact, at least in the Kazakhstan populations (Singh *et al*. 2010).

Other threats include habitat degradation, infrastructure, competition for grazing and disease. The only currently serious threat among these is competition for grazing in the Mongolian sub-population (Chimeddorj 2009). However, infrastructural development, such as canals, and habitat degradation through over-grazing by livestock were major issues in Soviet times, particularly in Russia (Lushchekina and Struchkov 2001; Sokolov and Zhirnov 1998; Robinson *et al*. 2003). Disease transmission between livestock and saigas, particularly foot-and-mouth, killed large numbers of saigas during epidemics (Sokolov and Zhirnov 1998; Morgan *et al*. 2006). Current densities of livestock and saigas are not high enough for habitat degradation or disease transmission to be major problems nowadays, particularly as livestock movements are currently constrained, such that they no longer undertake the long-distance migrations that used to mirror those of the saiga (Robinson and Milner-Gulland 2003; Kerven *et al*. 2006). In fact current pasture quality in much of the saiga range is reportedly good, due to a lack of grazing pressure. Major concerns in the near future include new infrastructural development, particularly for gas exploration in Uzbekistan (Bykova *et al*. 2009).

Conservation and management solutions

International concern for saigas began in the 1990s. Saiga horn was listed on Appendix II of CITES in 1995 due to concerns about high trade levels. A significant trade review led to the suspension of all international trade in saiga parts in 2001 (CITES 2004). The species was listed on Appendix II of the Convention on Migratory Species in 2002, and an MOU between the range states under CMS came into force in 2006 (CMS 2006a). This MOU included a commitment by signatories, including range states and international NGOs, to work towards implementing a medium term work programme over the next 5 years (CMS 2006b). This flurry of international concern, and particularly the CMS activities, has led to a strong international commitment to action towards conserving the species, which is helped by the existence of a saiga conservation network (www.saiga-conservation.com). Concerted action by governments and NGOs to improve law enforcement and public support for saiga conservation, combined with an improvement in rural livelihoods that has reduced the need to poach, have led to an apparent stabilization in numbers and signs of improved population structure in some populations (CMS 2006a). The only population that appears still to be declining is the transboundary one between Kazakhstan and Uzbekistan, for which management is much more challenging due both to its remoteness and the weaker jurisdictional authority over the population. A recent bilateral agreement on saiga conservation between the two countries represents a first step towards improving the situation.

A key challenge for the near future is to take advantage of the opportunity presented by the new political regimes in Central Asia to rethink land use planning, so that conservation at the ecosystem level can be integrated into sustainable land management while there is still flexibility (before livestock numbers increase again, and before major infrastructural developments are implemented). One example of this is the Altyn Dala Conservation Initiative, which is a collaboration between the Government of Kazakhstan and local and international NGOs to plan conservation on a landscape level within a large part of central Kazakhstan, with saigas as the umbrella species. A new protected area is in the process of being declared as part of this initiative. However, effective conservation planning is still hindered by a lack of understanding of saiga movement patterns in all parts of its range, its seasonal distribution and the drivers of migration (Singh *et al*., 2010). Without these, it is very hard to translate current observed locations of a low density, fragmented population into the long-term needs of a recovered population that can maintain intact migratory behaviour. With the Kazakhstan-based saiga populations in particular, we have the opportunity to safeguard a major migration of global significance, if action is taken promptly and with due regard to the species' requirements.

While the existing evidence is somewhat ambiguous, under some circumstances it appears as though long-distance migrants might be more vulnerable to threats such as habitat loss, over-exploitation and climate change. However, there are some circumstances where long-distance migrants might be more resilient to anthropogenic threats than short-distance migrants. For example, if migration distance is reflective of general vagility, long-distance migrants might be more capable of successfully responding to the relatively rapid forced range shifts that are predicted to accompany climate change (Walther *et al.* 2002; Malcolm *et al.* 2002). Greater mobility might also make them better equipped to move to new locations with higher habitat quality in response to habitat loss or degradation along traditional migration routes or in breeding or over-wintering ranges. Finally, long-distance migration might prevent or reduce the severity of over-exploitation by providing harvested populations with a temporary escape from exploitation pressure in some portion of their range. In comparison, migratory populations that remain in the same political jurisdiction may have no opportunity to escape (if only temporarily) from intensive harvesting pressure.

11.2.2.7 Responsiveness to resources and conditions during migration
The type of behaviour that migratory organisms exhibit while travelling between disjunct ranges could influence the extent to which individuals are responsive to, and capable of avoiding, threats that they are exposed to while migrating. A behaviour-oriented definition of migration has been proposed by Kennedy (1985), Dingle (1996) and Dingle and Drake (2007), in which the inhibition of responses to immediate experiences of resources and conditions is used to distinguish migratory movements from other types of movement behaviour. In contrast, exploratory behaviour that moves individuals beyond discrete ranges, but is also characterized by responsiveness to immediate surroundings, is called 'ranging' (Dingle 1996). Both types of behaviour can produce 'migratory' space use patterns (i.e., return-trip or back-and-forth movement between spatially separate ranges; Chapter 8, Baker 1978; Fryxell and Sinclair 1988). However, they have different implications for the extent to which

individuals are responsive to and capable of avoiding threats that they encounter while migrating. Specifically, 'migratory' behaviour (Dingle 1996; Kennedy 1985) is likely to be governed by a small number of large-scale decisions that are based on memory or genetic encoding and, thus, extend beyond local conditions and an individual's immediate perceptual range (Bolger *et al.* 2008). As a result, individuals that exhibit 'migratory' behaviour might be less capable of responding to and avoiding the effects of habitat loss/degradation or human exploitation along the migratory route. In contrast, if migration is produced by a series of local decisions made using information acquired from the resources and conditions while migrating (i.e., 'ranging' behaviour), then individuals might be more likely to react to and be resilient to threats that they encounter whilst migrating (Bolger *et al.* 2008).

11.3 Strategies for conserving and managing migratory species

11.3.1 Threat mitigation and sustainable management of extant migrations and migratory populations

A wide variety of strategies exist for conserving and managing migrations and migratory species. The effectiveness of these different approaches is dependent on the particular characteristics of the targeted migration. When the persistence of a migratory population and/or migratory behaviour is at risk, the nature of the major threats responsible can also affect the success of a particular conservation strategy.

11.3.1.1 Protected areas (static and dynamic)
The establishment of permanent reserves or protected areas, that exclude or reduce the intensity of threatening processes, is considered to be one of the most efficient and cost-effective strategies for conserving biodiversity (Balmford *et al.* 1995). Consequently, recent decades have seen major increases in the total area of the Earth's surface that falls under some sort of protected status (Terborgh and van Schaik 2002). However, the extent to which

they are successful at meeting conservation objectives varies (Ervin 2003).

Many species-specific conservation strategies involve delineating specific locations or 'protected areas' where the impacts of threatening processes are restricted or eliminated. However, given logistical constraints on size and number of reserves (Norton-Griffiths and Southey 1995), the inherent vagility of migratory populations can make designing effective reserve-based or location-specific strategies to conserve migratory behaviour and/or populations particularly challenging (Thirgood *et al.* 2004; Berger 2004; Kramer and Chapman 1999).

For migratory organisms, the effectiveness of fixed reserves is contingent on different aspects of their migratory behaviour. First, permanent reserves would be much more effective for protecting the habitat of organisms that exhibit fidelity to their ranges or migratory routes, compared with those that don't show repeated use of the same locations. Second, the vagility of migratory populations and socio-economic constraints on protected area size make it difficult to create protected areas that are large enough to encompass entire migratory trajectories. Short migration distances and aggregative behaviour are two characteristics that might render a migratory population more suited to fixed area-protection. For example, it might be possible to include the entire range of a short-distance migrant in a single reserve (Peres 2005). Additionally, if aggregative behaviour concentrates the activities of multiple individuals in specific locations, it would probably be much easier to delineate protected area boundaries and there might even be a reduction in the minimum reserve size needed to protect utilized habitat. For many long-distance migrants, the protection of an entire migratory trajectory within a single protected area is simply not feasible (Thirgood *et al.* 2004). Numerous bird species fall into this category and several authors have suggested that conservation strategies include a network of reserves that protect critical stopover areas along migration routes, as well as in breeding and wintering areas (Kirby *et al.* 2008; Mehlman *et al.* 2005; Fox and Madsen 1997). The potential effectiveness of such reserve networks for conserving long-distance migrants and migrations is higher for organisms that exhibit site fidelity and aggregative behaviour.

In addition to the impacts that different characteristics of migratory behaviour are likely to have on the conservation efficacy of protected areas, there are several other factors that might influence the success of such an approach. For example, fixed reserves are probably more effective at protecting migrations and migrants from some threats than they are from others. Specifically, static reserves can provide protection from habitat loss due to anthropogenic land conversion and/or development, as well as human exploitation, but they are less capable of protecting migrants from the impacts of climate change (Singh and Milner-Gulland in press). Additionally, given the existence of competing uses and limited availability of financial resources for establishing and maintaining protected areas, the creation of a single reserve or protected areas network, focused on the habitat requirements and locations used by one migratory species, could be difficult to justify from a biodiversity conservation perspective (Gerber *et al.* 2005). Thus, while such an approach might be effective for conserving a targeted population, practical considerations might make it infeasible unless the organism is an 'umbrella' species, one of multiple focal species (Fleishman *et al.* 2000; Roberge and Angelstam 2004), or is of significant economic or cultural value.

While static protected areas can be an effective approach for conserving some migrations and migratory organisms, reserves that are dynamic in space and/or time might be more effective in other circumstances. For migrants and non-migratory organisms that live in environments that undergo rapid successional change or feature a dynamic disturbance regime, fidelity to specific locations can be short-term, due to the ephemeral nature of suitable habitat. In such cases, reserves that are relocated in order to track spatial habitat shifts are likely to be more effective from a conservation perspective (Rayfield *et al.* 2008). Zakazniks are a form of dynamic reserve that was established during the Soviet era and still persist in the countries of the former USSR (Mallon and Kingswood 2001). Zakazniks are established at specific locations to promote high densities of certain wildlife species (e.g., via the imposition of hunting restrictions) and then removed from protection after approximately 5–10 years (Shahgedanova 2003; Weiner 2000), or even shorter periods. For example, poaching has caused populations of the migratory saiga antelope (*Saiga tatarica*) to decline throughout their range since the end of the Soviet period (Box 11.3). Seasonally shifting zakazniks, which protected the saiga at critical times during their annual migration, particularly during birth aggregations and the rut, were used during Soviet times (Bekenov *et al*. 1998) and expansion of this approach has recently been recommended (Mallon and Kingswood 2001; Robinson *et al*. 2008).

In addition to spatially dynamic protected areas, there are also reserves that are fixed in space, but only provide temporary protection for migratory organisms. For example, because mortality due to fisheries bycatch is a major threat to leatherback populations (Kaplan 2005), Shillinger *et al*. (2008) recommend that dynamic time/area closures be employed as a mechanism for conserving migratory leatherback turtles (*Dermochelys coriacea*). These consist of temporary closures or reductions in fishing activities along leatherback migration routes and foraging areas during the months that turtles are present (Shillinger *et al*. 2008; Lewison *et al*. 2004a).

11.3.1.2   *Extensive approaches*

While location-based approaches to conserving and managing migratory populations can be effective in some circumstances, there are alternative approaches that don't focus on delineating specific locations for protection or for implementing specific management prescriptions. Such extensive approaches might be more effective for organisms that face threats amenable to reduction or elimination at a broad scale (e.g., the general prohibition of certain kinds of hunting or fishing gear) or for migrants or migrations that have characteristics that make them difficult to protect using a reserve-based approach.

In some circumstances, protected areas might not be sufficient (Thirgood *et al*. 2004) or even appropriate for achieving conservation goals. In such cases, the mitigation of threats to migrants by imposing sustainable management practices on the broader landscape can complement or even take the place of reserves. There are several migratory characteristics that might enhance the value of such an approach. For example, long-distance migration and lack of

site fidelity are characteristics that probably increase the likelihood that migrants will spend time outside of reserve boundaries. Solitary behaviour during at least part of the migratory cycle might make it difficult to delineate a discrete protected area of sufficient size to make a significant contribution to conservation. Additionally, depending on the nature of the threats to migrants, the implementation of more sustainable land use or management practices on a broad scale might be more effective than the delineation of distinct protected areas.

For example, in Ontario, Canada, forest management practices are believed to have contributed to the degradation of habitat for woodland caribou (*Rangifer tarandus caribou*; Vors *et al*. 2007) and consequent patterns of range retraction and population decline (Schaefer 2003). In an effort to mitigate the impact of forestry, new guidelines have been developed to reduce the impact of industrial logging on caribou habitat. These guidelines focus on excluding forestry activities from documented calving sites and adjacent areas, and maintaining a shifting mosaic of large, contiguous and mature stands of conifer forest (i.e., winter habitat) throughout the managed landscape that surrounds an existing network of provincial and national protected areas (Racey *et al*. 1999). While woodland caribou migrate relatively short distances compared with other subspecies (Mallory and Hillis 1998), solitary behaviour during the summer (Darby and Pruitt 1984; Schaefer *et al*. 2001), lack of strong individual fidelity to locations used during winter (Ferguson and Elkie 2004), and the dynamic nature of suitable winter habitat, all favour a landscape wide-approach to conservation and management.

There are many additional examples of extensive conservation approaches that can complement or supplant protected areas. One example is the establishment of Wildlife Management Areas in the Serengeti to reduce poaching and promote sustainable harvesting of migratory and resident ungulates (Thirgood *et al*. 2004). Another is the construction of wildlife crossing structures (Gloyne and Clevenger 2001; Xia *et al*. 2007) or fish passages (Coutant and Whitney 1999; Kareiva *et al*. 2000), which can reduce the impacts of anthropogenic barriers (such as expressways and hydroelectric dams) on aquatic and terrestrial migrations.

11.3.1.3 Population management (limiting exploitation)

For migratory populations whose persistence is threatened by over-exploitation, general restrictions on harvesting can be an effective conservation mechanism. Mortality due to unintentional capture by commercial fisherman has been identified as a major threat to migratory sea turtles (Watson *et al.* 2005). Several different types of fishing gear, including trawls (Poiner and Harris 1996), gill nets (Julian and Beeson 1998) and long-lines (Lewison *et al.* 2004b), have been identified as having a negative impact. While establishing protected areas (e.g., dynamic time/area harvesting closure; Shillinger *et al.* 2007) represents one potential solution, switching to alternative gear (i.e., circle hooks and mackerel bait) can significantly reduce loggerhead and leatherback bycatch (Watson *et al.* 2005). For migrants that are threatened by direct over-exploitation, the imposition of more sustainable harvesting practices can serve as an important conservation strategy. However, such an approach may be very difficult to develop and implement for long-distance migrants that inhabit and travel through a variety of different political jurisdictions (Jarnemo 2008) or, as is the case for several marine species, the high seas where no single nation has jurisdiction.

11.3.1.4 Behaviour management (trophic interactions and predator mimicry)

Access to high quality forage and predator avoidance are two major mechanisms that are believed to underlie the evolution of migratory behaviour (Chapter 2; Fryxell and Sinclair 1988; Dingle 1996). Therefore, the manipulation of predation risk has the potential to alter migratory behaviour (Fryxell and Sinclair 1988; Alerstam *et al.* 2003). Patterns of range retraction and relatively high extinction rates amongst higher trophic level species such as large carnivores (Berger 1998; Laliberte and Ripple 2004) might lead to migration loss in their prey. While the impacts of carnivore loss on predator recognition, anti-predator vigilance and aggregative behaviour of their prey species has been demonstrated in some cases (Blumstein 2002; Berger 1998), there appears to be little research on the impact of predator loss on prey migration. One exception to this is the influence of predator presence on the diel migrations exhibited by planktonic copepods (Bollens and Frost 1989). After a week of separation from their planktivorous fish predators, copepods stopped migrating and only exposure to free-ranging predators was able to induce migration (Bollens and Frost 1989). For those populations for which migration loss may be tied to the loss of major predators, the re-introduction of such predators might be successful at restoring lost migrations. However, in some circumstances the resulting increase in predation-related mortality could have a negative impact on population viability, thus the potential implications of predator introductions for prey populations should be carefully considered a priori.

Management efforts focused on increasing the quantity of suitable forage available to migratory populations can also have an impact on migratory behaviour, intentional or not. For example, in temperate climates, supplemental feeding of cervid populations is a common strategy to reduce mortality during the winter season, when the availability of suitable forage is relatively low (Baker and Hobbs 1985; Ouellet *et al.* 2001; Peterson and Messmer 2006). Although food provision can have a positive impact on survival rates and productivity (Baker and Hobbs 1985; Peterson and Messmer 2006), it has also produced changes in the migration timing and the length of migratory routes used by mule deer (*Odocoileus hemionus*) in Utah (Wood and Wolfe 1988; Peterson and Messmer 2006).

Supplemental feeding has also been implemented as a direct strategy to arrest or alter migratory movements, in situations where traditional migratory patterns are believed to represent a significant human safety hazard (Gunderson *et al.* 2004; Andreassen *et al.* 2005). For example, in an effort to reduce the human safety hazard posed by moose migration in Norway, supplemental feeding has been used as a means of stopping migration or redirecting migratory routes (Andreassen *et al.* 2005). Similar strategies have also been implemented to control the location of feeding sites used by migratory pink-footed geese (*Anser brachyrhynchus*) in Denmark (Klaassen *et al.* 2008).

In addition to the direct manipulation of predator abundance to alter migratory behaviour, disturbances designed to simulate predator activity and invoke anti-predator responses have been used to alter different aspects of migration, such as the location of stopovers used during migration and the length of time that migrants spend at them (Kéry *et al.* 2006; Klaassen *et al.* 2006). For example, many migratory goose populations are subjected to 'scaring' activities designed to invoke a flight response along the entire length of their flyways, in an effort to minimize crop damage at stopover sites (Klaassen *et al.* 2008). Deterrents used to scare birds are highly varied and include auditory, visual and chemical deterrents, in addition to habitat modification and exclusion (Bishop *et al.* 2003). Many visual scaring tactics attempt to simulate the presence of a predator, with the goal of evoking fear and producing an avoidance response (Bishop *et al.* 2003). In a review of studies that evaluate the effectiveness of different techniques, Bishop *et al.* (2003) found that those that closely mimicked the predators of targeted species were relatively successful compared with the alternatives. However, effectiveness can be compromised by rapid habituation to the simulated threat (Marsh *et al.* 1992) or, in some cases, the provocation of an undesirable anti-predator response (e.g., swarming mock predators rather than leaving a specific area; Conover 1984). While it might be effective in meeting management objectives in some circumstances, an individual-based model developed by Klaassen *et al.* (2006) predicted that intensive scaring is likely to cause a decrease in food intake rates that, in turn, could have a negative impact on reproduction and survival. However, at least one attempt to evaluate whether intensive scaring has long-term fitness consequences found no significant effects (Kéry *et al.* 2006).

11.3.2 Restoration of lost migrations

Several different approaches have been used to restore migration pathways for populations that have lost them. These include restoration of damaged habitat along migration routes, removal of migration barriers and provision of passages that circumvent obstacles to migration.

11.3.2.1 Habitat restoration
If damage to critical habitat is the key factor underlying migration loss, efforts to restore or alter affected habitat might be sufficient to promote the re-establishment of lost migrations (although formal assessments of restoration effectiveness are limited; Block *et al.* 2001). Results from recent studies show that it is possible to restore degraded habitat to a condition that produces increased use by wildlife that previously avoided such areas. Habitat restoration in Jasper and Banff National Parks (Alberta, Canada) led to increased use by wolves (*Canis lupus;* Shepherd and Whittington 2006, Duke 2001) and the creation of a movement corridor for elk (*Cervus elaphus;* Shepherd and Whittington 2006), while the removal of ski trails and a tourist cabin from historic winter reindeer (*Rangifer tarandus*) range in Rondane (Norway), led to greater use by reindeer of this previously avoided area (Nellemann *et al.* 2009). Nellemann *et al.* (2009) argue that habitat restoration could be an effective approach for restoring lost migration routes amongst reindeer in Norway. However, this approach would require the rehabilitation of very large areas, which could be prohibitively expensive and logistically infeasible.

11.3.2.2 Barrier removal
In addition to the incremental degradation of important habitats, the migrations of both aquatic and terrestrial organisms are extremely vulnerable to anthropogenic barriers developed along migratory routes (Coutant and Whitney 1999; Kareiva *et al.* 2000; Gloyne and Clevenger 2001). Freshwater systems are particularly vulnerable to such blockages, as almost three quarters of North American and European rivers have regulated flows and more than 800 000 dams divert Earth's waterways (Giller 2005). Along with other aquatic barriers (e.g., culverts and road crossings), dams are major obstacles to migration and consequently, are major contributors to species loss and population decline amongst freshwater organisms (de Leaniz 2008; Kiffney *et al.* 2009). Dam removal is generally advocated for a variety of environmental restoration-based reasons (Stanley and Doyle 2003), including the restoration of aquatic migration routes (Stanley *et al.* 2007). However, the number of studies that have assessed its impact on migration restoration are relatively small (Bernhardt *et al.* 2005; Marks *et al.* 2009). The limited evidence that does exist

indicates that dam removals can be successful in restoring lost migrations (Schuman 1995; Stanley and Doyle 2003; Kiffney *et al*. 2009). Positive outcomes appear to be dependent on the attributes of the species targeted for restoration. For example, if a barrier has been in place long-term and a species is dependent on natal stream imprinting or conspecific pheromone cues to migrate, natural re-establishment of migration subsequent to barrier removal may be slow or non-existent (Vrieze and Sorensen 2001). Potential negative impacts associated with dam removal must also be considered (e.g., removal of barriers between native and hatchery raised populations or between native and invasive species; Stanley and Doyle 2003),

While there appear to be few terrestrial analogues for the absolute barriers to migration that can affect aquatic migrants, one example is the 'veterinary fences' constructed in Botswana in the 1950s to control the transmission of disease to cattle (Mbaiwa and Darkoh 2005). These fences span several hundreds of kilometres, block the migration of wildebeests, zebras (*Equus quagga*) and giraffes (*Giraffa camelopardalis*) and produce high mortalities. While restoration efforts are not yet widespread, the removals of two such fences have been followed by population increases amongst affected migrants (Mbaiwa and Mbaiwa 2006).

11.3.2.3 Barrier circumvention

In recent years, the negative impacts of dams on freshwater migrants have been mitigated by increased efforts to include fish passages in dam construction protocols (Box 11.4). Similar efforts have been made in terrestrial environments, where human-constructed linear features such as roads and railways represent the most common obstacle for land-bound migrants. While not impenetrable, these features can disrupt migratory behaviour (Xia *et al*. 2007) and reduce the size of migratory populations via an increase in collision-related mortalities (Kleist *et al*. 2007). Several different forms of wildlife passage have been built over or under these barriers as a means of reducing vehicular mortalities (of wildlife and humans) and facilitating animal movement between bisected areas (Clevenger and Waltho 2005). Assessments of the effectiveness of these structures have revealed a

mixed record of success in promoting movement and reducing vehicular mortalities (Kleist *et al*. 2007). For example, railway underpasses were successful at reducing Tibetan antelope *(Pantholops hodgsonii)* collision mortalities in China (Xia *et al*. 2007) but highway underpasses had little effect on moose collision mortalities in Sweden (Seiler *et al*. 2003). Effectiveness appears to be species-dependent and is affected by factors such as: passage type and dimensions; habitat suitability in the areas surrounding the crossing structure; and the presence of fencing around the linear feature (Ng *et al*. 2004; Clevenger and Waltho 2005; Kleist *et al*. 2007).

11.3.2.4 Re-introduction and translocation of migratory individuals

For several decades, re-introductions (via the release of translocated or captive-bred individuals) have served as a strategy for re-establishing populations in parts of their historic range where they are in decline or have been extirpated (Fischer and Lindenmayer 2000). The outcomes of these re-introduction efforts have been mixed (Griffith *et al*. 1989; Wolf *et al*. 1996), although it seems that the probability of success is higher if: Translocated individuals are obtained from a wild population (vs. captive born); relatively large numbers of animals are released (i.e., *n* >100); and the threat that resulted in the original decline is eliminated (Fischer and Lindenmayer 2000). For re-introductions to serve as an effective strategy for re-establishing lost migrations, they must not only succeed at creating self-sustaining populations in areas where migratory species have been lost, but they must also result in the return of migratory behaviour (Box 11.4). Whether or not re-introduction efforts manage to accomplish these goals depends on whether the approach chosen is appropriate for the circumstances of the specific situation (Fischer and Lindenmayer 2000). For example, while relatively few studies have explicitly evaluated this, success at restoring migratory behaviour may depend on the extent to which that behaviour is learned (e.g., transmitted from parents to offspring) or innate, with the advantage possibly going to those organisms with a strong innate component (e.g., the garden warbler *Sylvia borin*; Gwinner and Wiltschko 1978; but see Box 11.4).

Box 11.4 Conservation of aquatic migrants: anadromous and marine fishes

Biology

Marine environments are far from homogeneous, providing a wide range of habitat types and great contrast in temperature and productivity with latitude, proximity to landmasses and major ocean currents and depth. Many species undertake very large horizontal migrations between feeding areas (often located in zones of high productivity in upwellings or current boundary features, which may attract large numbers of predators) and breeding areas (usually located inshore in more sheltered areas that are removed from high levels of predation pressure). While breeding areas tend to be more static, feeding areas may themselves move over time, as the source of productivity changes with the seasons.

Many marine animals, in particular teleost fish and ommastrephid squid, are broadcast spawners, and may make use of ocean currents to distribute their eggs into environments favourable for juvenile survival, and to do this they migrate upstream to spawn. This behaviour is shown by all three major commercial ommastrephid squid (*Illex argentinus, Illex illecebrosus* and *Todarodes pacificus*; Boyle and Rodhouse, 2005). Although they do not have pelagic, drifting eggs, anadromous fish such as salmon adopt a similar strategy, actively migrating to breeding grounds (to which they exhibit strong site fidelity) in rivers to deposit eggs in favourable juvenile environments. These migrations are fundamental to the life-history strategy of the animal. When these migrations are interrupted, extinction of the affected population or stock can be expected.

Fishes that are truly oceanic, such as the Norwegian spring spawning herring (NSSH—*Clupea harengus*), also exhibit migrations that are driven by reproductive requirements and efforts to exploit food resources. The NSSH undertake an annual migration cycle that includes a spawning period on the Norwegian coast in early spring; followed by a spring-early summer phase, in which they feed on zooplankton (principally copepods and euphausids) in the Norwegian Sea and exhibit steady movement away from the coast; and a wintering area, either in a deep coastal Fjord or open ocean. Post-hatching larval drift is northwards along the coast as far north as the western Barents Sea and young herring (ages 1–3) remain in fjords and near the coast until they join the adult population.

Threats and vulnerabilities

Migration blockages are difficult to achieve in marine species unless spawning or feeding grounds are in coastal constrictions that can either be physically blocked or made uninhabitable by pollution and habitat loss. In contrast, experience has shown that anadromous fish are particularly vulnerable to blockages (e.g., by dams) or habitat destruction, which together with over-fishing, have contributed to significant population declines (Young 1999). Gustafson *et al.* (2007) showed recently that 29% of 1383 historic populations of US Pacific sockeye salmon (*Oncorhynchus nerka*), coho salmon (*Oncorhynchus kisutch*), chum salmon (*Oncorhynchus keta*), pink salmon (*Oncorhynchus gorbuscha*), and steelhead trout (*Oncorhynchus mykiss*) are now extinct (defined as having lost the original indigenous population or its anadromous component). Given the spawning site fidelity that salmon populations show, this loss translates into a similar level of ecological and genetic biodiversity loss (Gustafson *et al*. 2007).

Although truly oceanic fish species are unlikely to suffer the habitat degradation problems that have affected anadromous fish, they have suffered from over-exploitation, which in a number of cases has significantly altered population ranges and migratory patterns. The stock of Norwegian spring spawning herring was very large in the first half of the 20th century (i.e., 16 million tonnes of spawning stock biomass in 1945). Fishing pressure was and is particularly intense when fish aggregate at their spawning grounds (to which they return, year after year) and in their wintering areas. There was a rapid increase in fishing pressure in the 1950s and 1960s, caused by the entry of new vessels and the development of more efficient fishing gear such as nylon nets for purse seines and a hydraulic power block (Dragesund *et al*, 1997). This new gear caused an unsustainable increase in catches, particularly of juvenile fish. This in turn led to a collapse, starting in the mid-1950s with its lowest point in 1972 when the spawning stock was only 16 000 t (one thousandth of its peak recorded size).

The migration pattern of this species has been very dynamic. There have been four major shifts since 1950 (Holst *et al*. 2004), which corresponded with changes in population size. While these shifts may be partly a response to high population pressure on local resources, stock collapses have also served as a contributing factor. For

continues

Box 11.4 (*continued*)

example, as the stock contracted in the 1960s, so too did the feeding migration, partly in response to unfavourable feeding conditions north of Iceland. Spawning areas remained in most of the same places along the central Norwegian coast as they had previously, although a more northerly spawning component arose in the mid-1960s (Dragesund *et al.* 1997). After the collapse in the 1970s, a completely new migration pattern emerged in which all stock activities were centred in Norwegian waters and fjords.

Conservation and management solutions

The primary strategy for avoiding the loss of migratory behaviour and the extirpation of migratory populations, that can occur when the migratory routes of anadromous fishes are obstructed (Gustafson *et al.* 2007), is to protect or restore migration routes. Protection can be achieved by prohibiting the development of barriers or removing existing ones, avoiding habitat destruction at migration and spawning locations, and incorporating mechanisms (i.e., fishways) at existing barriers that facilitate fish passage (Coutant and Whitney 1999; Kareiva *et al.* 2000).

Additionally, although anthropogenic habitat loss and degradation have already resulted in local extinctions, which have reduced the biodiversity of salmon on the west coast of the USA, Gustafson *et al.* (2007) point to a number of studies that indicate that salmon retain the ability for robust evolutionary adaptability in the face of anthropogenic declines and climate change. A number of natural and artificial re-introductions of salmon from surviving populations into salmon-extinct tributaries have been successful (Young, 1999; Anderson and Quinn, 2007). These results indicate that efforts to restore declining and extirpated populations of anadromous fishes have a substantial chance of succeeding.

After its collapse due to over-exploitation in the 1950s and 1960s the Norwegian spring spawning herring stock began to recover in the mid-1980s. This recovery resulted from efforts to implement precautionary harvest restrictions aimed at reducing exploitation pressure which included measures to protect juveniles (ICES 2007). In the 1990s the distribution of the stock expanded as it grew in size until, in 1994, catches were made outside Norwegian waters for the first time in 26 years (Host *et al.* 2004).

The 2000s have seen a continued increase of stock size to 12 million tonnes (ICES 2008a) and an expansion of the feeding distribution to cover almost the entire area used for feeding in the 1950s, from northern Iceland through Jan Mayen Island almost to Bear Island. Over-wintering appears to be taking place once more in offshore areas (ICES 2008b). While the NSSH recovery is now considered complete, it occurred very slowly and may have been partially constrained by the post-collapse migration patterns adopted in the 1970s. Specifically, the Fjordal over-wintering sites used during this period might have increased the vulnerability of the stock to predation (Host *et al.* 2004). Therefore, the primary strategy for retaining large-scale fish migrations such as those described here (and also, for example for bluefin and other tuna; Block *et al.* 2005) is the same as the solution for managing stocks of fish generally, which is to keep them at a sufficient stock size that recruitment is not impaired.

In addition to the vulnerability of marine fishes to over-exploitation and the importance of sustainable harvesting strategies, the NSSH example demonstrates that migration patterns of oceanic species can be very flexible and may themselves be affected by stock collapses. Changes in migration pattern may have significant consequences for access of fleets and coastal fishers to a resource. For instance, while the NSSH was at a low level and restricted to the Norwegian Exclusive Economic Zone (EEZ), catches by other nations were not possible (with the exception of Russia, which fished the stock in its waters in the western Barents Sea). After the species expanded into international waters in 1994, other countries could participate once more (Norway, Russia, Faroes, Iceland and the EU). These changes in the countries engaged in the fishery created different incentives for cooperation and exploitation between the major players (Arnason *et al.* 2000; Hannesson 2004). Sustainable management of wide-ranging migratory fishes such as NSSH often requires the development of multilateral agreements (a process that can be complicated and time-consuming) and if stocks decline to levels that are low enough to disrupt migration patterns, management strategies will have to be revisited and redeveloped. This continuing revision of policy is likely to further complicate and even delay efforts to implement effective management approaches, which could have a negative impact on stock viability.

Despite the challenges associated with this approach, there are several instances where re-introductions have succeeded in restoring lost migrations. For example, migratory Atlantic salmon disappeared from the Rhine River in the 1950s following pollution-induced habitat degradation and canalization of the main course of the water way (Gerlier and Roche 1998). Following an improvement in water quality, salmon were re-introduced into the upper Rhine basin in 1992 and, by 1996, telemetry studies showed that marked individuals were attempting to migrate to spawning beds upstream and that, despite the presence of dams with ineffective fish passages, some individuals were successful in migrating up tributaries to spawn (Gerlier and Roche 1998). Efforts to re-introduce the white stork (*Ciconia ciconia*) to Switzerland after it was extirpated from the country in 1950 (as a consequence of habitat degradation and over-exploitation), were eventually successful at re-establishing a viable migratory population (Schaub *et al.* 2004). Re-introduction efforts began the year prior to extirpation. They involved the translocation of individuals from Algeria and their release in Switzerland after fledging. Owing to the poor return rate of these individuals, the strategy was altered so that translocated individuals were kept in captivity for four years before their release into the wild. While these efforts were successful at establishing a population of individuals who remained and began to breed in Switzerland, most birds failed to migrate and were dependent on artificial feeding programmes to survive the winter. However, by the early 1970s, the numbers of wild-born, unmanipulated birds had increased and so had the proportion of birds that fledged and migrated in a similar manner to other natural European populations. Thus, although re-introduction of white storks to Switzerland required decades of effort and major alterations in approach, those efforts appear to have been successful at re-establishing a lost migration.

11.3.2.5 *Teaching animals to follow lost or new migratory routes*

As mentioned above, it has proved particularly difficult to restore lost migrations amongst organisms that learn how and where to migrate by accompanying their parents on their first migratory journeys

(e.g., moose, Andersen 1991; trumpeter swans, Box 11.2). Re-introductions of captive-born or translocated individuals may be sufficient to re-establish extirpated populations, but those individuals are unlikely to re-establish migrations to and from a targeted location. From the 1990s onwards, researchers in North America and Europe have tried to develop migration training programmes designed to restore migration in birds for which the behaviour is learned (Ellis *et al.* 2003; Lumsden and Drever 2002). Migration training experiments have been implemented with a variety of large bird species, including Canada geese (*Branta canadensis*), sandhill cranes (*Grus canadensis*), whooping cranes (*Grus americana*) and Siberian cranes (*Grus leucogeranus*; Sladen *et al.* 2002; Lumsden and Drever 2002; Sorokin *et al.* 2002; Ellis *et al.* 2003; Sladen and Olsen 2005). Different protocols have been used in these experiments, but most share basic elements, such as: The incubation and hatching of offspring in summer habitat and the imprintation on either human captors or a guiding motorized vehicle; human initiation of aerial migration (e.g., using ultralight 'trikes', hang-gliders or trucks to guide flying birds); or guiding of birds along migratory routes from breeding sites to suitable over-winter locations (Ellis *et al.* 2003).

The success of these experiments has been variable, but the results indicate that all species are capable of learning to migrate (Ellis *et al.* 2003, SCFC undated a) and most are capable of making unassisted migratory journeys along learned routes in subsequent years (Ellis *et al.* 2003). Despite these successes, training approaches that involve human-led migrations have drawbacks that could limit their usefulness for the large-scale restoration of lost migrations. The main shortcoming involves the extensive level of resources (e.g., in terms of time, labour and flight costs) required to teach a single group of birds to migrate (Box 11.2). However, alternative approaches for guiding aerial migrations that might be more efficient and cost-effective (e.g., gas balloons, air ships) are in development and it is likely that the time needed for intensive training will be short, as trained birds should be capable of teaching subsequent generations how and when to migrate (Ellis *et al.* 2003; Sladen and Olsen 2005). Additionally, alternative training approaches that are less resource-intensive have their own draw-backs. For example, efforts to re-establish migratory populations by using similar organisms to cross-foster the young of targeted species and teach them to migrate have been successful, but they have also resulted in targeted species imprinting on, and attempting to breed with, individuals from the foster species, rather than conspecifics (Mahan and Simmers 1992; Reed 2004; SCFC undated b). Further, even if the imprinting problem is avoided, the need to place the targeted organism with a foster species that has very similar habitat requirements could compromise the long-term goal of population restoration, as the two species might eventually compete for resources at breeding or wintering areas or along the migration route.

In summary, 'migration training' appears to hold some promise for the restoration of lost migrations in birds that acquire migratory behaviour via social transmission of information. However, such approaches require intensive efforts and considerable resources (Sladen and Olsen 2005). Additionally, it is unclear how feasible human-led migration training might be for the re-establishment of migratory populations of non-avian taxa.

11.3.2.6 Pastoralism

Pastoralism and transhumance are forms of animal husbandry that are characterized by the strong dependence of people on the domestic livestock they tend (generally ungulates such as cattle, camels, goats and sheep) and the intermittent movement or migration of domestic herds (along with most or at least some of the humans that depend on them) in response to the shifting availability of resources (Marshall 1990). If pastoral or transhumant peoples and their livestock shift their locations in response to spatial variation in rainfall and vegetation growth then, in some respects, their movements mimic the migratory patterns of the wild ungulates with which they coexist (Chapter 10; Berkes 1999). Thus, mobile pastoral and transhumant lifestyles are potentially more consistent with the natural dynamics of the ecosystems where they are practised than permanent settlements and static agricultural practices (Rannestad *et al.* 2006; Okoti *et al.* 2004). As a result, it is possible that pastoral livestock might be performing some of the same ecological functions carried out by wild migratory

ungulates, which could be particularly important in situations where populations of undomesticated migrants have declined or been extirpated.

While there may be a case for viewing pastoral and transhumant livestock as the domestic analogues of their wild, migratory counterparts, the results of some studies indicate that pastoralism can have negative impacts on wildlife, due to competition and disease transmission (Caro *et al.* 1998; Prins, 1992; Niamir-Fuller 1998). Additionally, for several decades, African pastoralists and their livestock have been blamed for over-grazing and desertification. However, other studies have demonstrated that nomadic pastoralism has a much smaller ecological impact than sedentary livestock-based agriculture (Berkes 1999). In some cases, it can even enhance habitat suitability for wild migratory ungulates and accelerate nutrient cycling (Augustine *et al.* 2003). For example, Rannestad *et al.* (2006) found that, during the growing seasons, several wild ungulate species in the Lake Mburo ecosystem in Uganda appeared to preferentially select habitat in areas used for pastoral activities rather than that in the adjacent Lake Mburo National Park.

11.3.3 Variation between media

While migrations through two-dimensional environments (e.g., terrestrial migrations and those that occur in river systems) are more vulnerable to physical obstruction (see discussion in Section 11.2.1.5, above), they are also easier to protect than migrations through three-dimensional environments (e.g., aerial and many marine migrations).

Recent studies in North America suggest that wildlife corridors may be effective at preserving gene flow in terrestrial systems (Dixon *et al.* 2006). Similar corridors—for instance of preserved habitat or no-exploitation zones—may have to be much broader (in terms of space, the numbers of habitat types that require conservation action and the size of human use areas that are impacted) effectively to conserve wide-ranging aerial or oceanic animals, unless they exhibit fidelity to migration routes that are very narrow or focused at particular vulnerable points (e.g., green turtles; Box 11.1). If avian migrants use few intermediate habitats on their migrations, it may be sufficient to conserve only the migration route ends and possibly some key migratory stopo-vers (Mehlman *et al.* 2005), whereas for terrestrial animals, all intermediate habitats will need to be conserved.

The effectiveness of conservation efforts may also vary depending upon the jurisdiction that the migratory journey covers. For example, for most migratory birds, the vulnerable portions of their migration routes are over land and, therefore, conservation efforts need only deal with limited and clear national jurisdictional issues (although these may not be trivial problems to overcome). However, for oceanic seabirds such as albatross and petrels, much of their vulnerable migration takes place over the high seas, which have no national jurisdiction and no natural conservation or management bodies to regulate the activities of national fleets of merchant or fishing vessels operating in these waters (Birdlife International 2004). In such cases, effective conservation often requires the development of specific international agreements, which can be very difficult both to create and to enforce (Section 11.5).

11.4 Are migratory animals more or less difficult to manage than resident animals?

To a large extent, conservation and exploitation are two sides of the same coin and both goals share the challenges of managing migratory populations. For some migratory organisms, characteristics such as the tendency to form large aggregations, fidelity to distinct seasonal ranges and temporary refuge from site-specific threats, can make protection or exploitation efforts easier than they might be for non-migratory species. However, there are other aspects of the migration phenomenon that, on balance, make conservation and management of migrants much more complicated and challenging than efforts to conserve and manage non-migratory species.

Ecological theory is better developed for closed than open systems and, similarly, the allocation of conservation effort to the management of populations is largely focused on sedentary species (Klaassen *et al.* 2008). However, migrants operate at multiple spatio-temporal scales and their movements are wide-ranging and often multi-jurisdictional in nature. Thus, the successful conservation

or exploitation of migratory populations requires that we include the spatial dimension in our strategies that, in turn, calls for knowledge on how organisms respond to spatial heterogeneity on various scales. Hence, the cost of achieving the knowledge needed for modelling systems and advising managers (e.g., estimates of movement rates) is certainly higher for migratory than for sedentary populations.

Migrants differ from residents in at least two ways that are relevant for conservation and management; first in the extent of their life-time spatial distribution, and second in the spatial scale of the movement processes that generate that distribution. In terms of vulnerability, these characteristics, along with other aspects of the migratory life-style, make migration a double-edged sword. For instance, migratory organisms are by their nature able to cope with a wide range of environmental conditions and are considered to be relatively flexible and plastic, potentially making them more adaptable to certain threats (e.g., climate change; Chapter 2). Migration can also provide an escape from density-dependent effects, which can facilitate larger population sizes (vs. resident populations) and lower the risk associated with small population sizes. On the other hand, the large range sizes required for long-distance migration often result in increased exposure to extrinsic threats. Also, the migration process itself is potentially dangerous. For some species there is evidence that anthropogenic mortality is highest during migration (e.g., Trakimas and Sidaravičius 2008) and, for a number of birds, the migratory period features the greatest number of deaths from any cause (Sillett and Holmes 2002). Furthermore, many migratory organisms have evolved annual routines where the timing of events is crucial, which makes them vulnerable to climate change unless they can respond accordingly (Both *et al.* 2006; Chapter 6). The optimal timing of migration is also a non-trivial problem considering the suite of—sometimes opposite—selection pressures acting throughout the annual cycle (Hedenström *et al.* 2007; Jonzén *et al.* 2007).

The unique vulnerabilities of migrants to extrinsic influences have several implications for the development of effective conservation and management strategies. For instance, conservation approaches that focus on spatial control, such as reserves and other protected areas (Section 11.3.1.1), are far more difficult to design for wide-ranging migrants. Many migrants move out of protected areas and, therefore, need to be conserved in multiple-use land/sea-scapes where they interact with rural (and often impoverished) people. Thus, for efficient conservation of large-scale migrants, it may be necessary to take the whole migration route into account (Martin *et al.* 2007). However, it is only recently that conservation strategies for migratory species have been developed that consider how migratory animals are spatially connected between different periods of the annual cycle (e.g., Klaassen *et al.* 2008).

Efforts to develop management strategies that address the threats that migrants face across whole migration routes highlight a fundamental challenge that is much more common to migratory populations than it is to resident ones. In addition to moving outside of reserves, migrants frequently move across jurisdictional/political boundaries, which implies that they are typically the responsibility of more than one management authority (Jarnemo 2008). Thus, the successful conservation and management of migratory organisms often requires coordinated, or at least complementary, efforts amongst multiple jurisdictions. Achieving such inter-jurisdictional cooperation can be extremely challenging. For example, differences in opinion across political boundaries regarding development or activities (e.g., roads, dams, logging) that might obstruct a migrant's journey or destroy critical habitat, could serve as major impediments to developing coordinated conservation and management strategies.

In the last few decades, recognition of the importance of developing inter-jurisdictional conservation and management strategies for migratory populations has led to the development of several international treaties and agreements (Section 11.5). Some of these agreements are more comprehensive than others and, for all of them, compliance is difficult to enforce. However, given the prevalence of population declines amongst migratory organisms (Wilcove and Wikelski 2008) and recent increases in the loss of migratory behaviour (Berger 2004; Bolger *et al.* 2008), any serious effort to improve the management of migratory populations is likely to be a step in the right direction.

11.5 Institutional frameworks

Migratory organisms have wide-ranging 'lifetime tracks' (Dingle 1996) that often include movements across state, provincial and national boundaries. Despite the difficulties involved, a number of international multilateral treaties and agreements have been developed, which are relevant to the conservation of migratory species. These can be divided into agreements aimed at conservation, which have been used primarily to conserve endangered species of mammals, birds, reptiles and sharks:

- the Convention on the Conservation of Migratory Species of Wild Animals (also called the CMS or the Bonn Convention, 1983);
- the Convention on Biological Diversity (CBD);
- the Convention on International Trade in Endangered Species of Wild Fauna and Flora (CITES);
- the Convention on Wetlands of International Importance Especially as Waterfowl Habitat (Ramsar)

and those aimed primarily at both management and conservation of marine fish:

- the Convention for the Conservation of Antarctic Marine Living Resources (CCAMLR, 1982);
- the International Whaling Commission (IWC);
- the UN Convention on the Law of the Sea (UNCLOS, 1982) and its implementing agreement on highly migratory species, the UN Fish Stocks Agreement (UNFSA, 1995);
- the Regional Fisheries Management Organizations (RFMOs) for tuna and billfish (the Inter-American Tropical Tuna Agreement (IATTC); the International Convention for the Conservation of Atlantic Tunas (ICCAT); the Indian Ocean Tuna Commission (IOTC); the Convention for the Conservation of Southern Bluefin Tuna (CCSBT); and the Convention on the Conservation and Management of Highly Migratory Fish Stocks in the Western and Central Pacific Ocean (WCPFC))—the last of these was concluded using the framework of UNFSA;
- other RFMOs, such as NEAFC, the Northwest Atlantic Fisheries Organization (NAFO) and the Southeast Atlantic Fisheries Organization (SEAFO);

• the FAO's Code of Conduct on for Responsible Fisheries, and the International Plans of Action on Seabirds and Sharks developed under the Code.

The two agreements with the widest international application are the CMS and UNFSA. The former has 112 signatories (as of August 2009), UNFSA 75 (as of April 2009). UNCLOS, which contains some of the provisions and obligations of UNFSA, has 157 signatories.

Adopted in 1979, the CMS came into force in 1983, and operates under the auspices of the United Nations Environment Programme. The aim of the CMS is 'to conserve terrestrial, marine and avian migratory species throughout their range' (www. cms.int). The definition of 'migratory species' in the CMS text is 'the entire population or any geographically separate part of the population of any species or lower taxon of wild animals, a significant proportion of whose members cyclically and predictably cross one or more national jurisdictional boundaries. This has been broadly interpreted as any animal that is transboundary in its range. Hence species that it would be very hard to call migratory under any biologically accepted meaning of the term, such as gorillas (*Gorilla gorilla*), Bukhara deer (*Cervus elaphus bactrianus*) and Mediterranean monk seals (*Monachus monachus*), but which happen to have transboundary distributions, have MOUs or agreements under the treaty.

The CMS has two appendices. Appendix I lists migratory species 'threatened with extinction' and Appendix II lists migratory species that would 'significantly benefit from international cooperation'. Parties to CMS vote on amendments to the Appendices proposed by the Secretariat. Range states of species on the Appendices (those geopolitical units in which a species spends a part of its life cycle) are encouraged to implement global or regional conservation action plans. These may be legally binding treaties (termed Agreements by CMS) or non-legally binding Memoranda of Understanding (MoUs) between range states. Regular meetings with range states are held, and conservation plans drafted and adopted. Of the parties to the convention, most notably absent are Brazil, China, Russia and the USA, although some have signed species-specific MoUs.

Current instruments of the CMS (Agreements and MoUs) focus on birds, turtles and mammals, including the saiga antelope (Box 11.3). A shark agreement, however, is currently being put into place with the interest of 50 range states. It will potentially include seven species of shark, three of which are listed by CMS.

CMS has come under criticism in the past for its ambiguous text. For example, six marine turtle species are listed on CMS Appendix I, and CMS requires parties to *prohibit the taking of animals* that are Appendix I species, but exceptions are allowed under the following criteria:

(a) the taking is for scientific purposes;
(b) the taking is for the purpose of enhancing the propagation or survival of the affected species;
(c) the taking is to accommodate the needs of traditional subsistence users of such species; or
(d) extraordinary circumstances so require; provided that such exceptions are precise as to content and limited in space and time. Such taking should not operate to the disadvantage of the species.

The term 'traditional subsistence user' is undefined in the Convention text and has not been defined elsewhere by the CMS Secretariat. The UK, a party to CMS, allows the legal harvest of turtles for meat in several of its Overseas Territories (OTs) to which CMS is extended. As such, it could be argued that the UK is in breach of the Convention for allowing such harvests in OTs such as the Cayman Islands where turtle meat is not the main source of protein (Bell *et al.* 2006; Richardson *et al.* 2006). But the lack of definition of 'traditional subsistence' muddies the waters and leaves this text open to interpretation. That being said, the MoUs relating to turtles have led to widespread collaborations in the Indian Ocean and West Africa, which would have been unlikely to have formed without the efforts of the CMS Secretariat.

The only fish listed in Appendices 1 and 2 of the CMS are sharks (*Selachimorpha* spp.), sturgeon (*Acipenser* spp.) and the Mekong giant catfish (*Pangasianodon gigas*). No other migratory fish are currently listed as endangered or deserving of conservation through Agreements. Highly migratory fish species such as tuna (*Thunnus* spp.) and billfish (*Istiophoridae* spp.) are managed by the series of RFMOs mentioned above, most of which were concluded before UNFSA (1995) came into force. This management includes rational use in terms of the target species (the highly migratory fish). Although the tuna RFMOs have been criticized recently for their poor performance (e.g., ICCAT, Hurry *et al.*, 2008) and many stocks are overfished (technically, their population sizes are below those that generate the maximum sustainable yield), their conservation status is not generally threatened (though in the case of southern bluefin tuna (*Thunnus maccoyii*) and eastern Atlantic bluefin tuna (*Thunnus thynnus*) stocks are significantly depleted). Non-tuna RFMOs also interact with highly migratory stocks, such as the Norwegian spring spawning herring (Box 11.4), which is managed by NEAFC.

There has long been resistance to the use of 'conservation' instruments, such as CITES or the CMS, to regulate the management of exploited marine species. Some exploited species are currently listed in CITES Appendix II (e.g., basking shark, *Cetorhinus maximus*), but the proposal by the EU to list porbeagle (*Lamna nasus*) and spiny dogfish (*Squalus acanthias*) was rejected by the CITES parties in 2007, as an earlier proposal to list toothfish (*Dissostichus eleginoides*) also had been. They were rejected because of the inability, perceived by some Parties, of such instruments to effectively manage the commercial exploitation of marine species that are also subject to fisheries management jurisdiction.

In addition to having an obligation to manage their target stocks (some of which may be migratory) all RFMOs interact with non-target species through bycatch, incidental mortality of birds, mammals, sharks and turtles, and interactions with vulnerable benthic habitat. For some of the earlier international fisheries agreements, their obligations to deal with these issues may not have been as strong as for the most recent, such as WCPFC. Nevertheless, growing criticism both from within their membership and externally, which has often identified them as lacking in effective management of impacted ecosystems (Small 2005; Mooney-Seus and Rosenberg, 2007), has forced a change in the practices of RFMOs.

Most of them now have specific working groups and measures to consider their impact on other species, including migratory species.

As we have previously discussed, the key to the successful conservation and management of migratory species is cooperation between range states. Both the CMS and UNFSA, which are over-arching implementing agreements, deal with the overlap between their interests and those of other international organizations to which their signatories may also be Parties in a very general way. In other words, the agreement does not alter the rights and obligations of Parties that arise from other agreements (CMS Article XII and UNFSA Article 44). Thus neither agreement requires specific consideration of other international agreements and mechanisms. Nevertheless, such cooperation is increasingly recognized as important. Several Agreements and MOUs have been concluded under the CMS that involve species that may interact with RFMOs—both those listed above as dealing with highly migratory fish stocks and the many others that manage regional fisheries. Examples include Agreements on Cetaceans, Sharks (currently under negotiation) and Albatrosses and Petrels and MOUs on Marine Turtles (Atlantic Coast of Africa, Indian Ocean and South-East Asia) and Cetaceans of the Pacific Island States.

Formal recognition of the need for cooperative links between these instruments and RFMOs is only sometimes present. For instance, the MOU on Marine Turtles of the Indian Ocean and South-East Asia (IOSEA) recognizes in its text a number of international agreements, including CITES, the CBD and UNCLOS, but does not mention the IOTC with which cooperation is critical to avoid turtle bycatches during fishing. Cooperation between the IOTC and IOSEA is evident at a practical level, however, with the IOTC's Working Party on Ecosystems and Bycatch actively cooperating with IOSEA (IOTC 2008).

ACAP, the Agreement on the Conservation of Albatrosses and Petrels, is the most developed of the CMS Agreements in this regard. Its convention includes direct reference to relevant organizations and specifically requires, in Article XI, its Parties to: 'promote the objectives of this Agreement and develop and maintain coordinated and complementary working relationships with all relevant international, regional and sub-regional bodies, including those concerned with the conservation and management of seabirds and their habitats and other marine living resources, particularly with the Commission of CCAMLR and the Food and Agriculture Organization of the United Nations, and particularly in the context of the International Plan of Action for Reducing Incidental Catch of Seabirds in Longline Fisheries'. It also allows the Secretariat to enter into arrangements, with the approval of the Meeting of Parties, with other organizations and institutions as may be appropriate.

Cooperation with CCAMLR, which formally recognizes ACAP as an observer, is particularly strong since CCAMLR has a working group on the Incidental Mortality Associated with Fishing with significant membership overlap with ACAP. Cooperation with other organizations may not be as easy. Although Birdlife International is an observer to the WCPFC, ACAP is not; and although the fisheries of the CCSBT and IOTC have a direct effect on ACAP birds, these organizations mention only the FAO IPOA in terms of cooperation (e.g., the CCSBT Recommendation to Mitigate the Impact on Ecologically Related Species of Fishing for Southern Bluefin Tuna (2008) makes reference to the IPOA-SEABIRDS but not to the activities of ACAP).

In the sense that ACAP recognizes and seeks to create cooperative links with all existing formal and voluntary international initiatives and agreements, it can be seen as a model approach to the conservation of highly migratory species that should be followed by others.

11.6 Summary and conclusions

The extinction or local extirpation of any organism is lamentable from a biodiversity conservation perspective and, often, from an economic perspective (e.g., due to the loss of exploitation opportunities and/or ecosystem services). However, the loss of a migratory species, or the loss of migratory behaviour from a population, can have additional consequences. These include: The loss of energetic and material linkages between spatially distinct ecosystems; the loss of a

behavioural 'spectacle' that is of fundamental evolutionary and ecological interest (Dingle 1996; Dingle and Drake 2007); and the loss of a phenomenon with aesthetic appeal that can generate tangible socio-economic benefits (e.g., ecotourism focused on the migration of the Serengeti wildebeest; Haslam 2008). The negative impacts associated with migration loss should be of particular concern, as recent studies indicate that migration is a phenomenon at risk, due to local extirpation and population declines amongst migratory species, as well as unintentional and, in some cases, intentional (Andreassen *et al.* 2005) anthropogenic constraints on migratory behaviour (Wilcove and Wikelski 2008; Wilcove 2007; Berger 2004).

The extrinsic threats that migrants and migration face are numerous, however most can be assigned to one of four general categories, each of which is anthropogenic in nature. These four processes also have a negative impact on non-migratory species. However, habitat loss and degradation, over-exploitation and interrupted migration have all been associated with relatively unique consequences for migratory species and they have played a major role in observed migration losses (Wilcove and Wikelski 2008; Bolger *et al.* 2008). The impacts of global climate change on migratory species are just beginning to be documented. Yet researchers have already identified ways in which climate change is likely to have differential impacts on migrants, as compared with non-migratory species (e.g., the decoupling of migration timing from resource availability; Robinson *et al.* 2005; Kirby *et al.* 2008).

Each of the major threats that migratory organisms face is associated with its own general pattern of impacts. Yet determining the extent to which a particular migratory species or population is vulnerable to these impacts requires a detailed evaluation of their migratory attributes. This should include careful consideration of characteristics such as migration distance and behavioural plasticity and the determination of whether migrants possess certain traits that may predispose them to vulnerability (e.g., site fidelity, aggregative or solitary behaviour; Table 11.1) and, if so, identifying which stage(s) of the migratory journey they apply to. A detailed understanding of the migratory behaviour of a given organism is not just crucial for assessing its suscepti-bility to particular threats, it is also critical for devel-oping successful conservation and management strategies. A variety of different approaches (e.g., protected areas, barrier removal) are available for conserving, restoring and managing migratory pop-ulations, but the probability that each will be effec-tive is highly dependent on both the migratory attributes of a targeted population and the extrinsic threats that they face.

Finally, while some migrants might have charac-teristics that render them easier to protect or har-vest than non-migratory organisms, the difficulties associated with conserving or exploiting migratory species (vs. more sedentary ones) appear to out-weigh the advantages. Specifically, the multiple scales at which migrants move and select habitat and the long-distance movements that they under-take, are likely to expose them to a broader range of threats than those encountered by their non-migra-tory counterparts. Additionally, many migrants travel through and take up temporary residency in multiple jurisdictions, which makes it very difficult to apply conservation measures consistently or, for exploited species, to ensure that harvesting is sus-tainable across a migrant's entire range.

The numerous regulations and multilateral agre-ements devoted to migratory species are testimony to the difficulties associated with the conservation and management of such wide-ranging organisms. These institutional frameworks are not just indica-tors of the challenges associated with managing migrants, but they are often a critical component of sustainable exploitation and conservation success. This is because they formalize the inter-jurisdic-tional cooperation that is essential for the long-term persistence of both migrations and migratory pop-ulations. Existing legislation, conventions and MoUs concerning migratory species are not free of shortcomings. However, they can serve as valuable starting points for developing regulations and agreements that promote cohesive or complemen-tary conservation and management approaches throughout a migrant's range. While the develop-ment and consistent application of effective conser-vation strategies for migrants can be extremely difficult, recent increases in migration loss demon-strate the importance of facing and overcoming these difficulties so that further losses are avoided, and effective strategies for the conservation of migratory organisms become entrenched in man-agement practice.

CHAPTER 12

Conclusions

E.J. Milner-Gulland, John M. Fryxell, and Anthony R.E. Sinclair

12.1 Aim of the book

Our aim in this book has been to take a broad and integrated view of migration as a phenomenon, linking evolution with ecology and management, and theory with empirical research, and embracing all the major migratory taxa (including humans). This comparative and integrated approach, both within and between chapters, has promoted new insights and highlighted promising new avenues of research. For example, in Chapter 6, the parallels between taxa in their decision rules and cues has enabled the authors both to generalize about the relative importance of internal and external cues for different migratory stages, which is quite consistent between taxa, and also to highlight the key differences and the reasons behind these. Similarly, Chapter 8 makes clear that theory and empirical understanding can better inform each other if appropriate methods are used that allow questions to be formulated and tested in an integrated modelling framework.

This broad view, encompassing evolutionary and ecological processes, and extending beyond the target organism to consider the ecosystem-level dynamics of migration, is rarely found in migration research. In fact, Chapter 9 has had to rely on one of the few well-studied ecosystems that is formed and maintained by migration, the Serengeti, for its examples of the role of migration in ecosystems. Evolutionary thinking can illuminate currently observed ecological processes, as is made clear in Chapter 7, which deals with the interplay between uncertainty and predictability in driving the life history strategies of migrants and nomads. And our inclusion of human pastoralism (Chapter 10), with both its strong parallels and its fundamental differences from wild animal migration, also has the power to change how we think about both the function and the management of migratory systems more broadly.

Our intention was to look forwards rather than backwards; rather than simply reviewing the field of migration as it stands today, we rather highlight the lacunae in our knowledge and understanding and the exciting new possibilities that are now opening up, whether because of advances in our understanding of migration as a biological phenomenon or through the availability of a range of new technologies. Several broad themes have emerged that cut across chapters, systems and scales of enquiry. We consider them in turn.

12.2   What is migration?

It is easy, for a topic such as migration, to feel that the starting point should be a long discussion of the definition of the term, the exceptions to the definition and the exclusions that it implies. While such a discussion can be illuminating, it can also hamper research, preventing us from moving on to exploration of the processes underlying migration in all its forms. We have in this book explicitly taken a broad view of the phenomenon of migration and not sought to box it into a particular definition (see Chapter 1). The view taken of migration by our authors is relatively consistent throughout the book, despite this lack of guidance, suggesting that this approach has been a good one.

Key questions that our authors have pondered include whether migration is special in some way, or just another form of movement. Evolutionarily Stable Strategy (ESS) models developed in Chapter 3 suggest that the conditions favouring the evolution of migration are nearly ubiquitous. Virtually any system in which there is a combination of natural regulation, seasonality and habitat heterogeneity should favour some form of migration or another. On the other hand, these same models make clear that migration is rarely obligatory and the sole ESS—we should routinely expect to see mixtures of migrants and residents. By this criterion, migration is not particularly 'special', but rather one of the most logical outcomes imaginable. In Chapter 7, it is placed at one end of a continuum with nomadism at the other end, with the axis being the predictability and uncertainty of resource availability. The authors suggest that migration generally arises when resource availability in time and space is more predictable but there is severe scarcity at certain times and places. In this case, internal cues are of use in guiding the optimal exploitation of these resources. In contrast, the nomadic lifestyle evolves in response to unpredictable and uncertain resources, for which external cues are more useful. However, generalizations are inevitably proved to be unreliable. Movement patterns are highly plastic

in many species, such that, for example, the saiga antelope *Saiga tatarica* is migratory between ranges and nomadic within its summer range (Bekenov *et al.* 1998), a feature also shown by the Serengeti wildebeest *Connochaetes taurinus*, while in many species some individuals or populations are migratory and others are resident, with these roles potentially switching from year to year.

Chapter 2 gives a fascinating review of the variation in migratory strategies over a range of scales. The questions that can be asked about migration as a strategy vary depending on the spatial and temporal scale at which the system is studied and at which the processes happen, set against the range of scales at which environmental variation occurs. Chapter 7 gives an example from Australia, showing the way in which environmental variation can be characterized at different scales in different areas, and the different movement strategies that animals use to cope with this variation. Chapter 10 discusses the complexity of the constraints and opportunities that livestock herders face, and some of the differences and strong similarities with the movement decisions faced by wild animals at a range of temporal and spatial scales. At a coarse level, pastoralist movements appear very similar to wild ungulate migrations, driven by the distribution of resources in time and space, competition for resources and constraints on access (both environmental, such as water availability, and institutional, such as property rights).

A theme throughout the book is the importance of understanding the plasticity of migration as a strategy, in the face of environmental change. Migration is generally a response to relatively predictable and cyclic environmental change, but current anthropogenic change is rapid and deterministically trending, making it a challenge to migratory species. This change is occurring at a range of temporal and spatial scales; from long-term climate change to more immediate habitat destruction. Many of the chapters take up this theme, and ask how plastic the migratory syndrome is and thus how likely it is to respond at meaningful time scales to anthropogenic change. Chapter 2 gives the example of the loss of migration in populations of salmon and trout that have become land-locked, or its alteration into movements between freshwater habitats (Quinn 2005), while introduced rainbow trout *Onchorhynus mykiss* coming from possibly non-migratory ancestral stock have developed a migratory strategy in their new habitat (Pascual *et al.* 2001; Thrower *et al.* 2004; Ciancio *et al.* 2008).

Several authors have highlighted that multiple constraints and opportunities play into the development of migration as a component of an animal's life history strategy. This means that migration as a syndrome (incorporating the strategy and all the phenotypic and behavioural changes that go along with it) must be understood in the round rather than with the focus on single optimizations, a difficult task.

12.3   Crossing boundaries in our science

Something that has come out very strongly from both the individual chapters and from the collection as a whole is the importance of taking a cross-disciplinary approach if we are to make progress. Our authors' fundamental disciplines include genetics, evolutionary theory, behavioural, community and population ecology, physiology, animal mechanics, statistics, mathematical modelling, conservation, anthropology and more. The advantages of combining insights from a wide range of disciplines are clear throughout the book.

A cross-taxonomic approach is also fundamental to progress. Several of our authors have noted the preponderance of bird studies over those concerning other taxonomic groups. For example, Chapter 5 noted the lack of empirical studies on migratory energetics in taxa other than birds, in comparison with the theoretical models in Chapter 4, which although generally bird-derived, are more easily transferred to taxa using other media. A notable exception to this dearth of studies on the energetics of migration is the research on the monarch butterfly *Danaus plexippus*, summarized by Brower *et al.* (2006). Chapter 5 could then use similarities in the spatiotemporal patterns of fuel loading in the monarch and the thrush nightingale *Luscinia luscinia* to infer that a similar process controls fat storage in these two taxonomically distinct species.

This example indicates that looking across taxa can help us to formulate general hypotheses. Chapter 6 examined the different types of internal and external cues that are used in a range of taxa

and found a fair degree of consistency. Generally, external cues are used to start the preparation stage, internal cues indicate readiness to start and finish the migration, and external cues are used to guide the actual departure. However, these very rough generalizations are usually based upon observational studies; very little is known about the actual cues and decision rules that any animal uses. Even within a taxonomic group, a comparative approach is useful. Chapter 10 examined three very different human migratory systems and uncover broad similarities in the constraints that they face and how they manage these constraints, with finer scale differences. Understanding the cues and decision rules that are used should dramatically improve our ability to predict the consequences of externally imposed changes in the environment, but we are a long way from this understanding.

One area that could be revolutionized by the inclusion of migration is community ecology. As the authors of Chapter 7 highlighted, community ecology tends to be static and focused on interactions such as competition and predation in well-mixed, homogeneous environments. Migration forces us to dispense with our preconceptions, because ecological interactions should tend to be weaker in communities that are dominated by migrants, or at least temporally limited and variable. Chapter 9 has a particular focus on the role of migration in community ecology. The authors presented much suggestive evidence from the literature of the importance of migrations for ecosystem structure and dynamics, including substantial transfers of a wide range of biological material, such as organic matter, nitrogen, genes and diseases. However they found very little work that brought together the various elements and examined the cascading effects of these transfers on the system, either theoretical or empirical. This is a crucial hole in our understanding given how much disruption and collapse has already been recorded for migrations worldwide (Wilcove and Wikelski 2008; Harris *et al.* 2009), which must be having a substantial impact on migration-structured ecosystems.

However, our understanding of the chain of events in many migration–ecosystem interactions is sketchy and potentially misplaced. For example, the migratory sociable plover *Vanellus gregarius* is critically endangered and rapidly declining in number. It requires intensively grazed areas for ground nesting. Its breeding sites in Kazakhstan are in the range of two major migrations with which it may have co-evolved; the saiga antelope and pastoralists. In the 1990s, the pastoralist migrations ceased due to social disruption (Kerven 2003), and saiga numbers fell by over 90%, though its migrations continue (Singh *et al.* 2010). Watson *et al.* (2006) found that plovers now nest around settlements where grazing intensity remains high, but is continuous rather than migration-driven. Their results suggested that this continuous grazing caused egg mortality and potentially population declines. This was a compelling story, and conservation organizations linked the decline of the saiga through changes in vegetation to the decline in the sociable plover. Later studies revealed that recruitment was healthy, however and, in a further twist, satellite tracking has since revealed firstly the existence of previously unknown major colonies in Syria and Turkey, and secondly very high mortality of adult plovers from hunters at migration stopovers in the middle East, which may be driving recent declines (Kamp *et al.* 2009).

This case study illustrates the point made in Chapter 9 that rigorous empirical investigations of the population- and ecosystem-level effects of migration are urgently needed. In Chapter 3, we demonstrate that even the simplest models of migration can generate a range of surprising population dynamics, with migration sometimes being stabilizing, and other times not. The prediction that mixed ESS should predominate promises interesting community-level outcomes that have yet to be considered. The most thoroughly-studied example for which both empirical data and models exist, allowing exploration of the potential ecosystem-level effects of migration and its loss, is the Serengeti. The striking insight from these explorations is that migrants don't just passively respond to their environmental circumstances, as is often implicitly assumed, but that they also shape and modify their environment through their migratory behaviour. This may seem intuitively obvious once stated, but rarely are all the pieces of the jigsaw put together this compellingly. Pastoralists do exactly the same thing, with the addition of complex social

institutions that allow them to optimize their use of resources both in the shorter and longer terms, in the face of cultural and social constraints. As Chapter 10 points out, both pastoralists and wild migratory ungulates have very large environmental impacts, but those of pastoralists tend to be discussed in pejorative tones.

12.4 Scale and complexity

Several chapters illustrate the importance of scale for the study of migration, both in terms of the variation of resources in time and space (e.g. Chapter 7) and in terms of whether the object of study is changing in evolutionary or ecological time (Chapter 2). This comes on to the unit that we use to measure the benefits or costs of migration. Fitness is the ultimate unit, yet the physiologically-based studies described in Chapters 4 and 5 need to use more proximate measures such as rate of movement or energy gain/loss. If we are to relate these proxies to fitness we need to work at the level of the individual, and relate the decisions and constraints involved in migration to their overall life history strategy, recognizing that constraints at the ecological scale may not remain so at the evolutionary scale. Taking the example of migration energetics (Chapter 5), we need to look at the effects of changes in fuel load on fitness over the entire life history; the timing and quality of the staging refuelling stops are critical to maintaining a functioning migratory system, and can be very finely balanced. Chapter 5 gives the example of Delaware Bay, which is a single stopover in the spring migration of a population of red knots *Calidris canutus rufa*. Baker *et al.* (2004) showed that the dramatic decline of their food source (the eggs of spawning horseshoe crabs, *Limulus polyphemus*) due to over-harvesting and beach erosion meant that the knots couldn't gain mass as fast as before in order to reach their Arctic breeding areas in good condition. The knot population has declined dramatically, and Baker *et al.* (2004) suggest that this may be due to the deterioration of this one bay.

Following on from the need to understand the effects of changes in migration energetics within the entire life history, several authors noted the lack of data on survival throughout the life history, but particularly during migration. As survival is directly related to fitness, this is an important omission, and one that is highly conservation-relevant, but a difficult one to rectify as such data are not easy to obtain.

The theme of moving away from population-level data and analysis and the need to focus more on the individual level was echoed in many of our chapters. However Chapter 8 makes the important point that the group and population levels do tell us things that we are not able to study at the individual level. For example, within a group, only a small proportion of individuals may be actively migrating, following cues and making decisions based on environmental factors. The rest of the group may be following along, and therefore subject to different selection pressures and making different decisions based on different cues. Another reason for continued research at the group level is the existence of emergent properties at this level; for example the 'many wrongs' principle shows how groups may be more effective than individuals at navigation through the simple mechanism of pooling navigational errors (Simons *et al.* 2004). Chapter 8 makes the strong case that the best approaches combine individual and population-level data so that inference from one level can be used to inform model selection at the other level. Although population-level studies tend to suffer from being correlational rather than mechanistic, just focusing at the individual level loses information about large-scale collective processes, as well as the demographic context within which individual decisions must be executed. A further problem with a purely individual-based interpretation of migration is that nearly all empirical studies lack basic information on important landscape variables, such as food abundance and quality, disease prevalence, or predation risk. There is a limited amount that can be gleaned from habitat and spatial position as the sole determinants of migratory movement trajectories.

Chapter 2 highlights the need to look at further levels—genetic and phylogenetic—in order to improve our understanding of the linkages between ecological and evolutionary processes. Chapter 9 reminds us that the community level is also important, and highlights the multi-directional nature of feedbacks between migrants and their environment.

Chapter 7 focuses our minds on the statistical properties of the environment and on the environmental covariates that we need to include in our models if we are to understand how internal traits and behaviours interact with the animals' surroundings. The over-arching message is that we need to broaden our horizons, and to consider migration in context and at all relevant scales.

12.5 New approaches, new tools

The agenda set out in Section 12.4 is ambitious. It is only now that we are able to consider addressing these complexities in a practically feasible way, and to move beyond theory to an empirical understanding of how, why and when animal migration takes place. This expansion is due to the rapid and continuing advances in new technology in many of the areas relevant to migration studies.

Perhaps the most obvious of these is the massive advance in tracking technology over the last few years (Fig. 12.1). ARGOS technology first allowed satellite tracking of individual animals in the 1980s, and opened up a whole field, but was limited by location inaccuracies, weight of the transmitter and the medium in which animals moved (Rutz and Hays 2009). Weimerskirch (2010) recalls how the first ARGOS bird transmitter fitted in 1989 weighed 180 g, and the current equivalent weighs just 9 g, while Rutz and Hays (2009) remind us that 30 years ago, marine mammals were tracked by following balloons attached to the animal. In their paper, 'A horizon scan for global conservation issues in 2010', Sutherland *et al.* (2010) list mobile sensing technology as one of fourteen issues of potentially enormous future importance for conservation, and identify mobile phone apps as one particularly exciting aspect with the power to open up data logging and promote data sharing on a global scale.

These advances in tracking will allow research to focus more on the individual animal, rather than on collecting population-level information. This is vital for understanding the fitness consequences of migratory behaviour and environmental challenges, and for giving us a mechanistic rather than a correlational understanding of migratory behaviour. However, locational data are not enough:

(a)

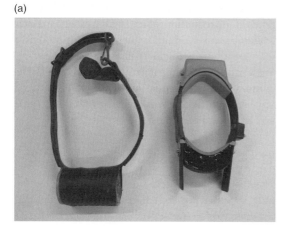

(b)

Figure 12.1   (a) Radiocollar deployed on adult moose in central Ontario, Canada in the late 1960s/early 1970s (left). Iridium GPS collar deployed on woodland caribou in north-western Ontario, Canada during the winter of 2010 (right). Photo: Ontario Ministry of Natural Resources. (b) Female Mormon cricket (*Anabrus simplex*) affixed with a 0.4 g radio transmitter, in 2005 (Photo: Darryl T. Gwynne).

Chapter 5 reminds us that we also need to obtain covariate data on the animal's behaviour, weight and physiological condition, and on environmental conditions such as wind speed, topography or temperature, enabling us to understand more clearly what they are doing and why in a particular location. This can then enable us to test fundamental components of the optimality models in Chapter 4, such as the range equation for how far an animal can travel and the role of drift in animal movement. New technology extends beyond positional devices, and includes devices that can help us to obtain these crucial additional data, such as accelerometers, heart rate loggers and depth loggers.

Although the revolution in biologging is extremely important, there are many other areas in which technological advances have opened up new fields of enquiry, or significantly decreased the difficulty of obtaining and processing large amounts of, or particular kinds of, data. Chapter 2 reminds us that advances in genomics and related subjects are opening the way to linking genetic variation to traits and fitness of individual organisms, and that this should lead to improved understanding of the underlying basis for the migratory syndrome and the interaction between genes and the environment in determining migratory behaviour. Molecular techniques are also able to inform ecological questions, through the use of genetic markers that give

information on population structure and connectivity (Sunnucks 2000). Non-invasive collection methods mean that these methods can be used even on threatened migratory species. For example Rudnick *et al.* (2008) used genetic analyses of shed feathers of the migratory Imperial eagle, *Aquila heliaca*, to analyse population structure and natal philopatry.

Modelling is also undergoing a revolution, as highlighted in Chapter 8. One of the major statistical problems that besets datasets typical of migration studies is non-identifiability. This is caused by multiple parameters playing off against each other so that it is not possible to separate out their confounding effects; this is prevalent in situations with a large number of parameters needing to be estimated within a model with a limited number of degrees of freedom. Datasets from migratory species may well be particularly prone to this, when typically few individuals are tagged, and their movements are affected by a plethora of internal and external cues and conditions. Alternatively they may only be observable at particular times (e.g. the breeding areas), such that inferences about demography are made on the basis of one or a few vital rates. One crucial requirement to address this problem is additional independent data that can provide an estimate for one or more of the unidentified parameters in the model. New technology such as biologgers can help in obtaining these data. However, new approaches in inferential modelling can also help. For example, Matthiopoulos *et al.* (2008) use a state space model to show that current imprecise estimates of trends in the UK grey seal *Halichoerus grypus* population, based on long-term pup counts, can only be improved with the addition of an independent estimate of population size, and demonstrate how such an estimate would narrow confidence intervals on the likely functional form of density dependence in this species. State space models allow both biological parameters (e.g. fecundity) and unobserved states (e.g. number of adult females) to be estimated in the same model based on noisy observational data (e.g. pup counts, with the addition of an independent total population size estimate). In essence, this approach bridges the traditional divide between empirically based statistical models and mechanistic models, by allowing both the parameters to be estimated and mechanistic models to be selected in the same framework.

One of the key reasons why it is now possible routinely to implement models such as these, as well as to cope with the increasing volume of data generated by biologger and genomic analyses, is the constantly increasing computer power at our disposal.

12.6 Moving towards a hypothesis-driven science of migration

One of the most intractable divisions in biology has been between theoreticians and empiricists. This has been as true in studies of migration as elsewhere and, apart from Chapter 8, the division is relatively clear even in the chapters of this book. Chapters 4 and 5 demonstrate that in the field of migration energetics, for example, theoretical models at the moment tend to be well ahead of the empirical evidence. As Chapter 4 shows, simple optimality reasoning has given us impressive insights that are a remarkably good fit in many cases to the limited empirical evidence that exists (mostly for birds).

However, the non-identifiability problem identified in Chapter 8 is a major stumbling block for further progress. Observed patterns in the empirical data can be generated in mechanistic models in a number of different ways, such that hypotheses about the underlying mechanisms cannot be distinguished from pattern alone. This observation leads to the strong recommendation by virtually all of our authors that we need to move from correlational studies towards proper hypothesis-testing. This will involve manipulative experiments, and moving from the laboratory into the wild, so that we can understand animal migratory behaviour in the context of realistically complex conditions. For example, Sword *et al.* (2005) translocated radio-tagged Mormon crickets *Anabrus simplex* from migratory bands to demonstrate the anti-predator benefits of banding together; 50–60% of their translocated crickets were depredated, compared with none of the crickets in the bands. Just as in this example, the new technologies identified in Section 12.5 will be invaluable tools allowing these experiments to be undertaken, allowing us to obtain the empirical evidence needed to test hypotheses generated by theoretical models.

More is needed, however. A generally valid criticism of the state of the art in migration studies today is that we are concerned either with pattern description or, separately, with theorizing and hypothesis formulation, rather than integration of prediction and testing. Fuller integration of theory and empirical study is needed so that one informs and inspires the other. This needs to go in both directions, with theory leading to empirical tests and unusual or counterintuitive observations leading to the development of new theory. This will require statistical analysis, mechanistic model building, a conceptual framework and ingenious experimental designs to collect the data. There are a range of methods available for all of this, and the use of multiple approaches for cross-validation of results generally sheds further light on the problem.

Climate change has come up throughout the book as a huge, globally significant process that will affect all migratory species. Climate change can be seen as an enormous manipulative experiment in the wild, albeit with limited scope for controls. If we seize the opportunity that it provides, and set up our questions in an appropriate way, we can learn a lot about the mechanisms underlying the migratory syndrome. This is already beginning, albeit currently mostly with correlational studies in birds. For example, Rubolini *et al.* (2007) show significant differences between bird species in their shifts in spring arrival date in Europe, which also varied by latitude and length of migration, with more southerly arrivals and shorter migrations showing a greater change in arrival date. The challenge now is to delve deeper, to understand the mechanisms behind these patterns.

12.7   Migration under threat

Climate change is just one of the threats to migrants. There is plenty of evidence from around the world that migrants and migration are at risk. Chapter 11 identified habitat loss and degradation, over-exploitation and interrupted migration (e.g. from barriers) as other particularly acute or prevalent threats to migratory species. Climate change itself is likely to have different effects on migrants than on non-migrants, by decoupling the timing of migration from resource availability, right along the migration route (Akola *et al.* 2004). And in fact there is evidence that not all migrants are under threat from climate change: Lepidopteran migrants appear to be benefiting from climate change such that the number of migrants is increasing in the UK (Sparks *et al.* 2007).

A key point that is made in Chapter 11 is that we need to consider threats to the phenomenon of migration separately from threats to the migratory species themselves. Harris *et al.* (2009) show that many ungulate migrations have vanished, but that this has not necessarily led to the extinction of the species themselves. This is not surprising given the plasticity and variation in movement behaviour predicted by ESS models (Chapter 3) and documented from empirical observations in other chapters (particularly Chapters 2 and 7). However, the large-scale effects of migration on ecosystem dynamics (Chapter 9), as well as the economic, aesthetic and scientific value of migration as a process, all demand that this phenomenon be conserved in its own right, not just as a means of conserving the species.

The degree to which migration is vulnerable depends on a range of attributes, and Chapter 11 demonstrates that in some circumstances migratory tendencies may buffer a species against threatening processes by enabling them to move to other areas or exploit spatial and temporal refuges from over-exploitation. For example Apostolaki *et al.* (2002) showed in a model that a marine reserve could be an effective buffer against over-exploitation even for migratory hake *Merluccius merluccius* stocks. However, migratory species are exposed to a wider range of threats than equivalent sedentary species, and are at risk if they show limited behavioural plasticity; hence it may be that the more robust strategy in the face of anthropogenic threats is nomadism (Chapter 7). Migration is also hard to conserve if the migrating individuals move through multiple jurisdictions. There are many international treaties that aim to tackle this problem at a regional or species-specific level, as well as the global Convention on Migratory Species, which is testament both to the difficulties inherent in managing migrants, and that there is at least some political will to do so.

One of the suggestions made in Chapter 11 for conserving some of the ecosystem services that wild migrants provide, if not the original migrations, is to make use of pastoralists, whose livestock herds could potentially perform some of the same functions of the original wild migrants. One of the great ironies, however, is that human pastoralist migratory behaviour is also under dire threat, with forced sedentarization, enclosure or agricultural development of their lands rife, leading to the loss of unique cultures, knowledge and the ecosystem services that these movements provide (Chapter 10). Maybe this is the basis for an alliance of migration biologists and anthropologists, both of whom recognize the importance of migratory movement as a phenomenon in its own right, intimately intertwined with the identity of the species or human cultural groups that display the behaviour.

The studies presented in this book suggest to us that we are on the threshold of a new era in the study of migration. Gone is rigidly discipline-bound, taxonomically limited, correlational or abstract research, dislocated from the real-world context and status of the species concerned. New understanding, new tools and a broader perspective among migration scientists mean that there is enormous potential for substantial academic advances in the field in the near future. But this is also a crucial time for the survival of migration as a phenomenon, and one at which it is vital that migration research is grounded in real-world problems and feels a sense of urgency to address them. The barriers between science and practice therefore also urgently need to be dissolved, and the rapid environmental changes that we are experiencing need to be seen as an opportunity to learn both about the drivers of migration and about effective approaches to conserving it. The Holy Grail is perhaps the development of a unifying framework for understanding migration; we are still a long way from this, as can be seen from the continuing disagreements about what types of movement the term encompasses. However, given the huge variation that has been observed in the syndrome and in the scales at which it needs to be addressed (from evolutionary, to ecological to physiological), maybe the concept of a unifying framework is actually a mirage.

References

Aarts, G., M., MacKenzie, B., McConnell, M., Fedak, and Matthiopoulos, J. (2008) Estimating space-use and habitat preference from wildlife telemetry data. Ecography, 31, 140–60.

Able, K.P. (1982) Skylight polarization patterns at dusk influence migratory orientation in birds. Nature, 299, 550–51.

Able, K.P. (1970) Radar study of altitude of nocturnal passerine migration. Bird-Banding, 41, 282–90.

Able, K.P. and Belthoff, J.R. (1998) Rapid 'evolution' of migratory behavior in the introduced house finch of eastern North America. Proceedings of the Royal Society B, Biological Sciences, 265, 2063–71.

Abraham, E.R. (2007) Sea-urchin feeding fronts. Ecological Complexity, 4, 161–68.

ACIA (2004) Impacts of a Warming Arctic—Arctic Climate Impact Assessment, Cambridge University Press: Cambridge.

Adams, C.C. (1918) Migration as a factor in evolution: its ecological dynamics. The American Naturalist, 52, 465–90.

Adams, M.E. (1982) The Baggara problem: attempts at modern change in Southern Darfur and Southern Kordofan (Sudan). Development and Change, 13, 259–89.

Adriaensen, F., Ulenaers, P., and Dhondt, A.A. (1993) Ringing recoveries and the increase in numbers of European great crested grebes (*Podiceps cristatus*). Ardea, 81, 59–70.

Ahlén, I., Hans, J. and Bach, L. (2009) Behavior of Scandinavian bats during migration and foraging at sea. Journal of Mammalogy, 90, 1318–23.

Ahola, M., Laaksonen, T., Sippola, K., Eeva, T., Rainio, K., and Lehikoinen, E. (2004) Variation in climate warming along the migration route uncouples arrival and breeding dates. Global Change Biology, 10, 1610–17.

Åkesson, S. and Hedenström, A. (2007) How migrants get there: migratory performance and orientation. BioScience, 57, 123–33.

Åkesson, S. and Hedenström, A. (2000) Wind selectivity of migratory flight departures in birds. Behavioral Ecology and Sociobiology, 47, 140–44.

Åkesson, S., Broderick, A. C., Glen, F., Godley, B. J., Luschi, P., Papi, F. and Hays, G. C. (2003) Navigation by green turtles: which strategy do displaced adults use to find Ascension Island? Oikos, 103, 363–72.

Albon, S.D. and Langvatn, R. (1992) Plant phenology and the benefits of migration in a temperate ungulate. Oikos, 65, 502–13.

Alerstam, T. and Hedenström, A. (1998) The development of bird migration theory. J. Avian. Biol., 29, 343–69.

Alerstam, T. and Pettersson, S.G. (1991) Orientation along great circles by migrating birds using a sun compass. Journal of Theoretical Biology, 152, 191–202.

Alerstam, T. (1978) Graphical illustration of pseudo-drift. Oikos, 30, 409–12.

Alerstam, T. (1979) Wind as selective agent in bird migration. Ornis Scandinavica, 10, 76–93.

Alerstam, T. (1990) Bird migration, Cambridge University Press: Cambridge.

Alerstam, T. (2001) Detours in bird migration. Journal of Theoretical Biology, 209, 319–31.

Alerstam, T. (2006) Conflicting evidence about long-distance animal navigation. Science, 313, 791–94.

Alerstam, T. and Enckell, P.H. (1979) Unpredictable habitats and evolution of bird migration. Oikos, 33, 228–32.

Alerstam, T. and Gudmundsson, G.A. (1999) Migration patterns of tundra birds: tracking radar observations along the Northeast Passage. Arctic, 52, 4, 346–71.

Alerstam, T. and Hedenström, A. (1998) The development of bird migration theory. Journal of Avian Biology, 29, 343–69.

Alerstam, T. and Lindström, Å. (1990) Optimal bird migration: the relative importance of time, energy, and safety. In E. Gwinner (ed.) Bird migration: Physiology and Ecophysiology, pp. 331–51, Springer: Berlin.

Alerstam, T., Hedenström, A., and Åkesson, S. (2003) Long-distance migration: evolution and determinants. Oikos, 103, 247–60.

Alerstam, T., Gudmundsson, G.A., Green, M., and Hedenström, A. (2001) Migration along orthodromic sun compass routes by arctic birds. Science, 291, 300–303.

Alexander, R.M. (2000) Walking and running strategies for humans and other mammals. In P. Domenici and R.W. Blake (eds) Biomechanics in animal behaviour, pp. 49–57, BIOS Scientific Publishers: Oxford.

Alexander, R.M. (2000) Principles of animal locomotion, Princeton University Press: Princeton.

Alimaev, I.I. (2003) Transhumant ecosystems: fluctuations in seasonal pasture productivity. In C. Kerven (ed.) Prospects for Pastoralism in Kazakhstan and Turkmenistan: From State Farms to Private Flocks, pp. 31–51, RoutledgeCurzon: London.

Allen, C.R. and Saunders, D.A. (2002) Variability between scales: predictors of nomadism in birds of an Australian Mediterranean-climate ecosystem. Ecosystems, 5, 348–59.

Allen, C.R. and Saunders, D.A. (2006) Multimodel inference and the understanding of complexity, discontinuity, and nomadism. Ecosystems, 9, 694–99.

Ambrosini, R., Møller, A.P., and Saino, N. (2009) A quantitative measure of migratory connectivity. Journal of Theoretical Biology, 257, 203–11.

Amstrup, S.C., McDonald, T.L, and Manly, B.F.J. (2005) Handbook of Capture-Recapture Analysis, Princeton University Press: Princeton, New Jersey.

Andersen, R. (1991) Habitat Deterioration and the Migratory Behaviour of Moose (*Alces alces* L.) in Norway. Journal of Applied Ecology, 28, 1, 102–108.

Anderson, J.H. and Quinn, T.P. (2007) Movements of adult coho salmon (*Oncorhynchus kisutch*) during colonization of newly accessible habitat. Canadian Journal of Fisheries and Aquatic Sciences, 64, 8, 1143–54.

Anderson, W.B. and Polis, G.A. (1999) Nutrient fluxes from water to land: seabirds affect plant nutrient status on Gulf of California islands. Oecologia, 118, 324–32.

Andersson, M. (1980) Nomadism and site tenacity as alternative reproductive tactics in birds. Journal of Animal Ecology, 49, 175–84.

Andreassen, H.P., Gundersen, H., and Storaas, T. (2005) The effect of scent-marking, forest-clearing and supplemental feeding on moose-train collisions. Journal of Wildlife Management, 69, 3, 1125–32.

Andriansen, H.K. (2003) The use and perception of mobility among Senegalese Fulani: New approaches to pastoral mobility, Kongevej Working Paper, Institute for International Studies: Copenhagen.

Anstey, M.L., Rogers, S.M.,. Ott, S.R, Burrows, M., and Simpson S.J. (2009) Serotonin mediates behavioral gregarization underlying swarm formation in desert locusts. Science, 323, 627–30.

Arendt, J. and Reznick, D. (2008) Convergence and parallelism reconsidered: what have we learned about the genetics of adaptation. Trends in Ecology and Evolution, 23, 26–32.

Arizaga, J., Barba, E., and Belda, E. J. (2008) Fuel management and stopover duration of blackcaps *Sylvia atricapilla* stopping over in northern Spain during autumn migration period. Bird Study, 55, 124–34.

Armstrong, R.B. and Laughlin, M.H. (1985) Metabolic indicators of fibre recruitment in mammalian muscles during locomotion. Journal of Experimental Biology, 115, 201–13.

Arnason, R., Magnússon, G., and Agnarsson, S. (2000) The Norwegian spring-spawning herring fishery: a stylized game model. Marine Resource Economics, 15, 293–319.

Apostolaki, P., Milner-Gulland, E.J., McAllister, M. and Kirkwood, G. (2002) Modelling effects of establishing marine reserves in nursery or spawning grounds. Canadian Journal of Fisheries and Aquatic Science, 59, 405–15.

Aristotle (350 BC) Historia Animalium.

Arnold, S.J. (1983) Morphology, performance and fitness. American Zoologist, 23, 347–61.

Auer, N.A. (1996) Importance of habitat and migration to sturgeons with emphasis on lake sturgeon. Canadian Journal of Fisheries and Aquatic Sciences, 53, Suppl. 1, 152–60.

Augustine, D.J. (2003) Long-term, livestock-mediated redistribution of nitrogen and phosphorus in an East African savanna. Journal of Applied Ecology, 40, 137–49.

Augustine, D.J., McNaughton, S.J., and Frank, D.A. (2003) Feedbacks between soil nutrients and large herbivores in a managed savanna ecosystem. Ecological Applications, 13, 1325–37.

Ayantunde, A.A. (1998) Influence of grazing regimes on cattle nutrition and performance and vegetation dynamics in Sahelian rangelands, University of Wageningen: Wageningen, The Netherlands.

Ayantunde, A.A., Fernández-Rivera, S., Hiernaux, P.H.Y., and van Keulen, H. (2001) Effect of timing and duration of grazing of growing cattle in the West African Sahel on diet selection, faecal output, eating time, forage intake and live-weight changes. Animal Science, 72, 117–28.

Bäckman, J. and Alerstam, T. (2003) Orientation scatter of free-flying nocturnal passerine migrants: components and causes. Animal Behaviour, 65, 987–96.

Bailey, D.W., J.E. Gross, Laca, E.A., Rittenhouse, L.R., Coughenour, M.B., Swift, D.M., and Sims, P.L. (1996) Mechanisms that result in large herbivore grazing distribution patterns. Journal of Range Management, 49, 5, 386–400.

Bairlein, F. (1998) The effect of diet composition on migratory fuelling in garden warblers *Sylvia borin*. Journal of Avian Biology, 29, 546–51.

Bairlein, F. and Gwinner, E. (1994) Nutritional mechanisms and temporal control of migratory energy accumulation in birds. Annual Review of Nutrition, 14, 187–215.

Baker, A.J., González, P.M., Piersma, T., Niles, L.J., de Lima Serrano do Nascimento, I., Atkinson, P.W., Clark, N.A., Minton, C.D.T., Peck, M.K., and Aarts, G. (2004) Rapid population decline in red knots: fitness consequences of decreased refueling rates and late arrival in Deaware Bay. Proceedings of the Royal Society B-Biological Science, 271, 875–82.

Baker, D.L. and Hobbs, N.T. (1985) Emergency feeding of mule deer during winter: tests of a supplemental ration. Journal of Wildlife Management, 49, 4, 934–42.

Baker, R.R. (1978) The Evolutionary Ecology of Animal Migration, Hodder and Stoughton: London.

Balmford, A., Leader-Williams, N. and Green, M.J.B. (1995) Parks or arks: where to conserve threatened mammals? Biodiversity and Conservation, 4, 595–607.

Banko, W. (1960) The Trumpeter Swan, North American Fauna, No. 63, U.S. Fish and Wildlife Service: Washington, DC.

Barnes, R.F.W. (1999) Is there a future for elephants in West Africa? Mammal Review, 29, 3, 175–99.

Barnett-Johnson, R., Ramos, F.C., Grimes, C.B., and MacFarlane, R.B. (2005) Validation of Sr isotopes in otoliths by laser ablation multicollector inductively coupled plasma mass spectrometry (LA-MC-ICPMS): opening avenues in fisheries science applications. Canadian Journal of Fisheries and Aquatic Sciences, 62, 2425–30.

Barta. Z, McNamara, J.M., Houston, A.I., Weber, T.P., Hedenström, A., and Féro, O. (2008) Optimal moult strategies in migratory birds. Philosophical Transactions of the Royal Society of London B, 363, 211–29.

Bartumeus, F. and Levin, S.A. (2008) Fractal reorientation clocks: Linking animal behavior to statistical patterns of search. Proceeding of the National Academy of Sciences USA, 105, 19072–77.

Bartumeus, F., Da Luz, M.G.E., Viswanathan, G.M., and Catalan, J. (2005) Animal search strategies: a quantitative random-walk analysis. Ecology, 86, 3078–87.

Baskin, L. (1991) Reindeer husbandry in the Soviet Union. In L. A. Renecker and R. J. Hudson (eds) Wildlife Production: Conservation and Sustainable Development, pp. 218–26, Agricultural and Forestry Experiment Station: Fairbanks, Alaska.

Baskin, Y. (1993) Trumpeter swans relearn migration. BioScience, 43, 2, 76–9.

Bassett, T.J. (1986) Fulani herd movements. The Geographical Review, 76, 233–48.

Bassett, T.M. and Turner, M.D. (2007) Sudden shift or migratory drift? Fulbe herd movements to the Sudano-Guinean region of West Africa. Human Ecology, 35, 33–49.

Batima, P. (2006) Climate change vulnerability and adaptation in the livestock sector of Mongolia: A final report submitted to Assessments of Impacts and Adaptations to Climate Change (AIACC), Project No. AS 06. International START Secretariat: Washington, D.C.

Bauchinger, U. and Klaassen, M. (2005) Longer days in spring than in autumn accelerate migration speed. J. Avian Biol., 36, 3–5.

Bauchinger, U. and McWilliams, S. (2009) Carbon turnover in tissues of a passerine bird: Allometry, isotopic clocks, and phenotypic flexibility in organ size. Physiological and Biochemical Zoology, 82, 787–97.

Bauer, S., Gienapp, P., and Madsen, J. (2008) The relevance of environmental conditions for departure decision changes en route in migrating geese. Ecology, 89, 1953–60.

Bauer, S., Barta, Z., Ens, B.J., Hays, G.C., McNamara, J.M., and Klaassen, M. (2009) Animal migration: linking models and data beyond taxonomic limits. Biology Letters, 5, 433–35.

Bayer, W. and Waters-Bayer, A. (1994) Forage alternatives from range and field: pastoral forage management and improvement in the African drylands. In I. Scoones (ed.) Living with Uncertainty: New Directions in Pastoral Development in Africa, pp. 58–78, Intermediate Technology Publications Ltd: London.

Bayly, N.J. (2006) Optimality in avian migratory fuelling behaviour: a study of a trans-Saharan migrant. Animal Behaviour, 71, 173–82.

Bayly, N.J. (2007) Extreme fattening by sedge warblers, Acrocephalus schoenobaenus, is not triggered by food availability alone. Animal Behaviour, 74, 471–79.

Bazazi, S., Buhl, J., Hale, J.J., Anstey, M.L., Sword, G.A., Simpson, S.J., and Couzin, I.D. (2008) Collective motion and cannibalism in locust migratory bands. Current Biology, 18, 735–39.

Bazazi S., Romanczuk, P., Thomas, S., Schimansky-Geier, L., Hale, J.J., Miller, G.A., Sword, G.A., Simpson, S.J., and Couzin, I.D. (in press) Nutritional state and collective motion: from individuals to mass migration. Proceedings of the Royal Society B, Biological Sciences.

Beach, H. (1981) Reindeer-herd management in transition: the case of Tuorpon Saameby in Northern Sweden, Uppsala Studies in Cultural Anthropology 3: Stockholm.

Beach H. and Stammler, F. (2006) Human–Animal relations in pastoralism. In F. Stammler and H. Beach (eds) People and Reindeer on the Move, Special Issue of the journal Nomadic Peoples,10, 2, pp. 5–29, Berghahn Publishers: Oxford.

Bearhop, S., Fiedler, W., Furness, R.W., Votier, S.C., Waldron, S., Newton, J., Bowen, G.J., Berthold, P., and Farnsworth, K. (2005) Assortative mating as a mechanism for rapid evolution of a migratory divide. Science, 310, 502–504.

Beauvilain, A. (1977) Les Peul du Dallol Bosso, Études Nigériennes, 42, Institut de recherche en sciences humaines: Niamey, Niger.

Bedard, J., Nadeau, A., and Gauthier, G. (1986) Effects of spring grazing by greater snow geese on hay production. Journal of Applied Ecology, 23, 65–75.

Bedunah, D. and Harris, R. (2005) Observations on changes in Kazak pastoral use in two townships in Western China: A loss of traditions. Nomadic Peoples, 9, 107–29.

Beekman J.H., Nolet B.A., and Klaassen, M. (2002) Skipping swans: Fuelling rates and wind conditions determine differential use of migratory stopover sites of Bewick's Swans *Cygnus bewickii*. Ardea, 90, 437–59.

Beekman, J.H., van Eerden, M.R. and Dirksen, S. (1991) Bewick's swans *Cygnus columbianus bewickii* utilising the changing resource of *Potamogeton pectinatus* during autumn in the Netherlands. Wildfowl, Suppl. 1, 238–48.

Behnke, R.H. (1994) Natural resource management in pastoral Africa. Development Policy Review, 12, 1, 5–27.

Behnke, R.H. (1999) Stock movement and range-management in a Himba community in north-western Namibia. In M. Niamir-Fuller (ed.) Managing Mobility in African Rangelands: the Legitimization of Transhumance, pp. 184–216, Intermediate Technology Publications: London.

Behnke, R.H. (2008) The drivers of fragmentation in arid and semi-arid landscapes. In Galvin, K.A., Reid, R.S., Behnke, R.H., and Hobbs, N.T. (eds) Fragmentation in Semi-Arid and Arid Landscapes: Consequences for Human and Natural Systems, pp. 305–40, Dordrecht, The Netherlands: Springer.

Behnke, R.H. and Scoones, I. (1993) Rethinking range ecology: implications for rangeland management in Africa. In R.H. Behnke, I. Scoones, and C. Kerven (1993) Range Ecology at Disequilibrium: New Models of Natural Variability and Pastoral Adaptation in African Savannas, pp. 1–30, Overseas Development Institute: London.

Behnke, R.H., Davidson, G., Jabbar, A., and Coughenour, M. (2008) Human and natural factors that influence livestock distributions and rangeland desertification in Turkmenistan. The Socio-Economic Causes and Consequences of Desertification and Central Asia, pp. 141–68, Springer Science + Business Media B.V.: Dordrecht, The Netherlands.

Bekenov, A.B., Grachev, Iu. A., and Milner-Gulland, E.J. (1998) The ecology and management of the saiga antelope in Kazakhstan. Mammal Review, 28, 1–52.

Bell, C.P. (2000) Process in the evolution of bird migration and pattern in avian ecogeography. Journal of Avian Biology, 31, 258–65.

Bell, C.D., Blumenthal, J.M., Austin, T.J., Solomon J.L., Ebanks-Petrie, G., Broderick, A.C., and Godley, B.J. (2006) Traditional Caymanian fishery may impede local marine turtle population recovery. Endangered Species Research, 2, 63–69.

Bell, C., Solomon, J.L., Blumenthal, J.M., Austin, T.J., Ebanks-Petrie, G., Broderick, A.C., and Godley, B.J. (2007) Monitoring and conservation of critically reduced marine turtle nesting populations: lessons from the Cayman Islands. Animal Conservation, 10, 39–47.

Bellrose, F.C. (1958) The orientation of displaced waterfowl in migration. Wilson Bull., 70, 20–40.

Bengis, R.G., Grant, R., and de Vos, V. (2003) Wildlife diseases and veterinary controls: a savannah ecosystem perspective. In J. Du Toit, K.H. Rogers, and H.C. Biggs (eds) The Kruger Experience, pp. 349–69, Island Press: Washington.

Benhamou, S. (2006) Detecting an orientation component in animal paths when the preferred direction is individual-dependent. Ecology, 87, 518–28.

Benoit, M. (1979) Le chemin des Peuls du Boobola: contribution à l'écologie du pastoralisme en Afrique des savanes. O.R.S.T.O.M.: Paris.

Berg, H. (1983) Random walks in biology, Princeton University Press: Princeton.

Berger, J. (1998) Future prey: some consequences of the loss and restoration of large carnivores. In T.M. Caro (ed.) Behavioral Ecology and Conservation Biology, pp. 80–100, Oxford University Press: New York.

Berger, J. (2004) The last mile: how to sustain long-distance migration in mammals. Conservation Biology, 18, 2, 320–31.

Berger, J., Cain, S.L., and Berger, K.M. (2006) Connecting the dots: an invariant migration corridor links the Holocene to the present. Biol. Letters, 2, 528–31.

Berkes, F. (1999) Sacred Ecology: Traditional Ecological Knowledge and Resource Management, Taylor and Francis: Philadelphia and London.

Berlioz, J. (1950) Carreteres generaux et origines des migrations. In P Grasse (ed.) Trate de zoologie, Vol. XV, Oiseaux, pp. 1074–88.

Bernhardt, E.S., Palmer, M.A., Allan, J.D., Alexander, G., Barnas, K., Brooks, S., Carr, J., Clayton, S., Dahm, C., Follstad-Shah, J., Galat, D., Gloss, S., Goodwin, P., Hart, D., Hassett, B., Jenkinson, R., Katz S., Kondolf, G.M., Lake, P.S., Lave, R., Meyer, J.L., O'Donnell, T.K., Pagano, L., Powell, B., and Sudduth, E. (2005) Synthesizing U.S. river restoration efforts. Science, 308, 636–37.

Berthold, P. (1984) The control of partial migration in birds: a review. Ring, 10, 253–65.

Berthold, P. (1991) Genetic control of migratory behaviour in birds. Trends in Ecology and Evolution, 6, 8, 254–57.

Berthold, P. (1996) Control of Bird Migration, Chapman and Hall: London.

Berthold, P. (1999) A comprehensive theory for the evolution, control and adaptability of avian migration. Ostrich, 70, 1–11.

Berthold, P. (2001) Bird Migration. A General Survey, Oxford University Press: Oxford.

Berthold, P. and Querner, U. (1981) Genetic basis of migratory behavior in European warblers. Science, 212, 77–79.

Berthold, P., van den Bossche, W., Fiedler, W., Gorney, E., Kaatz, M., Leshem, Y., Nowak, E., et al. (2001) The migration of the White Stork (Ciconia ciconia): a special case according to new data. Journal für Ornithologie, 142, 73–92.

Bevan, R.M., Butler, P.J., Woakes, A.J., and Prince, P.A. (1995) The energy expenditure of free-ranging black-browed albatrosses. Philosophical Transactions of the Royal Society B, 350, 119–31.

Biebach, H. (1983) Genetic determination of partial migration in the European robin (Erithacus rubecula). Auk, 100, 601–606.

Biebach, H. (1998) Phenotypic organ flexibility in Garden Warblers Sylvia borin during long-distance migration. Journal of Avian Biology, 29, 529–35.

Biebach, H., Friedrich, W., and Heine, G. (1986) Interaction of body mass, fat, foraging and stopover period in trans-Sahara migrating passerine birds. Oecologia, 69, 370–79.

Birdlife International (2004) Tracking ocean wanderers: the global distribution of albatross and petrels, Results from the Global Procellariiform tracking Workshop, 1–5 September, 2003, Birdlife International: Cambridge, UK.

Bishop, C.M. and Butler, P.J. (1995) Physiological modeling of oxygen-consumption in birds during flight. Journal of Experimental Biology, 198, 2153–63.

Bishop, J., McKay, H., Parrott, D., and Allan, J. (2003) Review of international research literature regarding the effectiveness of auditory bird scaring techniques and potential alternatives. UK Department for Environment, Food and Rural Affairs Report: London.

Bishop, C.M., Ward, S., Woakes, A.J., and Butler, P.J. (2002) The energetics of barnacle geese (Branta leucopsis) flying in captive and wild conditions. Comparative Biochemistry and Physiology a-Molecular and Integrative Physiology, 133, 225–37.

Bisson, I.A., Safi, K., and Holland, R.A. (2009) Evidence for repeated independent evolution of migration in the largest family of bats. PLoS ONE, 4, e7504.

Biuw, E., Boehme, L., Guinet, C., Hindell, M., Costa, D., Charrassin, J-B., Roquet, F., et al. (2007) Variations in behavior and condition of a Southern Ocean top predator in relation to in situ oceanographic conditions. Proceedings of the National Academy of Sciences of the United States of America, 104, 13705–10.

Bjorndal, K.A. (1997) Foraging ecology and nutrition of sea turtles. In P.L. Lutz and J.A. Musick (eds) The Biology of Sea Turtles, pp. 199–231, CRC Press: London.

Blackwell, P.G. (1997) Random diffusion models for animal movement. Ecological Modelling, 100, 87–102.

Blackwell, P.G. (2003) Bayesian inference for Markov processes with diffusion and discrete components. Biometrika, 90, 613–27.

Blaxter, K.L., Wainman, F.W., and Wilson, R. (1961) The regulation of food intake in sheep. Journal of Animal Production, 3, 51–61.

Bleich, V.C. and Pierce, B.M. (2001) Accidental mass mortality of migrating mule deer. Western North American Naturalist, 61, 1, 124–25.

Blem, C.R. (1980) The energetics of migration. In S.A. Gauthreaux (ed.) Animal Migration, Orientation and Navigation, pp. 175–224, Academic Press: London.

Block, B.A., Teo, S.L.H. Walli, A. Boustany, A. Stokesbury, M.J.W. Farwell, C.J. Weng, K.C., Dewar, H., and Williams, T.D. (2005) Electronic tagging and population structure of Atlantic bluefin tuna. Nature, 434, 1121–27.

Block, W.M., Franklin, A.B., Ward, J.P. Jr., Ganey, J.I., and White, G.C. (2001) Design and implementation of monitoring studies to evaluate the success of ecological restoration on wildlife. Restoration Ecology, 9, 3, 293–303.

Blumstein, D.T. (2002) Moving to suburbia: ontogenetic and evolutionary consequences of life on predator-free islands. Journal of Biogeography, 29, 685–92.

Bolger, D.T., Newmark, W.D., Morrison, T.A., and Doak, D.F. (2008) The need for integrative approaches to understand and conserve migratory ungulates. Ecology Letters, 11, 63–77.

Bolker, B.M. 2008, Ecological Models and Data in R, Princeton University Press: Princeton.

Bollens, S.M. and Frost, B.W. (1989) Predator-induced diel vertical migration in a planktonic copepod. Journal of Plankton Research, 11, 1047–65.

Bollens, S.M. and Frost, B.W. (1989) Zooplanktivorous fish and variable diel migration in the marine planktonic copepod Calanus pacificus. Limnology and Oceanography, 34, 1072–83.

Bonfiglioli, Angelo Maliki (1988) Dudal: Histoire de famille et histoire de troupeau chez un groupe de Wodaabe du Niger, Cambridge University Press: Cambridge.

Boone, R.B., Thirgood, S.J., and Hopcraft, J.G.C. (2006) Serengeti wildebeest migratory patterns modeled from rainfall and new vegetation growth. Ecology, 87, 1987–94.

Börger, L., Dalziel, B.D., and Fryxell, J.M. (2008) Are there general mechanisms of animal home range behaviour? A review and prospects for future research. Ecology Letters, 11, 637–50.

Both, C. and Visser, M.E. (2001) Adjustment to climate change is constrained by arrival date in a long-distance migrant bird. Nature, 411, 296–98.

Both, C., Bouwhuis, S., Lessells, C.M., and Visser, M.E. (2006) Climate change and population declines in a long-distance migratory bird. Nature, 441, 81–83.

Bowlin, M.S. and Wikelski, M.C. (2006) Calibration of heart rate and energy expenditure during flight and at rest in a passerine. Integrative and Comparative Biology, 46, E14–E14.

Bowlin, M. S. and Wikelski, M. (2006) Pointed wings, low wingloading and calm air reduce migratory flight costs in songbirds. PLoS ONE, 3, e2154.

Bowlin, M. S., Cochran, W.W., and Wikelski, M.C. (2005) Biotelemetry of New World thrushes during migration: Physiology, energetics and orientation in the wild. Integrative and Comparative Biology, 45, 295–304.

Boyd, I.L. (2004) Migration of marine mammals. In D. Werner (ed.) Biological Resources and Migration, pp 203–10, Springer-Verlag: Berlin.

Boyle, P, and Rodhouse, P. (2005) Cephalopods: ecology and fisheries, Blackwell: London.

Boyle, W.A. and Conway, C.J. (2007) Why migrate? A test of the evolutionary precursor hypothesis. The American Naturalist, 169, 344–59.

Bradshaw, G.A. and Spies, T.A. (1992) Characterizing canopy gap structure in forests using wavelet analysis. Journal of Ecology, 80, 205–15.

Bradshaw, W.E. and Holzapfel, C.M. (2006) Evolutionary response to rapid climate change. Science, 312, 1477–78.

Brattström, O., Wassenaar, L. I., Hobson, K. A., and Åkesson, S. (2008) Placing butterflies on the map—testing regional geographical resolution of three stable isotopes in Sweden using the monophagus peacock Inachis io. Ecography, 31, 490–98.

Breiman, L. (2001) Statistical modeling: the two cultures. Statistical Science, 16, 199–231.

Breman, H. and de Wit, C.T. (1983) Rangeland productivity and exploitation in the Sahel. Science, 221, 1341–47.

Broderick A.C., Coyne, M.C., Fuller, W.J., Glen, F., and Godley, B.J. (2007) Fidelity and over-wintering of sea turtles. Proceedings of the Royal Society of London Series B, 274, 1533–38.

Broderick, A.C., Frauenstein, R., Glen, F., Hays, G.C., Jackson, A.L., Pelembe, T., Ruxton, G.D., and Godley, B.J. (2006) Are green turtles globally endangered? Global Ecology and Biogeography, 15, 21–6. www.seaturtle.org/mtrg/pubs/Broderick_GEB_2006.pdf.

Brönmark, G., Skov, C., Broderson, J., Nilsson, P.A., and Hansson, L.-A. (2008) Seasonal migration determined by a trade-off between predator avoidance and growth. PLoS ONE, 3, e1957.

Brooks, T.M., Mittermeier, R.A., Mittermeier, C.G. da Fonseca, G.A.B., Rylands, A.B., Konstant, W.R., Flick, P., Pilgrim, J., Oldfield, S., Magin, G., and Hilton-Taylor, C. (2002) Habitat loss and extinction in the hotspots of biodiversity. Conservation Biology, 16, 4, 909–23.

Brower, L.P. and Malcolm, S.B. (1991) Animal Migrations: Endangered Phenomena. American Zoologist, 31, 1, 265–76.

Brower, L. P., Fink, L. S., and P. Walford (2006) Fueling the fall migration of the monarch butterfly. Integrative and Comparative Biology, 46, 1123–42.

Brown, C.G. (1992) Movement and migration patterns of mule deer in southeastern Idaho. Journal of Wildlife Management, 56, 246–53.

Brown, C.G., Waldron, S.A., and Longworth, J.W. (2008) Sustainable Development in Western China: Managing People, Livestock and Grasslands in Pastoral Areas, Edward Elgar: Cheltenham, UK.

Bruderer, B. and Salewski, V. (2008) Evolution of bird migration in a biogeographical context. Journal of Biogeography, 35, 1951–59.

Buckland, S.T., Anderson, D.R., Burnham, K.P., Laake, J.L., Borchers, D.L., and Thomas, L. (2001) Introduction to Distance Sampling: Estimating Abundance of Biological Populations, Oxford University Press: Oxford.

Buckland, S.T., Anderson, D. R., Burnham, K. P., Laake, J. L., Borchers, D. L. and Thomas L. (2004) Advanced Distance Sampling: Estimating Abundance of Biological Populations, Oxford University Press: Oxford.

Buhl, J., Sumpter, D.J.T., Couzin, I.D., Hale, J.J., Despland, E., Miller, E.R. and Simpson, S.J. (2006) From disorder to order in marching locusts. Science, 312, 1402–406.

Burnham, K.P. and D.R. Anderson. 2002, Model Selection and Multi-Model Inference, A Practical Information–Theoretic Approach, Springer Verlag: Berlin, Germany.

Butler, P.J. (1982) Respiration during flight and diving in birds. In A.D.F. Addink and N. Spronk (eds) Exogenous and Endogenous Influences on Metabolic and Neural Control, pp 103–14, Pergamon Press: London.

Butler, P.J. (1986) Exercise. In S. Nilsson and S. Holmgren (eds) Fish Physiology: Recent Advances, pp. 102–18, Croom Helm: London.

Butler, P.J. (1991) Exercise in birds. Journal of Experimental Biology, 160, 233–62.

Butler, P.J. and Bishop, C.M. (2000) Flight. In G. C. Whittow (ed.) Sturkie's Avian Physiology, 5th edn, Academic Press: New York, NY.

Butler, P.J. and Woakes, A.J. (1980) Heart-rate, respiratory frequency and wing beat frequency of free flying barnacle geese *Branta leucopsis*. Journal of Experimental Biology, 85, 213–26.

Butler, P.J., Woakes, A.J., and Bishop, C.M. (1998) Behaviour and physiology of Svalbard barnacle geese *Branta leucopsis* during their autumn migration. Journal of Avian Biology, 29, 536–45.

Bykova, E., Esipov, A., and Chernogaev, E. (2009) Will the saiga return to its traditional breeding grounds in Uzbekistan? Saiga News, 8, 13–5. [Online], Available: http://www.saiga-conservation.com/saiga_news. html. [April 2009].

Carmi, N., Pinshow, B., Porter, W.P., and Jaeger, J. (1992) Water and energy limitations on flight duration in small migrating birds. Auk, 109, 268–76.

Caro, T.M., Pelkey, N., Borner, M., Campbell, K.L.I., Woodworth, B.L., Farm, B.P., Kuwai, J.O., Huish, S.A., and Severre, E.L.M. (1998) Consequences of different forms of conservation for large mammals in Tanzania: preliminary analyses. African Journal of Ecology, 36, 303–20.

Carpenter, F.L., Hixon, M.A., Russell, R.W., Paton, D.C., and Temeles E. J. (1993) Interference asymmetries among age-sex classes of rufous hummingbirds during migratory stopovers. Behavioral Ecology and Sociobiology, 33, 297–304.

Carr, A. (1984) The sea turtle: so excellent a fishe. University of Texas Press: Austin.

Chaloupka, M., Bjorndal, K.A., Blalzs, G.H., Bolten, A.B., Ehrhart, L.M., Limpus, C.J., Suganuma, H., Troeng, S., and Yamaguchi, M. (2008) Encouraging outlook for recovery of a once severely exploited marine megaherbivore. Global Ecology and Biogeography, 17, 2, 297–304.

Chambers, L.E. (2008) Trends in timing of migration of south-western Australian birds and their relationship to climate. Emu, 108, 1–14.

Chan, K. (2001) Partial migration in Australian landbirds: a review. Emu, 101, 281–92.

Chapin III, F.S., Zavaleta, E.S. Eviner, V.T. Naylor, R. L. Vitousek, P.M. Reynolds, H.L. Hooper, D.U. Lavorel, S. Sala, O.E. Hobbie, S.E. Mack M.C., and Díaz, S. (2000) Consequences of changing biodiversity. Nature, 405, 234–42.

Chapman, J.W., Reynolds, D.R., Brooks, S.J., Smith, A.D., and Woiwod, I.P. (2006) Seasonal variation in the migration strategies of the green lacewing *Chrysoperla carnea* species complex. Ecological Entomology, 31, 378–88.

Chapman, J.W., Reynolds, D.R., Smith, A.D., Smith, E.T., and Woiwod, I.P. (2004) An aerial netting study of insects migrating at high altitude over England. Bulletin of Entomological Research, 94, 123–36.

Chapman, J.W., Reynolds, D.R., Hill, J.K., Sivell, D., Smith, A.D., and Woiwod, I.P. (2008a) A seasonal switch in compass orientation in a high-flying migrant moth. Current Biology, 18, R908–R909.

Chapman, J.W., Reynolds, D.R., Mouritsen, H., Hill, J.K., Riley, J.R., Sivell, D., Smith, A. D., and Woiwod, I.P. (2008b) Wind selection and drift compensation optimize migratory pathways in a high-flying moth. Current Biology, 18, 514–18.

Cheke, R.A. and Tratalos, J.A. (2007) Migration, patchiness, and population processes illustrated by two migrant pests. Bioscience, 57, 145–54.

Cherel, Y., Kernaléguen, L., Richard, P., and Guinet, C. (2009) Whisker isotopic signature depicts migration patterns and multi-year intra- and inter-individual foraging strategies in fur seals. Biology Letters, 5, 830–32.

Chesser, R.T. and Levey, D.J. (1998) Austral Migrants and the Evolution of Migration in New World Birds: Diet, Habitat, and Migration Revisited. The American Naturalist, 152, 311–19.

Chimeddorj, B. (2009) Public Awareness Needs Assessment for the Saiga Antelope Conservation Project in Mongolia. Saiga News, 8, 9–10. [Online], http://www.saiga-conservation.com/saiga_news.html [April 2009].

Christie, K.S. and Reimchen, T.E. (2005) Post-reproductive Pacific salmon, Oncorhynchus spp., as a major nutrient source for large aggregations of gulls, Larus spp. Canadian Field-Naturalist, 119, 202–207.

Ciancio, J.E., Pascual, M., Botto. M., Amaya, M., O'Neal, S., Riva Rossi, C., and Iribarne, O. (2008) Stable isotope profiles of partially migratory salmonid populations in Atlantic rivers of Patagonia. Journal of Fish Biology, 72, 1708–19.

Cimprich, D.A. and F.R. Moore (2006) Fat affects predator-avoidance behavior in gray catbirds (*Dumetella carolinensis*) during migratory stopover. Auk, 123, 1069–76.

Cimprich, D.A., Woodrey, M.S., and Moore, F.R. (2005) Passerine migrants respond to variation in predation risk during stopover. Animal Behaviour, 69, 1173–79.

Cissé, S. (1981) L'avenir du pastoralisme dans le delta central du Niger (Mali): Agriculture, élevage, ou agropastoralisme? Nomadic Peoples, 9, 16–21.

CITES (Convention on International Trade in Endangered Species of Wild Fauna and Flora) (2004) Conservation of Saiga tatarica. CoP 13, Document 32. Convention on International Trade in Endangered Species of Flora and Fauna. [Online], http://www.cites.org/eng/cop/13/doc/E13–32.pdf [April 2009].

Clark, C.W. and Mangel, M. (1986) The evolutionary advantages of group foraging. Theor. Pop. Biol., 30, 45–75.

Clark, C.W. and Mangel, M. (2000) Dynamic state variable models in ecology: methods, and applications, Oxford University Press: Oxford.

Clausen, P., Green, M., and Alerstam, T. (2003) Energy limitations for spring migration and breeding: the case of Brent geese Branta bernicla tracked by satellite telemetry to Svalbard and Greenland. Oikos, 103, 426–45.

Cleaveland, S., Packer, C., Hampson, K., Kaare, M., Kock, R., Craft, M., Lembo, T., Mlengeya, T., and Dobson, A. (2008) The multiple roles of infectious disease in the Serengeti ecosystem. In A.R.E. Sinclair, C. Packer, S.A.R. Mduma, and J.M. Fryxell (eds) Serengeti III: Human Impacts on Ecosystem Dynamics, pp. 209–40 Chicago University Press: Chicago.

Clevenger, A.P. and Waltho, N. (2005) Performance indices to identify attributes of highway crossing structures facilitating movement of large mammals. Biological Conservation, 121, 453–64.

Clewett, J.F., Clarkson, N.M., George, D.A., Ooi, S.H., Owens, D.T., Partridge, I.J., and Simpson, G.B. (2003) Rainman StreamFlow version 4.3: a comprehensive climate and streamflow analysis package on CD to assess seasonal forecasts and manage climate risk, QI03040, Department of Primary Industries: Queensland.

Clobert, J., Le Galliard, J. F., Cote, J., Meylan, S., and Massot M. (2009) Informed dispersal, heterogeneity in animal dispersal syndromes and the dynamics of spatially structured populations. Ecology Letters, 12, 197–209.

Clutton-Brock, T., Guiness, F.E., and Albon, S.D. (1982) Red Deer: Behavior and Ecology of Two Sexes, Wildlife Behavior and Ecology Series, The University of Chicago Press: Chicago.

CMS (Convention on Migratory Species) (2006a) Revised Overview Report of the First Meeting of the Signatories to the Memorandum of Understanding concerning Conservation, Restoration and Sustainable Use of the Saiga Antelope (Saiga tatarica tatarica) (CMS/SA- 1/ Report Annex 5), Convention on Migratory Species. [Online], Available: http://www.cms.int/species/ saiga/post_session/Annex_05_Revised_Overview_ Report_E.pdf

CMS (Convention on Migratory Species) (2006b) Medium term international work programme for the saiga antelope (2007–2011). (CMS/SA- 1/Report Annex 9): Convention on Migratory Species. [Online], Available: http://www.cms.int/species/saiga/post_session/ Annex_09_MediumTerm_Int_WrkProgm_E.pdf

Cochran, W.W. (1972) Long-distance tracking of birds. In S.R Galler, K. Schmidt-Koenig, G.J. Jacobs, and R.E Belleville (eds) Animal Orientation and Navigation, pp. 39–59, National Aeronautics and Space Adminstration: Washington, DC.

Cochran, W.W. and Kjos, C.G. (1985) Wind drift and migration of thrushes: A telemetry study. Illinois Natural History Survey Bulletin, 33, 297–330.

Cochran, W.W. and Wikelski, M. (2005) Individual migratory tactics of New World Catharus thrushes: Current knowledge and future tracking options from space. In R. Greenberg and P.P. Marra (eds) Birds of Two Worlds: The Ecology and Evolution of Migration, pp. 274–89, John Hopkins University Press: Baltimore.

Cochran, W.W., Mouritsen, H., and Wikelski, M. (2004) Migrating songbirds recalibrate their magnetic compass daily from twilight cues. Science, 304, 405–408.

Codling, E.A., Pitchford, J.W., and Simpson, S.D. (2007) Group navigation and the 'many-wrongs principle' in models of animal movement. Ecology, 88, 1864–70.

Cohen, D. (1967) Optimization of seasonal migratory behaviour. The American Naturalist, 101, 5–17.

Colin de Verdière, P. (1995) Étude compare de trois systèmes agropastoraux dans la région de Filingué—Niger. Doctorate thesis, Université de Hohenheim Allemagne et Institut National Agronomique: Paris-Grignon, France.

Conover, M.R. (1984) Comparative effectiveness of Avitrol, exploders and hawk-kites in reducing blackbird damage to corn. Journal of Wildlife Management, 48, 1, 109–16.

Conradt, L., Clutton-Brock, T. H., and Thomson, D. (1999) Habitat segregation in ungulates: are males forced into suboptimal foraging habitats through indirect competition by females? Oecologia, 119, 367–77.

Conradt, L., Krause, J., Couzin, I. D., and Roper, T. J. (2009) 'Leading according to need' in self-organizing groups. American Naturalist, 173, 304–12.

Cooke, S.J., Hinch, S.G., Wikelski, M., Andrews, R.D., Kuchel, L.J., Wolcott, T.G., and Butler P.J. (2004) Biotelemetry: a mechanistic approach to ecology. Trends in Ecology and Evolution, 19, 334–43.

Coppock, D.L., Ellis, J.E., and Swift, D.M. (1986) Livestock feeding ecology and resource utilization in a nomadic pastoral ecosystem. Journal of Applied Ecology, 23, 573–83.

Coppolillo, P.B. (2000) The landscape ecology of pastoral herding: spatial analysis of land use and livestock production in East Africa. Human Ecology, 28, 4, 527–60.

Coppolillo, P.B. (2001) Central-place analysis and modeling of landscape-scale resource use in an East African agropastoral system. Landscape Ecology, 16, 205–19.

Coughenour, M.B. (1991) Spatial components of plant-herbivore interactions in pastoral, ranching and native ungulate ecosystems. Journal of Range Management, 44, 530–42.

Coughenour, M.B., Coppock, D.L., and Ellis, J.E. (1990) Herbaceous forage variability in an arid pastoral region of Kenya: importance of topographic and rainfall gradients. Journal of Arid Environments, 19, 147–59.

Coulson, T., Guinness, F., Pemberton, J., and Clutton-Brock T. (2004) The demographic consequences of releasing a population of red deer from culling. Ecology, 85, 411–22.

Courchamp, F., Berec, L., and Gascoigne, J. (2008) Allee effects in ecology and conservation. Oxford University Press: Oxford.

Coutant, C.C. and Whitney, R.R. (1999) Fish behavior in relation to passage through hydropower turbines: a review. Transactions of the American Fisheries Society, 129, 351–80.

Couzin, I.D. and Krause, J. (2003) Self-organization and collective behavior in vertebrates. Advances in the Study of Behavior, 32, 32, 1–75.

Couzin, I.D. and Laidre, M.E. (2009) Fission–fusion populations. Current Biology, 19, R633–35.

Couzin, I.D., Krause, J., Franks, N.R., and Levin S.A. (2005) Effective leadership and decision-making in animal groups on the move. Nature, 433, 513–16.

Cox, D.R. (2006) Principles of Statistical Inference, Cambridge University Press: Cambridge, UK.

Cox, G.W. (1968) The role of competition in the evolution of migration. Evolution, 22, 180–92.

Cox, G.W. (1985) The evolution of avian migration systems between temperature and tropical regions of the New World. American Naturalist, 126, 451–74.

Cox, T.M., Lewison, R.I., Zydelis, R., Crowder L.B., Safina, C., and Read, A.J. (2007) Comparing effectiveness of experimental and implemented bycatch reduction measures: the ideal and the real. Conservation Biology, 21, 1155–64.

Craig, A.S. and Herman, L.I. (1997) Sex differences in site fidelity and migration of humpback whales (*Megaptera novaeangliae*) to the Hawaiian Islands. Canadian Journal of Zoology, 75, 1923–33.

Crick, H.Q.P. (2004) The impact of climate change on birds. Ibis, 146, 48–56.

Cumming, H.G. and Beange, D.B. (1987) Dispersion and movements of Woodland Caribou near Lake Nipigon, Ontario. Journal of Wildlife Management, 51, 1, 69–79.

Cunnison, I. (1966) Baggara Arabs: Power and the Lineage in a Sudanese Nomad Tribe, Clarendon Press: Oxford.

Curio, E., Ernst, U., and Vieth, W. (1978) Cultural transmission of enemy recognition: one function of mobbing. Science, 202, 899–901.

Dalziel, B.D., Morales, J.M., and Fryxell, J.M. (2008) Fitting probability distributions to animal movement trajectories: using artificial neural networks to link distance, resources, and memory. American Naturalist, 172, 248–58.

Darby, W.R. and Pruitt, W.O. (1984) Habitat use, movements and grouping behavior of Woodland Caribou, *Rangifer tarandus caribou*, in southeastern Manitoba. Canadian Field-Naturalist, 98, 2, 184–90.

Davies, S. (1984) Nomadism as a response to desert conditions in Australia. Journal of Arid Environments, 7, 183–95.

Davis, M.B. and Shaw, R.G. (2001) Range shifts and adaptive responses to quaternary climate change. Science, 292, 673–79.

Dawson, A. (2008) Control of the annual cycle in birds: endocrine constraints and plasticity in response to ecological variability. Philosophical Transactions of the Royal Society of London, Series B, 363, 1621–33.

De Boer, W.F. and Prins, H.H.T. (1989) Decisions of cattle herdsmen in Burkina Faso and optimal foraging models. Human Ecology, 17, 445–64.

de Bruijn, M. and van Djik, H. (1995) Arid Ways: Cultural Understandings of Insecurity in Fulbe Society, Central Mali, Thela Publishers: Amsterdam.

de Leaniz, C.G. (2008) Weir removal in salmonid streams: implications, challenges and practicalities. Hydrobiologia, 609, 83–96.

Dean, W.R.J. (2004) Nomadic Desert Birds, Springer Verlag, London.

Dean, W.R.J., Barnard, P., and Anderson, M.D. (2009) When to stay, when to go: trade-offs for southern African arid-zone birds in times of drought. South African Journal of Science, 105, 24–28.

Deinum, B. (1984) Chemical composition and nutritive value of herbage in relation to climate. In H. Riley, and A.O. Skjelvag (eds) The Impact of Climate on Grass Production and Quality. Proceedings of the 10th General Meeting of the European Grassland Federation, As, Norway, pp. 338–50.

Del Hoyo, J., Elliott, A., and Sargatal, J. (1994) Handbook of the Birds of the World. Volume 2: New World Vultures to Guineafowl, Lynx Edicions: Barcelona.

D'Eon, R.G. and Delparte, D. (2005) Effects of radio-collar position and orientation on GPS radio-collar performance, and the implications of PDOP in data screening. Journal of Applied Ecology, 42, 383–88.

Deppe, J.L. and Rotenberry J.T. (2008) Scale-dependent habitat use by fall migratory birds: vegetation structure, floristics, and geography. Ecological Monographs, 78, 461–487.

Déregnacourt, S., Guyomarc'h, J., and Belhamra, M. (2005) Comparison of migratory tendency in European quail Coturnix c. coturnix domestic Japanese quail Coturnix c. japonnica and their hybrids. Ibis, 147, 25–36.

Dierschke, V., Delingat, J., and Schmaljohann, H. (2003) Time allocation in migrating northern wheatears (Oenanthe oenanthe) during stopover: is refuelling limited by food availability or metabolically? Journal für Ornithologie, 144, 33–44.

Dierschke, V., Mendel, B., and Schmaljohann, H. (2005) Differential timing of spring migration in northern wheatears *Oenanthe oenanthe*: hurried males or weak females? Behavioral Ecology and Sociobiology, 57, 470–80.

Dingle, H. (1980) Ecology and evolution of migration. In S.A. Gauthreaux (ed.) Animal Migration, Orientation, and Navigation, pp. 1–100, Academic Press: London.

Dingle, H. (1989) The evolution and significance of migratory flight. In G.J. Goldsworthy (ed.) Insect Flight, pp. 99–114, CRC Press: Boca Raton.

Dingle, H. (1996) Migration: the Biology of Life on the Move, Oxford University Press: New York.

Dingle, H. (2006) Animal migration: is there a common migratory syndrome? Journal of Ornithology, 147, 212–20.

Dingle, H. and Drake, V.A. (2007) What is migration? Bioscience, 57, 113–21.

Dingle, H., Rochester, W.A., and Zalucki, M.P. (2000) Relationships among climate, latitude and migration: Australian butterflies are not temperate-zone birds. Oecologia, 124, 196–207.

Dixon, J.D., Oli, M.K., Wooten, M.C., Eason, T.H., McCown, J.W., and Paetkau, D. (2006) Effectiveness of a regional corridor in connecting two Florida black bear populations. Conservation Biology, 20, 155–62.

Dodson, J.J. (1997) Fish migration: an evolutionary perspective. In J.-G. Godin (ed.) Behavioural Ecology of Teleost Fishes, pp 10–36, Oxford University Press: Oxford.

Doebeli, M. (1995) Dispersal and dynamics. Theoretical Population Biology, 47, 82–106.

Doebeli, M. and Ruxton, G. (1997) Evolution of dispersal rates in metapopulation models: branching and cyclic dynamics in phenotype space. Evolution, 51, 1730–41.

Dolman, P.M. and Sutherland, W.J. (1995) The response of bird populations to habitat loss. Ibis, 137, s1, S38–S46.

Dorst, J. (1961) The Migrations of Birds, Heinemann: London.

Dragesund, O., Johannessen, A., and Ulltang, Ø. (1997) Variation in migration and abundance of Norwegian spring spawning herring (*Clupea harengus* L.). Sarsia, 82, 2, 97–105.

Drake, V.A. and Gatehouse, A.G. (eds) (1995) Insect Migration: Tracking Resources Through Space and Time, Cambridge University Press: Cambridge.

Drent, R.J., Fox, A.D., and Stahl, J. (2006) Travelling to breed. Journal of Ornithology, 147, 122–34.

Dudley, R. and Srygley, R.B. (2008) Airspeed adjustment and lipid reserves in migratory Neotropical butterflies. Func. Ecol., 22, 264–70.

Dudley, R., Srygley, R.B., Oliveira, E.G., and DeVries, P.J. (2002) Flight speeds, lipid reserves, and predation of the migratory neotropical moth *Urania fulgens* (Uraniidae). Biotropica, 34, 452–58.

Dugger, K.M., Faaborg, J., Arendt, W.J., and Hobson, K.A. (2004) Understanding survival and abundance of overwintering warblers: does rainfall matter? The Condor, 106, 744–60.

Duke, D.L. (2001) Wildlife use of corridors in the Central Canadian Rockies: multivariate use of habitat characteristics and trends in corridor use. Thesis, University of Alberta: Edmonton, Alberta, Canada.

Dulvy, N.K., Sadovy, Y., and Reynolds, J.D. (2003) Extinction vulnerability in marine populations. Fish and Fisheries, 4, 25–64.

Durant, S.M., Caro, T.N., Collins, D.A., Alawi, R.M. and Fitzgibbon, C.D. (1988) Migration patterns of Thomsons gazelles and cheetahs on the Serengeti Plains. African Journal of Ecology, 26, 257–68.

Duriez, O., Bauer, S., Destin, A., Madsen, J., Nolet, B.A., Stillman, R.A., and Klaassen, M. (2009) What decision rules might pink-footed geese use to depart on migration? An individual-based model. Behavioral Ecology, 20, 560–69.

Dusenberry, D.B. (2001) Physical constraints in sensory ecology. In F.G. Barth and A. Schmid. (eds) Ecology of Sensing, pp. 1–17, Springer: Berlin.

Dwyer, M. and Istomin, K. (2008) Theories of nomadic movement: a new theoretical approach for understanding the movement decisions of Nenets and Komi reindeer herders. Human Ecology, 17, 445–64.

Ebbesson, L.O.E., Ebbesson, S.O.E., Nilsen, T.O., Stefansson, S.O., and Holmqvist, B. (2007) Exposure to continuous light disrupts retinal innervation during the pre-optic nucleus during parr-smolt transformation in Atlantic salmon. Aquaculture, 273, 345–49.

Edwards, A.M. (2008) Using likelihood to test for Lévy flight search patterns and for general power-law distributions in nature. Journal of Animal Ecology, 77, 1212–22.

Edwards, A.M., R.A. Phillips, N.W. Watkins, M.P. Freeman, E.J. Murphy, V. Afanasyev, S.V. Buldyrev, M.G.E. da Luz, E.P. Raposo, H.E. Stanley, and G.M. Viswanathan. 2007. Revisiting Levy flight search patterns of wandering albatrosses, bumblebees and deer. Nature 449:1044–1048.

Egevang, C., Stenhouse, I.J., Phillips, R.A., Petersen, A., Fox, J.W., and Silk, J.R.D. (2010) Tracking of Arctic terns *Sterna paradisaea* reveals longest animal migration. Proc. Natl. Acad. Sci., 107, 2078–2081, doi/10.1073/pnas.0909493107

Eichorn, G., Drent, R.H., Stahl, J., Leito, A., and Alerstam, T. (2009) Skippig the Baltic: the emergence of a dichotomy of alternative spring migration strategies in Russian barnacle geese. Journal of Animal Ecology, 78, 63–72.

Ellegren, H. (1991) Stopover ecology of autumn migrating bluethroats *Luscinia svecica svecica* in relation to age and sex. Ornis Scandinavica, 22, 340–48.

Elliot, J.J. and Arbib, R.S.J. (1953) Origin and status of the house finch in the eastern United States, Auk, 70, 31–37.

Ellis, D.H., Sladen, W.J.L., Lishman, W.A., Clegg, K.R., Duff, J.W., Gee, G.F., and Lewis, J.C. (2003) Motorized migrations: the future or mere fantasy? BioScience, 53, 3, 260–64.

Ellis, J.E. and Swift, D.M. (1988) Stability in African pastoral ecosystems: alternate paradigms and implications for development. Journal of Range Management, 41, 450–59.

Enloe, J.G. and David, F. (1995) Rangifer herd behaviour: Seasonality of hunting in the Magdalenian of the Paris Basin. In L.J. Jackson and P. T. Thacker (eds) Caribou and Reindeer Hunters of the Northern Hemisphere, pp. 47–63. Avebury Press: Aldershot.

Erdenebaatar, B. (2003) Mongolia case study 1: Studies on long-distance transhumant grazing systems in Uvs and Khuvsgul aimags of Mongolia, 1999–2000. In J.M. Suttie and S.G. Reynolds (eds) Transhumant Grazing Systems in Temperate Asia, pp. 31–68, Food and Agriculture Organization of the United Nations (FAO): Rome.

Erni, B., Liechti, F., and Bruderer, B. (2002) Stopover strategies in passerine bird migration: A simulation study. Journal of Theoretical Biology, 219, 479–93.

Ervin, J. (2003) WWF: Rapid Assessment and Prioritization of Protected Area Management (RAPPAM) Methodology, WWF: Gland, Switzerland.

Evans, P.R. (1966) Migration and orientation of passerine night migrants in northeast England. Journal of Zoology, 150, 319–69.

Farnsworth, K.D. and Beecham, J.A. (1997) Beyond the ideal free distribution: more general models of predator distribution. Journal of Theoretical Biology, 187, 389–96.

Faugère, O., Dockes, A.C., Perrot, C., and Faugère, B. (1990a) L'élevage traditionnel des petits ruminants au Senegal. I. Pratiques de conduite et d'exploitation des animaux chez les éleveurs de la région de Kolda. Revue elevage médicine vétérinaire pays tropicaux, 43, 2, 249–59.

Fedak, M., Lovell, P., McConnell, B., and Hunter, C. (2002) Overcoming the constraints of long range radio telemetry from animals: getting more useful data from smaller packages. Integrative and Comparative Biology, 42, 3–10.

Fedorovich, B.A. (1973) Natural conditions of the arid zones of the USSR and ways of development of their livestock sectors. In Essays on the Agricultural History of the Peoples of Central Asia and Kazakhstan, pp. 207–22, Nauka: Leningrad.

Ferguson, M.M. and Duckworth, G.A. (1997) The status and distribution of lake sturgeon, *Acipenser fulvescens*, in the Canadian provinces of Manitoba, Ontario and Quebec: a genetic perspective. Environmental Biology of Fishes, 48, 299–309.

Ferguson, S.H. and Elkie, P.C. (2004) Seasonal movement patterns of woodland caribou (*Rangifer tarandus caribou*). Journal of Zoology, 262, 125–34.

Fernandez-Gimenez, M.E. (1997) Landscapes, Livestock, and Livelihoods: Social, Ecological, and Land-Use Change among the Nomadic Pastoralists of Mongolia, University of California: Berkeley, Berkeley, CA.

Fernandez-Gimenez, M.E. (1999) Sustaining the steppes: a geographical history of pastoral land use in Mongolia. The Geographical Review, 89, 315–42.

Fernandez-Gimenez, M.E. (2002) Spatial and social boundaries and the paradox of pastoral land tenure: a case study from postsocialist Mongolia. Human Ecology, 30, 49–78.

Fernandez-Gimenez, M.E. and Allen-Diaz, B. (2001) Vegetation change along gradients from water sources in three grazed Mongolian ecosystems. Plant Ecology, 157, 101–18.

Fernandez-Gimenez, M.E. and Batbuyan, B. (2004) Law and disorder: Local implementation of Mongolia's Land Law. Development and Change, 35, 141–65.

Fernandez-Gimenez, M.E. and Le Febre, S. (2006) Mobility in pastoral systems: dynamic flux or downward trend? International Journal of Sustainable Development and World Ecology, 13, 1–22.

Fernandez-Gimenez, M.E., Batbuyan, B., and Oyungerel, J. (2007) Climate, economy and land policy: effects on pastoral mobility patterns in Mongolia. In X. Sun and N. Naito (eds) Mobility, Flexibility, and Potential of Nomadic Pastoralism in Eurasia and East Africa, pp. 3–24, Graduate School of Asian and African Area Studies: Kyoto University, Kyoto, Japan.

Fernandez-Gimenez, M.E., Kamimura, A., and Batbuyan, B. (2008) Implementing Mongolia's Land Law: Progress and Issues, Final report to the Central Asian Legal Exchange, Central Asian Legal Exchange: Nagoya University: Nagoya, Japan.

Ferraroli, S., Georges, J., Gaspar, P., and Le Maho, Y. (2004) Where leatherback turtles meet fisheries. Nature, 429, 521–22.

Fieberg, J. and Delgiudice, G. (2008) Exploring migration data using interval-censored time-to-event models. Journal of Wildlife Management, 72, 1211–19.

Fiedler, W. (2003) Recent changes in migratory behaviour of birds: a compilation of field observations and ringing data. In P. Berthold, E. Gwinner, and E. Sonnenschein (eds) Avian Migration, pp. 21–38, Springer Verlag: Berlin.

Finke, P. (2000) Changing property rights systems in Mongolia, Max Planck Institute for Social Anthropology, Working Paper No. 3: Halle, Germany.

Finstad G.L., Kielland K.K., and Schneider, W.S. (2006) Reindeer herding in transition: historical and modern day challenges for Alaskan reindeer herders. In F. Stammler, and H. Beach (eds) People and Reindeer on the Move, Special Issue of the journal Nomadic Peoples, 10, 2, 31–49, Oxford: Berghahn Publishers.

Fischer, J. and Lindenmayer, D.B. (2000) An assessment of the published results of animal relocations. Biological Conservation, 96, 1–11.

Fisher, M.J., Rao, I.M., Thomas, R.J., and Lascano, C.E. (1996) Grasslands in the well-watered tropical lowlands. In J.Hodgson and A.W. Illius (eds.) The Ecology and Management of Grazing Systems, pp. 393–428, CAB International: Wallingford, UK.

Fleishman, E., Murphy, D.D., and Brussard, P.F. (2000) A new method for selection of umbrella species for conservation planning. Ecological Applications, 10, 2, 569–79.

Fleming, T. H. and Eby, P. (2003) Ecology of bat migration. In T.H. Kunz and M.B. Fenton (eds) Bat Ecology, pp. 156–208, Chicago University Press: Chicago.

Forbes, B. and Stammler, F. (2009) Arctic climate change discourse: the contrasting politics of research agendas in the West and Russia. Polar Research. Vol. 28, Climate Change Vulnerability and Adaptation in the Arctic, pp. 28–42.

Fox, A.D. and Madsen, J. (1997) Behavioural and distributional effects of hunting disturbance on waterbirds in Europe: implications for refuge design. The Journal of Applied Ecology, 34, 1, 1–13.

Frank, D.A., Inouye, R.S., Huntly, N., Minshall, G.W., and Anderson J.E. (1994) The biogeochemistry of a north-temperate grassland with native ungulates—nitrogen dynamics in Yellowstone National Park. Biogeochemistry, 26, 163–88.

Fransson, T. (1998) Patterns of migratory fuelling in whitethroats Sylvia communis in relation to departure. Journal of Avian Biology, 29, 569–73.

Fransson, T. and Weber, T. P. (1997) Migratory fuelling blackcaps (Sylvia atricapilla) under perceived risk of predation. Behavioral Ecology and Sociobiology, 41, 75–80.

Fransson, T., Barboutis, C., Mellroth, R., and Akriotis, T. (2008) When and where to fuel before crossing the Sahara desert—extended stopover and migratory fuelling in first-year garden warblers Sylvia borin. Journal of Avian Biology, 39, 133–38.

Fransson, T., Jakobsson, S., Johansson, P., Kullberg, C., Lind, J., and Vallin, A. (2001) Bird migration—Magnetic cues trigger extensive refuelling. Nature, 414, 35–36.

Freitag-Ronaldson, S. and Foxcroft, L.C. (2003) Anthropogenic influences at the ecosystem level. In

J. Du Toit, K.H. Rogers, and H.C. Biggs (eds) The Kruger Experience, pp. 391–421, Island Press: Washington.

Fretwell, D.D. and Lucas, H.L. 1970. On territorial behaviour and other factors influencing habitat distribution in birds. Acta Biotheoretica, 19, 16–36.

Frey, K.E. and Smith, L.C. (2003) Recent temperature and precipitation increases in West Siberia and their association with the Arctic Oscillation. Polar Research, 22, 287–300.

Fry, M. (2004) The status of the Saiga antelope in the Ustiurt region of western Kazakhstan. MSc thesis, Imperial College: London. [Online], Available: http://www.iccs.org.uk/thesis.htm [April 2009].

Fryxell, J.M. (1991) Forage quality and aggregation by large herbivores, The American Naturalist, 138, 478–98.

Fryxell, J.M. (1995) Aggregation and migration by grazing ungulates in relation to resources and predators. In A.R.E. Sinclair and P. Arcese (eds) Serengeti II: Dynamics, Management, and Conservation of an Ecosystem, pp. 257–73, Chicago University Press: Chicago.

Fryxell, J.M. and Sinclair, A.R.E. (1988a) Seasonal migration by white-eared kob in relation to resources. African Journal of Ecology, 26, 17–31.

Fryxell, J.M. and Sinclair, A.R.E. (1988b) Causes and consequences of migration by large herbivores. Trends in Ecology and Evolution, 3, 237–41.

Fryxell, J. Greever, J., and Sinclair, A.R.E. (1988) Why are migratory ungulates so abundant? American Naturalist, 131, 781–98.

Fryxell, J.M., Wilmshurst, J.F., and Sinclair, A.R.E. (2004) Predictive models of movement by Serengeti grazers. Ecology, 85, 2429–35.

Fryxell, J.M., Wilmshurst, J.F., Sinclair, A.R.E., Haydon, D.T., Holt, R.D., and Abrams, P.A. (2005) Landscape scale, heterogeneity, and the viability of Serengeti grazers. Ecology Letters, 8, 328–35.

Fryxell, J.M.,M. Hazell, L. Borger, B.D. Dalziel, D.T. Haydon, J.M. Morales, T. McIntosh, and R.C. Rosatte. (2008) Multiple movement modes by large herbivores at multiple spatiotemporal scales. Proceedings of the National Academy of Sciences of the United States of America, 105, 19114–19.

Gagliardo, A., Ioale, P., Savini, M., and Wild, M. (2008) Navigational abilities of homing pigeons deprived of olfactory or trigeminally mediated magnetic information when young. Journal of Experimental Biology, 211, 2046–51.

Gallais, J. (1975) Paysans et pasteurs du Gourma. La condition sahélienne, CNRS: Paris.

Galvin, K.A., Reid, R.S., Behnke, R.H., and Hobbs, N.T. (eds) (2008) Fragmentation in Semi-Arid and Arid Landscapes: Consequences of Human and Natural Systems, Springer: Dordrecht, The Netherlands.

Gannes, L.Z. (2002) Mass change pattern of blackcaps refueling during spring migration: Evidence for physiological limitations to food assimilation. Condor, 104, 231–39.

Garin, P., Faye, A. Lericollais, A., and Sissokho, M. (1990) Évolution du rôle du bétail dans la gestion de la fertilité des terroirs Sereer au Sénégal. Les Cahiers de la Recherche Développement, 26, 65–84.

Gatehouse, A.G. (1987) Migration: a behavioral process with ecological consequences? Antenna, 11, 10–12.

Gauthreaux, S. (1982) The ecology and evolution of avian migration systems. Avian Biology, 6, 93–168.

Gaylard, A., Owen-Smith, N., and Redfern J. (2003) Surface water availability: implications for heterogeneity and ecosystem processes. In J. Du Toit, K. H. Rogers, and H. C. Biggs (eds) The Kruger Experience, pp. 171–88, Island Press: Washington.

Geist, V. (1999) Deer of the World: Their Evolution, Behaviour, and Ecology, Swan Hill Press: Shrewsbury.

Geller, M. K. and Vorzhonov, V. V. (1975) Migratsii i sezonnoe razmeshchenie dikikh severnykh olenei taimyrskoi populiatsii. In E. E. Syroechkovski (ed.) Dikii Severnyi Olen' v SSSR, pp. 80–88, Central Board of Game Management and Nature Reserves of the RSFSR: Moscow.

Gelman, A. and J. Hill (2007) Data Analysis using Regression and Multilevel/Hierarchical Models, Cambridge University Press: Cambridge.

Georgiadis, N.J. and McNaughton, S.J. (1990) Elemental and fibre contents of savanna grasses: variation with grazing, soil type, season and species. Journal of Applied Ecology, 27, 623–34.

Gerber, L.R., Heppell, S.S., Ballantyne, F., and Sala, E. (2005) The role of dispersal and demography in determining the efficacy of marine reserves. Canadian Journal of Fisheries and Aquatic Sciences, 62, 863–71.

Gereta, E., Mwangomo, E., and Wolanski, E. (2009) Ecohydrology as a tool for the survival of the threatened Serengeti ecosystem. Ecohydrology and hydrobiology, 9, 115–24.

Gerlier, M. and Roche, P. (1998) A radio telemetry study of the migration of Atlantic salmon (Salmo salar L.) and sea trout (Salmo trutta trutta L.) in the upper Rhine. Hydrobiologia, 371/372, 283–93.

Gibo, D.L. and Pallett, M.J. (1979) Soaring flight of monarch butterflies, Danaus plexippus (Lepidoptera, Danaidae), during the late summer migration in Southern Ontario. Canadian Journal of Zoology, 57, 1393–1401.

Gilbert, M., Slingenbergh, J., and Xiao, X. (2008) Climate change and avian influenza. Revue Scientifique Et Technique-Office International Des Epizooties, 27, 459–66.

Gill, R.E., Tibbitts, T.L., Douglas, D.C., Handel, C.M., Mulcahy, D.M., Gottschalck, J.C., Warnock, N., McCaffery, B.J., Battley, P.F., and Piersma T. (2009) Extreme endurance flights by landbirds crossing the Pacific Ocean: ecological corridor rather than barrier? Proceedings of the Royal Society B-Biological Sciences, 276, 447–58.

Giller, P.S. (2005) River restoration: seeking ecological standards. Editor's introduction. Journal of Applied Ecology, 42, 201–207.

Gilly, W.F., Markaida, U., Baxter, C.H., et al. (2006) Vertical and horizontal migrations by the jumbo squid Dosidicus gigas revealed by electronic tagging. Marine Ecology-Progress Series, 324, 1–17.

Gilyazov, A. and Sparks, T.H. (2002) Change in the timing of migration of common birds at the Lapland Nature Reserve (Kola Peninsula, Russia) during 1931–1999. Avian Ecology and Behaviour, 8, 35–47.

Gloyne, C.C. and Clevenger, A.P. (2001) Cougar (Puma concolor) use of wildlife crossing structures on the Trans-Canada highway in Banff National Park, Alberta. Wildlife Biology, 7, 2, 117–24.

Godley, B.J., Broderick, A.C., and Hays, G.C. (2001) Nesting of green turtles (Chelonia mydas) at Ascension Island, South Atlantic. Biological Conservation, 97, 151–58.

Godley, B.J., Lima, E.H.S.M., Åkesson, S., Broderick, A. C., Glen, F., Godfrey, M.H., Luschi, P. and Hays, G.C. (2003) Movement patterns of green turtles in Brazilian coastal waters described by satellite tracking and flipper tagging. Marine Ecology Progress Series, 253, 279–88.

Goldstein, D.L. and Pinshow, B. (2006) Taking physiology to the field: using physiological approaches to answer questions about animals in their environments. Physiological and Biochemical Zoology, 79, 237–41.

Goodchild, M.F., Friedl, M.A., and Case, T.J. (eds) (2001) Spatial Uncertainty in Ecology, Springer-Verlag: New-York,

Goodhue, R.E. and McCarthy, N. (1999) Fuzzy access: modelling grazing rights in sub-Saharan Africa. In N. McCarthy, D. Swallow, M. Kirk and P. Hazell (eds) Property Rights, Risk, and Livestock Development in Africa, pp. 191–210, International Food Policy Research Institute: Washington, DC.

Gordo, O. (2007) Why are bird migration dates shifting? A review of weather and climate effects on avian migratory phenology. Climate Research, 35, 37–58.

Granberg, L., Soini K., and Kantanen. J. (eds) (2009) Sakha Ynaga—Cattle of the Yakuts. Annales Academiae Scientiarum Fennicae Humaniora 355: Helsinki. Submitted.

Gray, P. (2006) 'The Last Kulak' and other stories of post-privatization life in Chukotka's tundra, in People and

Reindeer on the Move. In F. Stammler, and H. Beach (eds) Special Issue of the journal Nomadic Peoples, 10, 2, pp 50–67, Berghahn Publishers: Oxford.

Green, A.A. (1988) Elephants of the Pendjari-Singou-Mekrou region, West Africa. Mammalia, 52, 4, 557–65.

Green, J.A., Halsey, L.G., Wilson, R.P., and Frappell, P.B. (2009) Estimating energy expenditure of animals using the accelerometry technique: activity, inactivity and comparison with the heart-rate technique. Journal of Experimental Biology, 212, 471–82.

Green, M. and Alerstam, T. (2000) Flight speeds and climb rates of Brent geese: mass-dependent differences between spring and autumn migration. Journal of Avian Biology, 31, 215–25.

Green, M., Alerstam, T., Clausen, P., Drent, R., and Ebbinge, R.S. (2002) Dark-bellied Brent geese Branta bernicla bernicla, as recorded by satellite telemetry, do not minimize flight distance during spring migration. Ibis, 144, 106–21.

Green, M., Alerstam, T., Gudmundsson, G.A., Hedenstrom, A., and Piersma, T. (2004) Do Arctic waders use adaptive wind drift? Journal of Avian Biology, 35, 305–15.

Greenwood, P.J. (1980) Mating systems, philopatry, and dispersal in birds and mammals. Animal Behavior, 28, 1140–62.

Griffioen, P.A. and Clarke, M.F. (2002) Large-scale bird-movement patterns evident in eastern Australian atlas data. Emu, 102, 97–125.

Griffith, B., Scott, J.M., Carpenter, J.W., and Reed C. (1989) Translocation as a species conservation tool: status and strategy. Science, 245, 477–82.

Grimm, V.,E. Revilla, U. Berger, F. Jeltsch, W.M. Mooij, S. F. Railsback, H.H. Thulke, J. Weiner, T. Wiegand, and D.L. DeAngelis (2005) Pattern-oriented modeling of agent-based complex systems: lessons from ecology. Science, 310, 987–91.

Grinnel, J. (1931) Some angles in the problem of bird migration. Auk, 48, 22–32.

Griswold, C.K., Taylor, C.M., and Norris D.R. (2010) The evolution of migration in a seasonal environment. Proceedings of the Royal Society of London (B), doi:10.1.1098/rspb.2010.0550.

Gross, M.R. (1987) Evolution of diadromy in fishes. American Fisheries Society Symposium, 1, 14–25.

Grouzis, M. (1988) Structure, productivité, et dynamique des systèmes écologiques Sahéliens (Mare d'Oursi, Burkina Faso), Collection Études et Thèses, Éditions de l'ORSTOM: Paris.

Grunbaum, D. (1998a) Schooling as a strategy for taxis in a noisy environment. Evolutionary Ecology, 12, 503–22.

Grunbaum, D. (1998b) Using spatially explicit models to characterize foraging performance in heterogeneous landscapes. American Naturalist, 151, 97–115.

Gschweng, M., Kalko, E.K.V., Querner, U., Fiedler, W., and Berthold, P. (2008) All across Africa: highly individual migration routes of Eleonora's falcon. Proceedings of the Royal Society B-Biological Sciences, 275, 2887–96.

Gudmundsson, G.A. and Alerstam, T. (1998) Optimal map projections for analysing long-distance migration routes. J Avian Biol., 29, 597–605.

Gudmundsson, G.A., Linddström, Å., and Alerstam, T. (1991) Optimal fat loads and long-distance flights by migrating knots Calidris canutus, sanderling C. alba and turnstones Arenaria interpres. Ibis, 133, 140–52.

Guillemain, M., Bertout, J.M., Christensen, T.K., Poysa, H., Vaananen, V.M., Triplet, P., Schricke V., et al. (2010) How many juvenile Teal Anas crecca reach the wintering grounds? Flyway-scale survival rate inferred from wing age-ratios. Journal of Ornithology, 151, 51–60.

Gulliver, P.H. (1975) Nomadic movements: causes and implications. In T. Monod (ed.) Pastoralism in Tropical Africa, pp. 369–86, OUP: London.

Gunderson, H., Andreassen, H.P., and Storaas, T. (2004) Supplemental feeding of migratory moose Alces alces: forest damage at two spatial scales. Wildlife Biology, 10, 213–23.

Gunn, A. (1998) Caribou and muskox harvesting in the Northwest Territories. In E.J. Milner-Gulland and R. Mace (eds) Conservation of Biological Resources, pp. 314–30, Blackwell Science: Oxford.

Gustafson, T., Lindkvist, B., Gotborn, L., and Gyllin, R. (1977) Altitudes and flight times for swifts Apus apus L. Ornis Scandinavica, 8, 87–95.

Gustafson, R.G., Waples, R.S., Myers, J.M., Weitkamp, L.A., Bryant, G.J., Johnson, O.W., and Hard, J.J. (2007) Pacific salmon extinctions: quantifying lost and remaining diversity. Conservation Biology, 21, 4, 1009–20.

Gwinner, E. (1969) Untersuchungen zur Jahresperiodik von Laubsängern. Die Entwicklung des Gefieders, des Gewichts und der Zugunruhe bei Jungvögeln der Arten Phylloscopus bonelli, Ph. sibilatrix und Ph. collybita. Journal of Ornithology, 110, 1–21.

Gwinner, E. (1977) Circannual rhythms in bird migration. Annual Review of Ecology and Systematics, 8, 381–405.

Gwinner, E. (1990) Bird Migration: Physiology and Ecophysiology. Springer-Verlag: Berlin.

Gwinner, E. (1996) Circadian and circannual programmes in avian migration. The Journal of Experimental Biology, 199, 39–48.

Gwinner, E. and Wiltschko, W. (1978) Endogenously controlled changes in migratory direction of the Garden Warbler, Sylvia borin. Journal of Comparative Physiology, 125, 267–73.

Gwinner, E. and Wiltschko, W. (1980) Circannual changes in migratory orientation of the Garden Warbler, *Sylvia borin*. Behavioral Ecology and Sociobiology, 7, 73–78.

Habeck, J. O. (2005) What it means to be a herdsman: The Practice and Image of Reindeer Husbandry among the Komi of Northern Russia, Vol. 5, Halle Studies in the Anthropology of Eurasia, Lit: Münster

Habou, A., and Danguioua, A. (1991) Transfert du capital – Bétail au Niger (des pasteurs aux autres catégories socio-professionnelles), Secrétariat Permanent du Comite National du Code Rural: Niamey, Niger.

Hahn, S., Bauer, S., and Liechti F. (2009) The natural link between Europe and Africa—2.1 billion birds on migration. Oikos, 118, 624–26.

Haining, R. (2003) Spatial Data Analysis: Theory and Practice, Cambridge University Press: Cambridge.

Hamelin, F. and Lewis, M. (2010) A differential game theoretical analysis of mechanistic models for territoriality, Journal of Mathematical Biology, 61, 665–694.

Hamilton, W.D. (1971) Geometry for the selfish herd. Journal of Theoretical Ecology, 31, 295–311.

Hancock, P.A. and Hutchinson, M.F. (2006) Spatial Interpolation of large climate data sets using thin plate smoothing splines. Environmental Modelling and Software, 21, 1684–94.

Hancock, P.A. and Milner-Gulland, E. J. (2006) Optimal movement strategies for social foragers in unpredictable environments. Ecology, 87, 2094–2102.

Hannesson, R. (2004) Sharing the Herring: Fish Migrations, Strategic Advantage, and Climate Change, SNF Working Paper No. 27/2004. [Online], Available: http://bora.nhh.no:8080/bitstream/2330/382/1/A2704.pdf [January 2009].

Hanrahan T.P., Dauble, D.D., and Geist, D.R. (2004) An estimate of chinook salmon (*Oncorhynchus tshawytscha*) spawning habitat and redd capacity upstream of a migration barrier in the upper Columbia River. Canadian Journal of Fisheries and Aquatic Sciences, 61, 1, 23–33.

Hanski, I. (1998) Metapopulation dynamics. Nature, 396, 41–49.

Hansson, L. A. and Hylander, S. (2009) Size-structured risk assessments govern *Daphnia* migration. Proceedings of the Royal Society B-Biological Sciences, 276, 331–36.

Haro, A., Richkus, W., Whalen, K., Hoar, A., Busch, W.-D., Lary, S., Brush, T., and Dixon, D. (2000) Population decline of the American eel: implications for research and management. Fisheries, 25, 9, 7–16.

Harrington, R., Owen-Smith, N., Viljoen, P.C., Biggs, H.C., Mason, D.R., and Funston, P. (1999) Establishing the causes of the roan antelope decline in the Kruger National Park, South Africa. Biological Conservation, 90, 69–78.

Harris, G., Thirgood, S., Grant, J., Hopcraft, C., Cromsigt, J.P.G.M., and Berger, J. (2009) Global decline in aggregated migrations of large terrestrial mammals. Endangered Species Research, 7, 55–76.

Harvey, P.H. and Pagel, M.D. (1991) The comparative method in evolutionary biology, Oxford University Press: Oxford.

Hary, I., Schwartgz, H-J., Peilert, V.H.C., and Mosler C. (1996) Land degradation in African pastoral systems and the destocking controversy. Ecological Modelling, 89, 227–33.

Hasegawa, H. and DeGange, A.R. (1982) The short-tailed albatross, *Diomedea albatrus*, its status, distribution and natural history. American Birds, 36, 806–14.

Haslam, C. (2008) Nobody to watch wildebeest migration in Kenya. The Sunday Times, 8 June [Online], Available: http://www.timesonline.co.uk/tol/travel/news/article4076148.ece. [November 2008].

Hastie, T., Tibshirani, R., and Friedman, J. (2001) The elements of statistical learning: data mining, inference, and prediction: Springer Series in Statistics, Springer: New York.

Hawkes, L.A., Broderick, A.C., Coyne, M.S., Godfrey, M.H., and Godley, B.J. (2007) Only some like it hot— quantifying the environmental niche of the loggerhead sea turtle. Diversity and Distributions, 13, 447–57.

Hawkes, L.A., Broderick, A.C., Coyne, M.S., Godfrey, M.S., Lopez-Jurado, L.F., Lopez-Suarez, P., Merino, S.E., Varo-Cruz, N., and Godley, B.J. (2006) Phenotypically linked dichotomy in sea turtle foraging requires multiple conservation approaches. Current Biology, 16, 990–95.

Hay, R.K.M. and Heide, O.M. (1984) The response of high-latitude Norwegian grass cultivars to long photoperiods and cool temperatures. In H. Riley and A.O Skjelvag (eds) The Impact of Climate on Grass Production and Quality, Proceedings of the 10th General Meeting of the European Grassland Federation, As, Norway, pp. 46–50.

Haydon, D.T., Morales, J.M., Yott, A., Jenkins, D.A., Rosatte, R., and Fryxell, J. (2008) Socially-informed random walks: Incorporating group dynamics into models of population spread and growth. Proceedings of the Royal Society B-Biological Sciences, 275, 1101–1109.

Hays, G.C. (2008) Sea turtles: A review of some key recent discoveries and remaining questions. Journal of Experimental Marine Biology and Ecology, 356, 1–7.

Hays, G.C., Houghton, J.D.R., and Myers, A.E. (2004) Endanqered species—Pan-Atlantic leatherback turtle movements. Nature, 429, 522.

Healey, M. C. and Groot, C. (1987) Marine migration and orientation of ocean-type Chinook and sockeye salmon. American Fish Society Symposium, 1, 298–312.

Heath, M.R., Boyle, P.R., Gislason, A., Gurney, W.S.C., Hay, S.J., Head, E.J.H., Holmes, S., Ingvarsdottir, A., Jonasdottir, S.H., Lindeque, P., Pollard, R.T., Rasmussen, J., Richards, K., Richardson, K., Smerdon, G. and Speirs, D. (2004) Comparative ecology of over-wintering Calanus finmarchicus in the northern North Atlantic, and implications for life-cycle patterns. ICES Journal of Marine Science, 61, 698–708.

Hebblewhite, M. and Merrill, E. (2007) Multiscale wolf predation risk for elk: does migration reduce risk? Oecologia, 152, 377–87.

Hebblewhite, W. and Merrill, E.H. (2009) Trade-offs between predation risk and forage differ between migrant strategies in a migratory ungulate. Ecology, 90, 3445–54.

Hebblewhite, M., Merrill, E. and McDermid, G. (2008) A multi-scale test of the forage maturation hypothesis in a partially migratory ungulate population. Ecological Monographs, 78, 141–66.

Hebblewhite, M., White, C.A., Nietvelt, C.G., McKenzie, J.A., Hurd, T.E., Fryxell, J.M., Bayley, S.E. and Paquet, P. (2005) Human activity mediates a trophic cascade caused by wolves. Ecology, 86, 2135–44.

Hedenström, A. (1992) Flight performance in relation to fuel load in birds. Journal of Theoretical Biology, 158, 535–37.

Hedenström, A. (1993) Migration by soaring or flapping flight in birds: the relative importance of energy cost and speed. Philosophical Transactions of the Royal Society of London B Biological Sciences, 342, 353–61.

Hedenström, A. (2001) Twenty-three testable predictions about bird flight. In P. Berthold, E. Gwinner, and E. Sonnenschein (eds) Avian Migration, pp. 563–82, Springer-Verlag: Berlin.

Hedenström, A. (2003a) Optimal migration strategies in animals that run: a range equation and its consequences. Animal Behaviour, 66, 631–36.

Hedenström, A. (2003b) Scaling migration speed in animals that run, swim and fly. J. Zool., 259, 155–60.

Hedenström, A. (2006) Scaling of migration and the annual cycle of birds. Ardea, 94, 399–408.

Hedenström, A. (2008) Adaptations to migration in birds: behavioural strategies, morphology and scaling effects. Philosophical Transactions of the Royal Society B, 363, 287–99. doi:10.1098/rstb.2007.2140.

Hedenström, A. (2009) Optimal migration strategies in bats. Journal of Mammalogy, 90, 1298–1309.

Hedenström, A. and Alerstam, T. (1992) Climbing performance of migrating birds as a basis for estimating limits for fuel-carrying capacity and muscle work. Journal of Experimental Biology, 164, 19–38.

Hedenström, A. and Alerstam, T. (1996) Skylark optimal flight speeds for flying nowhere and somewhere. Behavioral Ecology, 7, 121–26.

Hedenström, A. and Alerstam, T. (1997) Optimum fuel loads in migratory birds: distinguishing between time and energy minimization. Journal of Theoretical Biology, 189, 227–34.

Hedenström, A. and Pettersson, J. (1986) Differences in fat deposits and wing pointedness between male and female willow warblers caught on spring migration at Ottenby, SE Sweden. Ornis Scandinavica, 17, 182–85.

Hedenström, A. and Pettersson, J. (1987) Migration routes and wintering areas of willow warblers Phylloscopus trochilus (L) ringed in Fennoscandia. Ornis Fennica, 64, 137–43.

Hedenström, A., Johansson, L.C., and Spedding, G.R. (2009) Bird or bat: comparing airframe design and flight performance. Bioinspiration and Biomimetics, 4, 1.

Hedenström, A., Barta, Z., Helm, B., Houston, A.I., McNamara, J.M., and Jonzén, N. (2007) Migration speed and scheduling of annual events by migrating birds in relation to climate change. Climate Research, 35, 79–91.

Heino, M. and Hanski, I. (2001) Evolution of migration rate in a spatially realistic metapopulation model. American Naturalist, 157, 495–511.

Helbig, A. J. (1991) Inheritance of migratory direction in a bird species—A cross-breeding experiment with SE-migrating and SW-migrating blackcaps (Sylvia atricapilla). Behavioral Ecology and Sociobiology, 28, 9–12.

Helbig, A.J. (2003) Evolution of migration: a phylogenetic and biogeographic perspective. In P Berthold, E Gwinner and E Sonnenschein (eds) Avian Migration, pp. 3–20. Springer-Verlag: Heidelberg.

Helfield, J.M. and Naiman R.J. (2001) Effects of salmon-derived nitrogen on riparian forest growth and implications for stream productivity. Ecology, 82, 2403–2409.

Helm, B., Gwinner, E., and Trost, L. (2005) Flexible seasonal timing and migratory behavior: results from Stonechat breeding programs. Annals of the New York Academy of Sciences, 1046, 216–27.

Helms, C.W. (1963) The annual cycle and zugunruhe in birds. Proceedings of the International Ornithological Congress, 13, 925–39.

Hiernaux, P. (1998) Effects of grazing on plant species composition and spatial distribution in rangelands of the Sahel. Plant Ecology, 138, 2, 191–202.

Hiernaux, P. and L. Diarra. (1984) Savanna burning, a controversial technique for rangeland management in the Niger flood plains of central Mali. In J.C. Tothill and J.J. Mott (eds) Ecology and Management of the World's Savannas, pp. 238–43, Australian Academy of Sciences: Canberra.

Hiernaux, P. and Turner, M.D. (1996) The effect of the timing and frequency of clipping on nutrient uptake and production of Sahelian annual rangelands. Journal of Applied Ecology, 33, 387–99.

Hilborn, R. and Mangel, M. (1997) The Ecological Detective: Confronting Models with Data. Monographs in Population Biology, Princeton University Press: Princeton.

Hind, A., Gurney, W.S.C., Heath, M., and Bryant, A.D. (2000) Overwintering strategies in Calanus finmarchicus. Marine Ecology Progress Series, 193, 95–107.

Hirche, H.J. (1996) Diapause in the marine copepod, Calanus finmarchicus—a review. Ophelia, 44, 129–43.

Hjeljord, O. (2001) Dispersal and migration in northern forest deer—are there unifying concepts? Alces, 37, 2, 353–70.

Hoar, W. S. (1988) The physiology of smolting salmonids. In W. S. Hoar and D. J. Randall (eds) Fish Physiology, pp. 275–343, Academic Press: London.

Hobbs, N.T. and Swift, D.M. (1988) Grazing in herds: when are nutritional benefits realized? American Naturalist, 131, 760–64.

Hobson, K.A. and D.R. Norris. (2008) Animal migration: a context for using new techniques and approaches. In K.A. Hobson and L. I. Wassenaar (eds) Tracking Animal Migration with Stable Isotopes, pp. 1–19 Terrestrial Ecology Series, Elsevier: Amsterdam.

Hobson, K.A. and Wassenaar, L.I. (2008), Tracking Animal Migration with Stable Isotopes: Terrestrial Ecology Series, Elsevier.

Holdo, R.M. (2007) Elephants, fire, and frost can determine community structure and composition in Kalahari woodlands. Ecological Applications, 17, 558–68.

Holdo, R.M., Holt, R.D., and Fryxell, J.M. (2009a) Grazers, browsers, and fire influence the extent and spatial pattern of tree cover in the Serengeti. Ecological Applications, 19, 95–109.

Holdo, R.M., Holt, R.D., and Fryxell, J.M. (2009b) Opposing rainfall and nutrient gradients best explain the wildebeest migration in the Serengeti. The American Naturalist, 173, 431–45.

Holdo, R.M., Holt, R. Coughenour, D.M.B., and Ritchie, M.E. (2007) Plant productivity and soil nitrogen as a function of grazing, migration and fire in an African savanna. Journal of Ecology, 95, 115–28.

Holdo, R.M., Sinclair, A.R.E., Metzger, K.L., Bolker, B.M., Dobson, A.P., Ritchie, M.E., and Holt, R.D. (2009c) A disease-mediated trophic cascade in the Serengeti: implications for ecosystem structure and C stocks. PLOS Biology, 7, e1000210.

Holland, R.A., Borissov, I., and Siemers, B.M. (2010) A nocturnal mammal, the greater mouse-eared bat, calibrates a magnetic compass by the sun. Proceedings of the National Academy of Sciences, 107, 15, 6941–45.

Holland, R.A., Wikelski, M., and Wilcove, D.S. (2006b) How and why do insects migrate? Science, 313, 794–96.

Holland, R. A., Wikelski, M., Kummeth, F., and Bosque, C. (2009) The secret life of oilbirds: new insights into the movement ecology of a unique avian frugivore. PLoS ONE, 4, 12, e8264. doi:10.1371/journal.pone.0008264.

Holland, R.A., Thorup, K., Vonhof, M.J., Cochran, W.W., and Wikelski, M. (2006a) Navigation—bat orientation using Earth's magnetic field. Nature, 444, 702.

Holst, J.C., Rottingen, I., and Melle, W. (2004) The herring. In H.R. Skjoldal (ed.) The Norwegian Sea Ecosystem, pp. 203–26, Tapir Academic Press: Trondheim.

Holt, R.D. (1997) On the evolutionary stability of sink populations. Evolutionary Ecology, 11, 723–31.

Holt, R.D. (2004) Implications of system openness for local community structure and ecosystem function. In G. E. Polis, M. E. Power, and G. R. Huxel (eds) Food Webs at the Landscape Level, pp. 96–114, University of Chicago Press: Chicago.

Holt, R.D. (2008) Theoretical perspectives on resource pulses. Ecology, 89, 671–81.

Holt, R.D. and McPeek, M.A. (1996) Chaotic population dynamics favors the evolution of dispersal. American Naturalist, 148, 709–18.

Holyoak, M., Leibold, M.A., Mouquet, N.M., Holt, R.D., and Hoopes M.F. (2005) Metacommunities: a framework for large-scale community ecology. In M. Holyoak, M.A. Leibold, and R.D. Holt (eds) Metacommunities: Spatial Dynamics and Ecological Communities, pp. 1–32 University of Chicago Press: Chicago.

Horne, J.S., Garton, E.O., Krone, S.M., and Lewis, J.S. (2007) Analyzing animal movements using Brownian bridges. Ecology, 88, 2354–63.

Houston, A.I. (1998) Models of optimal avian migration: state, time and predation. Journal of Avian Biology, 29, 395–404.

Houston, A.I. and McNamara, J.M. (1999) Models of Adaptive Behaviour: An Approach Based on State, Cambridge University Press: Cambridge.

Houston, A.I., Stephens, P.A., Boyd, I.L., Harding, K.C., and McNamara, J.M. (2007) Capital or income breeding? A theoretical model of female reproductive strategies. Behavioral Ecology, 18, 241–50.

Hume, I.D. and Biebach, H. (1996) Digestive tract function in the long-distance migratory garden warbler, Sylvia borin. Journal of Comparative Physiology B, 166, 388–95.

Humphrey, C. and Sneath, D. (1999) The End of Nomadism? Society, State and the Environment in Inner Asia, Duke University Press: Durham, NC.

Hunsaker, C.T., Goodchild, M.F., Friedl, M.A., and Case, T.J. (2001) Spatial Uncertainty in Ecology, Springer Verlag: New-York.

Hunter, E., Metcalfe, J.D., O'Brien, C.M., Arnold, G.P., and Reynolds, J.D. (2004) Vertical activity patterns of free-swimming adult plaice in the southern North Sea. Marine Ecology-Progress Series, 279, 261–73.

Hurford, A. (2009) GPS measurement error gives rise to spurious 180° turning angles and strong directional biases in animal movement data. PLoS ONE, 4, e5632.

Hurry, G.D., Hayashi, M., and Maguire, J.J. (2008) Report of the independent review of ICCAT, ICCAT document PLE-106/2008. [Online], Available: http://www.iccat.int/com2008/ENG/PLE-106.pdf.

Huse, G. and Giske, J. (1998) Ecology in Mare Pentium: an individual-based spatio-temporal model for fish with adapted behaviour. Fisheries Research, 37, 163–78.

Huse, G., Railsback, S., and Fernö, A. (2002) Modelling changes in migration pattern of herring: collective behaviour and numerical domination. Journal of Fish Biology, 60, 571–82.

Huse, G., Strand, E., and Giske, J. (1999) Implementing behaviour in individual-based models using neural networks and genetic algorithms. Evolutionary Ecology, 13, 469–83.

Hutterer, R., Ivanova, T., Meyer-Cords, C., and Rodrigues, L. (2005) Bat migrations in Europe: a review of banding data and literature, Federal Agency for Nature Conservation: Bonn, Germany.

Hvenegaard, G.T., Butler, J.R., and Krystofiak, D.K. (1989) Economic values of bird watching at Point Pelee National Park, Canada. Wildlife Society Bulletin, 17, 4, 526–31.

ICES (The International Council for the Exploration of the Sea) (2007) Report of the Working Group on Northern Pelagic and Blue Whiting Fisheries, 27 August–1 September 2007, Vigo, Spain, [Online], Available: http://www.ices.dk/reports/ACOM/2007/WGNPBW/ACFM2907.pdf. [February 2009].

ICES (The International Council for the Exploration of the Sea) (2008a) Norwegian spring-spawning herring. Report of the Advisory Committee on Fisheries Management, [Online], Available: http://www.ices.dk/committe/acom/comwork/report/2008/2008/her-noss.pdf. [December 2008].

ICES (The International Council for the Exploration of the Sea) (2008b) Report of the Working Group on Widely Distributed Stocks (WGWIDE), 2–11 September 2008, ICES CM 2008/ACOM, 13.

Illius, A. and Gordon, I. (1992) Modeling the nutritional ecology of ungulate herbivores: evolution of body size and competitive interactions. Oecologia, 89, 428–34.

Illius A. and O'Connor, T. (1999) On the relevance of non-equilibrium concepts to semi-arid grazing systems. Ecological Applications, 8, 798–813.

Inglis, J.M. (1976) Wet season movements of individual wildebeests of the Serengeti migratory herd. East African Wildlife Journal, 14, 17–34.

Ingold, T. (1980) Hunters, Pastoralists and Ranchers, Cambridge University Press: Cambridge.

Inouye, D.W., Barr, B., Armitage, K.B., and Inouye, B.D. (2000) Climate change is affecting altitudinal migrants and hibernating species. Proceedings of the National Academy of the Sciences, 97, 4, 1630–33.

IOTC (Indian Ocean Tuna Commission) (2008) Report of the Fourth Session of the IOTC Working Party on Ecosystems and Bycatch, 20–22 October 2008, Bangkok, Thailand.

IPCC (Intergovernmental Panel on Climate Change) (2001) Climate Change 2001: The Scientific Basis. Contribution of Working Group I to the Third Assessment Report of the IPCC, J.T. Houghton, Y. Ding, D.J. Griggs, M. Noguer, P.J. van der Linden, X. Dai, K. Maskell and C.A. Johnson (eds) Cambridge University Press: Cambridge.

Ito, T., Miura, N., Lhagvasuren, B., Enkhbileg, D., Takatsuki, S., Tsunekawa, A., and Jiang, Z. (2005) Preliminary evidence of a barrier effect of a railroad on the migration of Mongolian gazelles. Conservation Biology, 19, 945–48.

Ito, T.Y., Miura, N., Lhagvasuren, B., Enkhbileg, D., Takatsuki, S., Tsunekawa, A., and Jiang, Z. (2006) Satellite tracking of Mongolian gazelles (*Procapra gutturosa*) and habitat shifts in their seasonal ranges. Journal of Zoology, 269, 291–98.

Izhaki, I. and Safriel, U.N. (1985) Why do fleshy-fruit plants of the Mediterranean scrub intercept fall- but not spring-passage of seed-dispersing migratory birds? Oecologia, 67, 40–43.

Jaeger, M.E., Zale, A.V., McMahon, T.E., and Schmitz, B.J. (2005) Seasonal movements, habitat use, aggregation, exploitation, and entrainment of saugers in the lower Yellowstone River: an empirical assessment of factors affecting population recovery. North American Journal of Fisheries Management, 25, 1550–68.

Jamemo, A. (2008) Seasonal migration of male red deer (*Cervus elaphus*) in southern Sweden and consequences for management. European Journal of Wildlife Research, 54, 327–33.

Jefferies, R.L., Jano, A.P., and Abraham, K.F. (2006) A biotic agent promotes large-scale catastrophic change in the coastal marshes of Hudson Bay. Journal of Ecology, 94, 234–42.

Jenni, L. and Jenni-Eiermann, S. (1992) Metabolic patterns of feeding, overnight fasted and flying night migrants during autumn migration. Ornis Scandinavica, 23, 251–59.

Jenni, L. and Jenni-Eiermann, S. (1998) Fuel supply and metabolic constraints in migrating birds. Journal of Avian Biology, 29, 521–28.

Jenni, L. and Kéry, M. (2003) Timing of autumn bird migration under climate change: advances in long-distance migrants, delays in short-distance migrants. Proceedings of the Royal Society of London Series B, 270, 1523, 1467–71.

Jenni-Eiermann, S., Jenni, L., Kvist, A., Lindström, A., Piersma, T., and Visser, G.H. (2002) Fuel use and metabolic response to endurance exercise: a wind tunnel study of a long-distance migrant shorebird. Journal of Experimental Biology, 205, 2453–60.

Jiang, Z., Takatsuki, S., Li, J., Wang, W.,Gao, Z., and Ma, J. (2002) Seasonal variation in foods and digestion of Mongolian gazelles in China. Journal of Wildlife Management, 66, 1, 40–47.

Johnson, C.G. (1969) Migration and Dispersal of Insects by Flight, Methuen: London.

Johnson, C.J., Parker, K.L., Heard, D.C. and Gillingham, M.P. (2002) Movement parameters of ungulates and scale-specific responses to the environment. Journal of Animal Ecology, 71, 225–35.

Jones, J., Barg, J.J., Sillett, T.S., Veit, M.L., and Robertson, R.J. (2004) Minimum estimates of survival and population growth for cerulean warblers (Dendroica cerulea) breeding in Ontario, Canada. Auk, 121, 15–22.

Jonsen, I.D., Flenming, J.M., and Myers, R.A. (2005) Robust state-space modeling of animal movement data. Ecology, 86, 2874–80.

Jonsen, I.D., Myers, R.A., and James, M.C. (2006) Robust hierarchical state-space models reveal diel variation in travel rates of migrating leatherback turtles. Journal of Animal Ecology, 75, 1046–57.

Jonsson, B. and Jonsson, N. (1993) Partial migration: niche shift versus sexual maturation in fishes. Reviews in Fish Biology and Fisheries, 3, 4, 348–65.

Jonsson, B. and N. Jonsson (2003) Migratory Atlantic salmon as vectors for the transfer of energy and nutrients between freshwater and marine environments. Freshwater Biology, 48, 21–27.

Jonzén, N., Hendenström, A., and Lundberg, P. (2007a) Climate change and the optimal arrival of migratory birds. Proceedings of the Royal Society B, Biological Sciences, 274, 269–74.

Jonzén, N., Ergon, T., Lindén, A., and Stenseth, N.C. (eds) (2007b) Bird migration and climate. Climate Research Special, 17, 1–180.

Jonzén, N., Lindén, A., Ergon, T., Knudsen, E., Vik, J.O., Rubolini, D., Piacentini, D., Brinch, C., Spina, F., Karlsson, L., Stervander, M., Andersson, A., Waldenström, J., Lehikoinen, A., Edvardsen, E., Solvang,

R., and Stenseth, N.C. (2006) Rapid advance of spring arrival dates in long-distance migratory birds. Science, 312, 1959–61.

Jordano, P. (1987) Frugivory, external morphology and digestive system in Mediterranean sylviid warblers Sylvia spp. Ibis, 129, 175–89.

Jorgensen, S.J., Reeb, C.A., Chapple, T.K., Anderson, S., Perle, C., van Sommeran, S.R., Fritz-Cope, C., Brown, A.C.; Klimley, A.P., and Block, B.A. (2010) Philopatry and migration of Pacific white sharks. Proceedings of the Royal Society B-Biological Sciences, 277, 679–88.

Joseph, I., Lessa, E.P., and Christidis, L. (1999) Phylogeny and biogeography in the evolution of migration: shorebirds of the Charadrius complex. Journal of Biogeography, 26, 329–42.

Julian, F. and Beeson, M. (1998) Estimates of marine mammal, turtle, and seabird mortality for two California gillnet fisheries: 1990–1995. Fisheries Bulletin, 96, 271–84.

Kaartvedt, S. (1996) Habitat preferences during overwintering and timing of seasonal vertical migration of Calanus finmarchicus. Ophelia, 44, 145–56.

Kaimal, B., Johnson, R., and Hannigan, R. (2009) Distinguishing breeding populations of mallards (Anas platyrhynchos) using trace elements. Journal of Geochemical Exploration, 102, 44–48.

Kaitala, A., Kaitala, V., and Lundberg P. (1993) A theory of partial migration. American Naturalist, 142, 59–81.

Kamp, J., Sheldon, R.D., Koshkin, M.A., Donald, P.F., and Biedermann, R. (2009) Post-Soviet steppe management causes pronounced synanthropy in the globally threatened Sociable Lapwing Vanellus gregarius. Ibis, 151, 452–63.

Kaplan, I.C. (2005) A risk assessment for Pacific leatherback turtles (Dermochelys coriacea). Canadian Journal of Fisheries and Aquatic Science, 62, 1710–19.

Karasov, W.H. (1990) Digestion in birds: chemical and physiological determinants and ecological implications. Studies in Avian Biology, 13, 391–415.

Karasov, W.H. and Pinshow, B. (2000) Test for physiological limitation to nutrient assimilation in a long-distance passerine migrant at a springtime stopover site. Physiological and Biochemical Zoology, 73, 335–43.

Kareiva, P., Marvier, M., and McClure, M. (2000) Recovery and management options for spring/summer chinook salmon in the Columbia River Basin. Science, 290, 5493, 977–79.

Keast, A. (1968) Seasonal movements in the Australian honeyeaters (Meliphagidae) and their ecological significance. Emu, 67, 159–209.

Keeton, W. T. (1980) Avian orientation and navigation: new developments in an old mystery. pp. 137–58, Acta

XVII Congr Intern Ornithol I, Deutsche Ornithologen-Gesellschaft: Berlin.

Kelly, J.F., DeLay, L.S., and Finch, D.M. (2002) Density-dependent mass gain by Wilson's Warblers during stopover. Auk, 119, 210–13.

Kennedy, J.S. (1961) A turning point in the study of insect migration. Nature, 189, 785–91.

Kennedy, J.S. (1985) Migration, behavioural and ecological. In M.A. Rankin (ed.) Migration: Mechanisms and Adaptive Significance, pp. 5–26, Contributions to Marine Science, Marine Science Institute, The University of Texas: Austin.

Kenward, R.E. (2001) A Manual for Wildlife Radiotracking, Academic Press: London.

Kerlinger, P. and Moore, F.R. (1989) Atmospheric structure and avian migration. Current Ornithology, 6, 109–42.

Kerven, C.K. (ed.) (2003) Prospects for Pastoralism in Kazakhstan and Turkmenistan: From State Farms to Private Flocks, Routledge Curzon: London.

Kerven, C.K. (2004) The influence of cold temperatures and snowstorms on rangelands and livestock in northern Asia. In S. Vetter (ed.) Rangelands at Equilibrium and Non-Equilibrium: Recent Developments in the Debate around Rangeland Ecology and Management, pp. 41–55, Programme for Land and Agrarian Studies, University of the Western Cape: Cape Town.

Kerven, C.K., Shanbaev, K, Alimaev, I., Smailov, A., and Smailov, K. (2008) Livestock mobility and degradation in Kazakhstan's semi-arid rangelands. In R. Behnke (ed.) The Socio-Economic Causes and Consequences of Desertification in Central Asia, pp. 113–40, Springer: Dordrecht, The Netherlands.

Kerven, C.K., Alimaev, I.I., Behnke, R., Davidson, G., Malmakov, N., Smailov, A., and Wright, I. (2006) Fragmenting pastoral mobility: changing grazing patterns in post-Soviet Kazakhstan. In D.J. Bedunah, E.D. McArthur, and M. Fernandez-Gimenez, (comps) Rangelands of Central Asia: Proceedings of the Conference on Transformations, Issues, and Future Challenges, 27 January 2004, Salt Lake City, UT. Proceeding RMRS-P-39. U.S. Department of Agriculture, Forest Service, Rocky Mountain Research Station: Fort Collins, CO.

Kéry, M., Madsen, J., and Lebreton J.-D. (2006) Survival of Svalbard pink-footed geese Anser brachyrhynchus in relation to winter climate, density and land-use. Journal of Animal Ecology, 75, 1172–81.

Kiffney, P.M., Pess, G.R., Anderson, J.H., Faulds, P., Burton K., and Riley, S.C. (2009) Changes in fish communities following recolonization of the Cedar River, WA, USA by Pacific Salmon after 103 years of local extirpation. River Research and Applications, 25, 438–52.

King, A.A. and Schaffer, W.M. (1999) The rainbow bridge: Hamiltonian limits and resonance in predator-prey dynamics. Journal of Mathematical Biology, 39, 439–69.

King, J.R. and Mewaldt, L.R. (1981) Variation of bodyweight in gambel white-crowned sparrows in winter and spring—Latitudinal and photoperiodic correlates. Auk, 98, 752–64.

Kirby, J.S., Stattersfield, A.J., Butchart, S.H.M., Evans, M.I., Grimmett, R.F.A., Jones, V.R., O'Sullivan, J., Tucker, G.M., and Newton, I. (2008) Key conservation issues for migratory land and waterbird species on the world's major flyways. Bird Conservation International, 18, S49–S73.

Kirilenko, A.P. and Solomon, A.M. (1998) Modeling dynamic vegetation response to rapid climate change using bioclimatic classification. Climatic Change, 38, 15–49.

Kitti, H., Gunslay, N., and Forbes, B.C. (2006) Defining the Quality of Reindeer Pastures: The perspective of Sami Reindeer Herders. In B.C. Forbes, M. Bölter, L. Müller-Wille, J. Hukkinen, F. Müller, N. Gunslay, and Y. Konstantinov (eds) Reindeer Management in Northernmost Europe, Vol. 184, Ecological Studies, pp. 141–65, Springer: Berlin, Heidelberg.

Klaassen, M., Kvist, A., and Lindström, A. (2000) Flight costs and fuel composition of a bird migrating in a wind tunnel. Condor, 102, 444–51.

Klaassen, M., Bauer, S., Madsen, J., and Possingham, H. (2008) Optimal management of a goose flyway: migrant management at minimum cost. Journal of Applied Ecology, 45, 1446–52.

Klaassen, M., Bauer, S, Madsen J., and Tombre, I. (2006) Modelling behavioural and fitness consequences of disturbance for geese along their spring flyway. Journal of Applied Ecology, 43, 92–100.

Klaassen, M., Lindström, A., Meltofte, H., and Piersma, T. (2001) Ornithology—Arctic waders are not capital breeders. Nature, 413, 794.

Klaassen, M., Beekman, J.H., Kontiokorpi, J., Mulder, R.J.W., and Nolet, B.A. (2004) Migrating swans profit from favourable changes in wind conditions at low altitude. Journal of Ornithology, 145, 142–51.

Klaassen, R.H.G. and Nolet, B.A. (2008) Persistence of spatial variance and spatial pattern in the abundance of a submerged plant. Ecology, 89, 2973–79.

Kleist, A.M., Lancia, R.A., and Doerr, P.D. (2007) Using Video Surveillance to Estimate Wildlife Use of a Highway Underpass. The Journal of Wildlife Management, 71, 8, 2792–800.

Klokov, K.B. (1997) Northern reindeer of Taimyr Okrug as the focus of economic activity: Contemporary problems of reindeer husbandry and the wild reindeer hunt. Polar Geography, 21, 233–71.

Knight, T.M., McCoy, M.W., Chase, J.M., McCoy, K.A, and Holt, R.D. (2005) Trophic cascades across ecosystems. Nature, 437, 880–83.

Koehler, A.V., Pearce, J.M., Flint, P.L., Franson, J.C., and Ip, H.S. (2008) Genetic evidence of intercontinental movement of avian influenza in a migratory bird: the northern pintail (Anas acuta) Molecular Ecology, 17, 4754–62.

Koenig, W.D. and Knops, J.M.H. (1998) Scale of mast seeding and tree-ring growth. Nature, 396, 225–26.

Kölzsch, A., and Blasius, B. (2008) Theoretical approaches to bird migration. European Physical Journal-Special Topics 157:191–208.

Kokko, H. (1999) Competition for early arrival in migratory birds. Journal of Animal Ecology, 68, 940–50.

Kokko, H. and Lundberg, P. (2001) Dispersal, migration, and offspring retention in saturated habitats. The American Naturalist, 157, 2, 188–202.

Koops, M.A. (2004) Reliability and the value of information. Animal Behaviour, 67, 103–11.

Korpimäki, E. (1986) Gradients in population fluctuations of Tengmalm's owl Aegolius funereus in Europe. Oecologia, 69, 195–201.

Kram, R. and Taylor, C.R. (1990) Energetics of running: a new perspective. Nature, 346, 265–266.

Kramer, D.L. and Chapman, M.R. (1999) Implications of fish home range size and relocation for marine reserve function. Environmental Biology of Fishes, 55, 65–79.

Kramer, G. (1959) Recent experiments on bird orientation. Ibis, 101, 399–416.

Krupnik, I. (1993) Arctic Adaptations: Native Whalers and Reindeer Herders of Northern Eurasia, University Press of New England: Hanover, New Hampshire.

Kuha, J. (2004) AIC and BIC—Comparisons of assumptions and performance. Sociological Methods and Research, 33, 188–229.

Kühl, A., Balinova, N., Bykova, E., Esipov, A., Arylov, I.A., Lushchekina, A.A., and Milner-Gulland, E.J. (2009) The role of saiga poaching in rural communities: linkages between attitudes, socio-economic circumstances and behaviour. Biological Conservation, 142, 1442–1449.

Kullberg, C., Fransson, T., and Jakobsson, S. (1996) Impaired predator evasion in fat blackcaps (Sylvia atricapilla) Proceedings of the Royal Society of London Series B Biological Sciences, 263, 1671–75.

Kullberg, C., Henshaw, I., Jakobsson, S., Johansson, P., and Fransson, T. (2007) Fuelling decisions in migratory birds: geomagnetic cues override the seasonal effect. Proceedings of the Royal Society B-Biological Sciences, 274, 2145–51.

Kumpula, T. (2006) Very High Resolution Remote Sensing Data in Reindeer Pasture Inventory in Northern

Fennoscandia In B.C. Forbes, M. Bölter, L. Müller-Wille, J. Hukkinen, F. Müller, N. Gunslay, and Y. Konstantinov, Reindeer Management in Northernmost Europe, Vol. 184, Ecological Studies, pp. 167–85, Springer: Berlin, Heidelberg.

Kunz, T.H., Wrazen, J.A., and Burnett, C.D. (1998) Changes in body mass and fat reserves in pre-hibernating little brown bats (Myotis lucifugus). Ecoscience, 5, 8–17.

Kvist, A. and Lindström, A. (2000) Maximum daily energy intake: it takes time to lift the metabolic ceiling. Physiological and Biochemical Zoology, 73, 30–36.

Kvist, A. and Lindström, A. (2003) Gluttony in migratory waders—unprecedented energy assimilation rates in vertebrates. Oikos, 103, 397–402.

Kvist, A., Klaassen, M., and Lindström, Å. (1998) Energy expenditure in relation to flight speed: what is the power of mass loss rate estimates? J. Avian Biol., 29, 485–98.

Kwan, D. (1994) Fat reserves and reproduction in the green turtle, Chelonia mydas. Wildlife Research, 21, 257–66.

Lack, D. (1954) The Natural Regulation of Animal Numbers, Oxford University Press: London.

Lack, D. (1968) Bird migration and natural selection. Oikos, 19, 1–9.

Laliberte, A.S. and Ripple, W.J. (2004) Range Contractions of North American Carnivores and Ungulates. BioScience, 54, 2, 123–38.

Lande, R., Engen, S., and Sæther, B.-E. (2003) Stochastic Population Dynamics in Ecology and Conservation, Oxford Series in Ecology and Evolution, Oxford University Press: Oxford.

Langvatn, R. and Albon, S.D. (1986) Geographic clines in body weight of Norwegian red deer: a novel explanation of Bergmann's rule? Holartic Ecology, 9, 285–93.

Langvatn, R. and Hanley, T.A. (1993) Feeding-patch choice by red deer in relation to foraging efficiency. Oecologia, 95, 164–70.

Lank, D. B., Butler, R.W., Ireland, J. and Ydenberg, R. C. (2003) Effects of predation danger on migration strategies of sandpipers. Oikos, 103, 303–320.

Lank, D.B., Pither, J., Chipley, D., Ydenberg, R.C., Kyser, T.K., and Norris, D.R. (2007) Trace element profiles as unique identifiers of western sandpiper (Calidris mauri) populations. Canadian Journal of Zoology, 85, 579–83.

Laris, P. (2002) Burning the seasonal mosaic: preventative burning strategies in the wooded savanna of southern Mali. Human Ecology, 30 (2), 155–86.

Le Boeuf, B.J., Crocker, D.E., Costa, D.P., Blackwell, S.B., Webb, P.M., and Houser, D.S. (2000) Foraging Ecology of Northern Elephant Seals. Ecological Monographs, 70, 353–82.

Le Houérou, H.N. (1989) The Grazing Land Ecosystems of the African Sahel, Ecological Studies 75, Springer-Verlag: Berlin, New York.

Le Pendu, Y. and Ciofolo, I. (1999) Seasonal movements of giraffes in Niger. Journal of Tropical Ecology, 15, 3), 341–53.

Lehikoinen, E., Sparks, T.H., and Zalakevivus, M. (2004) Arrival and departure dates. In: A.P. Møller, W. Fiedler and P. Berthold (eds) Birds and Climate Change. Advances in Ecological Research 35, pp. 1–31, Elsevier: Amsterdam, The Netherlands.

Leimgruber, P., McShea, W.J., Brookes, C.J., Bolor-Erdene, Wemmer, L.C., and Larson, C. (2001) Spatial patterns in relative primary productivity and gazelle migration in the Eastern Steppes of Mongolia. Biological Conservation, 102, 205–12.

Leon, J. (2009) Evaluating the Use of Local Knowledge in Species Distribution Studies: A Case Study of Saiga Antelope in Kalmykia, Russia, MSc thesis, Imperial College: London. Available at www.iccs.org.uk.

Lepczyk, C.A., Murray, K.G., Winnett, M.K., Bartell, P., Geyer, E., and Work, T. (2000) Seasonal fruit preferences for lipids and sugars by American robins. Auk, 117, 709–17.

Levey, D.J. and Stiles, F.G. (1992) Evolutionary precursors of long-distance migration: resource availability and movement patterns in Neotropical landbirds. The American Naturalist, 140, 447–76.

Levins, R. and Culver, D. (1971) Regional coexistence of species and competition between rare species (mathematical model/habitable patches). Proceedings of the National Academy of Sciences of the United States of America, 68, 1246ff.

Lewison, R.L. and Crowder, L.B. (2007) Putting Longline Bycatch of Sea Turtles into Perspective. Conservation Biology, 21, 79–86.

Lewison, R.L., Freeman, S.A., and Crowder, L.B. (2004a) Quantifying the effects of fisheries on threatened species: the impact of pelagic longlines on loggerhead and leatherback sea turtles. Ecology, 7, 221–31.

Lewison, R.L., Crowder, L.B., Read, A.J., and Freeman, S.A. (2004b) Understanding impacts of fisheries bycatch on marine megafauna. Trends in Ecology and Evolution, 19, 11, 599–604.

Liechti, F. (2006) Birds: blowin' by the wind? Journal of Ornithology, 147, 202–11.

Liechti, F. and Bruderer, B. (1998) The relevance of wind for optimal migration theory. Journal of Avian Biology, 29, 561–68.

Liechti, F. and Schaller, E. (1999) The use of low-level jets by migrating birds. Naturwissenschaften, 86, 549–51.

Liechti, F., Klaassen, M., and Bruderer, B. (2000) Predicting migratory flight altitudes by physiological migration models. Auk, 117, 205–14.

Lincoln, F.C. (1939) The Migration of American Birds, Doubleday, Doran, and Co.: New York.

Lind, J., Fransson, T., Jakobsson, S., and Kullberg, C. (1999) Reduced take-off ability in robins (Erithacus rubecula) due to migratory fuel load. Behavioral Ecology and Sociobiology, 46, 65–70.

Lindström, A. (1990) The role of predation risk in stopover habitat selection in migrating Bramblings Fringilla montifringilla. Behavioral Ecology, 1, 102–106.

Lindström, A. (1991) Maximum fat deposition rates in migrating birds. Ornis Scandinavica, 22, 12–19.

Lindström, A. and Alerstam, T. (1992) Optimal fat loads in migrating birds: a test of the time-minimization hypothesis. American Naturalist, 140, 477–91.

Lindström, A., Daan, S., and Visser, H.G., (1994) The conflict between moult and migratory fat deposition: a photoperiodic experiment with bluethroats. Animal Behaviour, 48, 1173–81.

Lindström, A., Hasselquist, D., Bensch, S., and Mats, G. (1990) Asymmetric contests over resources for survival and migration: a field experiment with bluethroats. Animal Behaviour, 40, 453–61.

Linnaeus, C. (1757) Migrationes Avium.

Löft, E.R., Menke, J.W., and Burton, T.S. (1984) Seasonal movements and summer habits of female black-tailed deer. Journal of Wildlife Management, 48, 1317–25.

Lohmann, K.J. and Lohmann, C.M.F. (1996) Orientation and open-sea navigation in sea turtles. Journal of Experimental Biology, 199, 73–81.

Lohmann, K.J., Hester, J.T. and Lohmann, C.M.F. (1999) Long-distance navigation in sea turtles. Ethology, Ecology and Evolution, 11, 1–23.

Lohmann, K.J., Lohmann, C.M.F., and Endres, C.S. (2008a) The sensory ecology of ocean navigation. Journal of Experimental Biology, 211, 1719–28.

Lohmann, K.J., Luschi, P., and Hays, G.C. (2008b) Goal navigation and island-finding in sea turtles. Journal of Experimental Marine Biology and Ecology, 356, 83–95.

Lohmann, K.J., Putman, N.F., and Lohmann, C.M.F. (2008) Geomagnetic imprinting: a unifying hypothesis of long-distance natal homing in salmon and sea turtles. Proceedings of the National Academy of Sciences of the United States of America, 105, 49, 19096–101.

Lohmann, K.J., Cain, S.D., Dodge, S.A., and Lohmann, C.M.F. (2001) Regional magnetic fields as navigational markers for sea turtles. Science, 294, 364–66.

Loonen, M., Zijlstra, M., and Vaneerden, M.R. (1991) Timing of wing molt in greylag geese Anser anser in relation to the availability of their food plants. Ardea, 79:252–59.

Loreau, M., Mouquet, N.M., and Holt, R.D. (2005) From metacommunities to metaecosystems. In M. Holyoak, M.A. Leibold, and R.D. Holt (eds) Metacommunities: Spatial Dynamics and Ecological Communities, pp. 418–38, University of Chicago Press: Chicago.

Louchart, A. (2008) Emergence of long distance bird migrations: a new model integrating global climate changes. Naturwissenschaften, 95, 1109–19.

Lovejoy, N.R., Mullen, S.P., Sword, G.A., Chapman, R.F., and Harrison, R.G. (2006) Ancient trans-Atlantic flight explains locust biogeography: molecular phylogenetics of Schistocerca. Proceedings of the Royal Society B, Biological Sciences, 273, 767–74.

Lumsden, H.G. (1984) The pre-settlement breeding distribution of Trumpeter (*Cygnus buccinator*) and Tundra swans (*Cygnus columbianus*) in Eastern Canada. Canadian Field-Naturalist, 98, 415–24.

Lumsden, H.G. and Drever, M.C. (2002) Overview of the Trumpeter Swan Reintroduction Program in Ontario, 1982–2000. Waterbirds: The International Journal of Waterbird Biology, 25, Special Publication 1, Proceedings of the Fourth International Swan Symposium, 2001, pp. 301–12.

Lundberg, J. and Moberg, F. (2003) Mobile link organisms and ecosystem functioning: Implications for ecosystem resilience and management. Ecosystems, 6, 87–98.

Lundberg, P. (1987) Partial bird migration and evolutionary stable strategies. Journal of Theoretical Biology, 125, 351–60.

Lundberg, P. (1988) The evolution of partial migration in birds. Trends in Ecology and Evolution, 3, 172–75.

Luschi, P., Hays, G.C., and Papi, F. (2003) A review of long-distance movements by marine turtles, and the possible role of ocean currents. Oikos, 103, 293–302.

Luschi, P., Hays, G.C., del Seppia, C., Marsh, R., and Papi, F. (1998) The navigational feats of green sea turtles migrating from Ascension Island investigated by satellite telemetry. Proceedings of the Royal Society of London Series B-Biological Sciences, 265, 2279–84.

Luschi, P., Åkesson, S., Broderick, A.C., Glen, F., Godley, B.J., Papi, F. and Hays, G.C. (2001) Testing the navigational abilities of ocean migrants: displacement experiments on green sea turtles (*Chelonia mydas*) Behavioral Ecology and Sociobiology, 50, 528–34.

Lushchekina, A.A. and Struchkov, A. (2001) The saiga antelope in Europe: Once again at the brink? The Open Country, 3, 11–24.

Lyons, J.E., Collazo, J.A., and Guglielmo, C.G. (2008) Plasma metabolites and migration physiology of semipalmated sandpipers: refueling performance at five latitudes. Oecologia, 155, 417–27.

Mace, G.M. and Reynolds, J.D. (2001) Exploitation as a conservation issue. In J.D. Reynolds, G.M. Mace, K.H. Redford, and J.G. Robinson (eds) Conservation of Exploited Species, pp. 3–15, Cambridge University Press: Cambridge.

Maddock, L. (1979) The 'migration' and grazing succession. In A.R.E. Sinclair and M. Norton-Griffiths (eds) Serengeti: Dynamics of an Ecosystem, pp. 104–29, University of Chicago Press: Chicago.

Magnus, O. (1555) Historia de Gentibus Septenionalibus.

Mahan, T.A. and Simmers, B.S. (1992) Social preference of four cross-foster reared Sandhill Cranes. Proceedings of the North American Crane Workshop, 6, 114–119.

Mahoney, S.P. and Schaefer, J.A. (2002) Hydroelectric development and the disruption of migration in caribou. Biological Conservation, 107, 147–53.

Mahoney, S.P. and Schaefer, J.A. (2002) Long-term changes in demography and migration of Newfoundland caribou. Journal of Mammalogy, 83, 957–63.

Maj, E. (2009) La vache sedentaire, le renne et le cheval nomades chez les Evenes et las Iakoutes des monts de Verkhihansk (Republique Sakha, Iakoutie). Fondation Fyssen—Annales, 23, 36–48.

Malcolm, J.R., Markham, A., Neilson, R.P., and Garaci, M. (2002) Estimated migration rates under scenarios of global climate change. Journal of Biogeography, 29, 835–49.

Malechek, J.C. (1984) Impacts of grazing intensity and specialized grazing systems on livestock response. In Developing Strategies for Rangeland Management, pp. 1129–58, National Academy of Sciences and Westview Press: Boulder, Colorado and London.

Mallon, D.P. and Kingswood, S.C. (compilers) (2001) Antelopes, Part 4: North Africa, the Middle East, and Asia, Global Survey and Regional Action Plans, SSC Antelope Specialist Group, IUCN:Gland, Switzerland and Cambridge, UK.

Mallory, F.F. and Hillis, T.L. (1998) Demographic characteristics of circumpolar caribou populations: ecotypes, ecological constraints, releases, and population dynamics. Rangifer, Special Issue, 10, 49–60.

Mandel, J.T., Bildstein, K.L., Bohrer, G., and Winkler, D.W. (2008) The movement ecology of migration in Turkey Vultures. Proceedings of the National Academy of Sciences of the United States of America, 105, 19102–19107.

Mangel, M. (1994) Climate change and salmonid life history variation. Deep Sea Research II, 41, 75–106.

Mangel, M. and Clark, C.W. (1988) Dynamic Modeling in Behavioral Ecology, Princeton University Press: Princeton, New Jersey.

Mangel. M. and Satterthwaite, W.H. (2008) Combining proximate and ultimate approaches to understand life history variation in salmonids with application to fisheries, conservation, and aquaculture. Bulletin of Marine Science, 83, 107–30.

Marcovaldi, M.Â., Baptistotte, C., De Castilhos, J.C., Gallo, B.M.G., Lima, E.H.S.M., Sanches, T.M., and Vieitas, C.F. (1998) Activities by Project TAMAR in Brazilian sea turtle feeding grounds. Marine Turtle Newsletter, 80L, 5–7.

Marks, J.C., Haden, G.A., O'Neill, M and Pace, C. (in press) Effects of flow restoration and exotic species removal on recovery of native fish: lessons from a dam decommissioning. Restoration Ecology, http://dx.doi.org/10.1111/j.1526-100X.2009.00574.x

Marra, P.P., Francis, C.M., Mulvihill, R.S., and Moore, F.R. (2005) The influence of climate on the timing and rate of spring bird migration. Oecologia, 142, 307–15.

Marsh, R.E., Erickson, W.A., and Salmon, T.P. (1992) Scarecrows and Predator Models for Frightening Birds from Specific Areas. In J. E. Borrecco and R. E. Marsh (eds) Proceedings of the Fifteenth Vertebrate Pest Conference, University of Nebraska, Lincoln, University of California: Davis.

Marshall, F. (1990) Origins of specialized pastoral production in East Africa. American Anthropologist, 92, 873–94.

Martin, T.G., Chadès, I., Arcese, P., Marra, P.P., Possingham, H.P., and Norris, D.R. (2007) Optimal Conservation of Migratory Species. Public Library of Science One, 2, E751.

Marty, A. (1993) La gestion des terroirs et les éleveurs: Un outil d'exclusion ou de négociation? Revue Tiers Monde, 34, 134, 329–44.

Matthiopoulos, J., Harwood, J., and Thomas, L. (2005) Metapopulation consequences of site fidelity for colonially breeding mammals and birds. Journal of Animal Ecology, 74, 716–27.

Matthiopoulos, J., Thomas, L., McConnell, B., Duck, C., Thompson D., Pomeroy, P., Harwood, J., Milner-Gulland, E.J., Wolf, N., and Mangel, M. (2008) Putting Long-Term, Population Monitoring Data to Good Use: The Causes of Density Dependence in UK grey seals. SCOS Briefing paper 06/08.

May, R.M. (1976) Simple models with very complicated dynamics. Nature, 261, 459–67.

Maynard-Smith, J. (1982) Evolution and the Theory of Games, Cambridge University Press: Cambridge.

Mayr, E. (1961) Cause and Effect in Biology. Science, 134, 1501–1506.

Mayr, E. and Meise, W. (1930) Theoretisches zur Geschichte des Vogulzuges. Der Vogelzug (Berlin), 1, 149–72.

Mbaiwa, J. E. and Darkoh, M.B.K (2005) Sustainable development and natural resource competition and conflicts in the Okavango Delta, Botswana. Botswana Notes and Records, 37, 40–60.

Mbaiwa, J.E. and Mbaiwa, O.I. (2006) The effects of veterinary fences on wildlife populations in Okavango Delta, Botswana. International Journal of Wilderness, 12, 3, 17–23.

McCabe, J.T. (1983) Land use among the pastoral Turkana. Rural Africana, 15–16, 109–26.

McCabe, J.T. (1994) Mobility and land use among African pastoralists: old conceptual problems and new interpretations. In E. Fratkin, K.A. Galvin, and E.A. Roth (eds) African Pastoralist Systems: An Integrated Approach, pp. 69–89, Lynne Rienner Publishers: Boulder.

McClelland, G.B. (2004) Fat to the fire: the regulation of lipid oxidation with exercise and environmental stress. Comparative Biochemistry and Physiology, Part B, 139, 443–60.

McClenachan, L., Jackson, J.B.C., and Newman, M.J.H. (2006) Conservation Implications of Historic Sea Turtle Nesting Beach Loss. Frontiers in Ecology and the Environment, 4, 6, 290–96.

McConville, A.J., Grachev, I.A., Keane, A., Coulson, T., Bekenov, A., and Milner-Gulland, E.J. (2009) Reconstructing the observation process to correct for changing detection probability of a critically endangered species. Endangered Species Research, 6, 231–37.

McCormick, S.D., Hansen, L.P., Quinn, T.P., and Saunders, R.L. (1998) Movement, migration, and smolting of Atlantic salmon (Salmo salar). Canadian Journal of Fisheries and Aquatic Sciences, 55, Suppl. 1, 77–92.

McEneaney, T. (2001) Yellowstone Bird Report, 2000. National Park Service, Yellowstone Center for Resources, Yellowstone National Park: Wyoming, YCR–NR–2001–01.

McGuire, L.P. and Guglielmo, C.G. (2009) What can birds tell us about the migration physiology of bats? Journal of Mammalogy, 90, 1290–97.

McKelvey, K.S. and Noon, B.R. (2001) Incorporating uncertainties in animal location and map classification into habitat relationships modeling. In C. T. Hunsaker, M. F. Goodchild, M. A. Friedl, and T. J. Case (eds) Spatial Uncertainty in Ecology, pp. 72–90, Springer-Verlag: New-York.

McKenzie, J. (2001) The selective advantage of urban habitat use by elk in Banff National Park M.Sc. Thesis, University of Guelph: Ontario.

McKinnon, L., Smith, P.A., Nol, E., Martin, J.L., Doyle, F.I., Abraham, K.F., Gilchrist, H.G., Morrison, R.I.G., and Bety, J. (2010) Lower predation risk for migratory birds at high latitudes. Science, 327, 326–27.

McLellan, B.N. and Shackleton, D.M. (1988) Grizzly bears and resource-extraction industries: effects of roads on behavior, habitat use, and demography. Journal of Applied Ecology, 25, 451–60.

McNamara, J.M. and Houston, A.I. (2008) Optimal annual routines: behaviour in the context of physiology and ecology. Philosophical Transactions of the Royal Society B, 363, 301–19.

McNaughton, S.J. (1976) Serengeti migratory wildebeest: facilitation of energy flow by grazing. Science, 191, 92–4.

McNaughton, S.J. (1984) Grazing lawns: animals in herds, plant form, and coevolution. American Naturalist, 124, 863–86.

McPeek, M.A. and Holt, R.D. (1992) The evolution of dispersal in spatially and temporally varying environments. American Naturalist, 140, 1010–27.

McQuire, L.P. and Guglielmo, C.G. (2009) What can birds tell us about the migration physiology of bats? Journal of Mammalogy, 90, 1290–97.

McWilliams, S.R., Kearney, S.B., and Karasov, W.H. (2002) Diet preferences of warblers for specific fatty acids in relation to nutritional requirements and digestive capabilities. Journal of Avian Biology, 33, 167–74.

Mduma, S.A.R., Sinclair, A.R.E., and Hilborn, R. (1999) Food regulates the Serengeti wildebeest: a 40-year record. Journal of Animal Ecology, 68, 1101–22.

Mehlman, D.W., Mabey, S.E., Ewert, D.N., Duncan, C., Abel, B., Cimprich, D., Sutter, R., and Woodfrey, M. (2005) Conserving stop-over sites for forest-dwelling migratory landbirds. Auk, 122, 1–11.

Meitner, C.J., Brower, L.P., and Davis, A.K. (2004) Migration patterns and environmental effects on stopover of monarch butterflies (Lepidoptera, Nymphalidae) at Peninsular Point, Michigan. Environmental Entomology, 33, 249–56.

Menu, S., Gauthier, G., and Reed, A. (2005) Survival of young greater snow geese (Chen caerulescens atlantica) during fall migration. The Auk 122, 479–96.

Mikkola, K. (2003) Red Admirals Vanessa atalanta (Lepidoptera: Nymphalidae) select northern winds on southward migration. Entomol Fenn, 14, 15–24.

Milá, B., Smith, B.S., and Wayne, R.K. (2006) Postglacial population expansion drives the evolution of long-distance migration in a songbird. Evolution, 60, 2403–409.

Mills, M.G.L. and Funston, P. (2003) Large carnivores and savannah heterogeneity In J. Du Toit, K.H. Rogers, and H.C. Biggs (eds) The Kruger Experience, pp. 370–88, Island Press: Washington.

Millspaugh, J.J. and Marzluff, J.M. (2001), Radio Tracking and Animal Populations, Academic Press: San Diego.

Milner-Gulland, E.J. (2001) A dynamic game model for the decision to join an aggregation. Ecological Modeling, 145, 85–99.

Milner-Gulland, E.J., Bukreeva, O.M., Coulson, T.N., Lushchekina, A.A., Kholodova, M.V., Bekenov, A.B., and Grachev, Iu.A. (2003) Reproductive collapse in saiga antelope harems. Nature, 422, 135.

Milner-Gulland, E.J., Kholodova, M.V., Bekenov, A., Bukreeva, O.M., Grachev, Iu.A., Amgalan, L., and Lushchekina, A.A. (2001) Dramatic declines in Saiga antelope populations. Oryx, 35, 340345.

Minetti, A.E. (1995) Optimum gradient of mountain paths. Journal of Applied Physiology, 79, 1698–1703.

Mitchell, C.D. (1994) Trumpeter swan (Cygnus buccinator). In A. Poole (ed.) The Birds of North America, Online, [Online], Available: http://bna.birds.cornell.edu.sub-zero.lib.uoguelph.ca/bna/species/105 [December 2008], Cornell Lab of Ornithology, Ithaca.

Mitchell-Olds, T., Feder, M., and Wray, G. (2008) Evolutionary and ecological functional genomics. Heredity, 100, 101–102.

Mitcheson, Y.S.D., Cornish, A, Domeier, M, Colin, P.L., Russell, M., and Lindeman, K.C. (2008) A Global Baseline for Spawning Aggregations of Reef Fishes. Conservation Biology, 22, 5, 1233–44.

Moil, R.J., Millspaugh, J.J. Beringer, J. Sartwell, J., and He, Z.H. (2007) A new 'view' of ecology and conservation through animal-borne video systems. Trends in Ecology and Evolution, 22, 660–68.

Møller, A.P., Rubolini, D., and Lehikoinen, E. (2008) Populations of migratory bird species that did not show a phenological response to climate change are declining. Proceedings of the National Academy of Sciences of the United States of America, 105, 42, 16195–200.

Mooney-Seus, M.L. and Rosenberg, A.A. (2007) Regional Fisheries Management Organizations (RFMOs): Progress in Adopting Precautionary Approach and Ecosystem-Based Management [Online], Available: at:http://www.illegal-fishing.info/uploads/RFMO-report-FortHill-0207.pdf [January 2009].

Moorcroft, P.R. and Lewis, M.A. (2006), Mechanistic Home Range Analysis: Monographs in Population Biology, Princeton University Press: Princeton.

Moore, A., Freake, S.M., and Thomas, I.M. (1990) Magnetic particles in the lateral line of the Atlantic salmon (Salmo salar L). Philosophical Transactions of the Royal Society of London, 329, 11–15.

Moore, F.R. and Yong, W. (1991) Evidence of food-based competition among passerine migrants during stopover. Behavioral Ecology and Sociobiology, 28, 85–90.

Morales, J.M., D.T. Haydon, J. Frair, K.E. Holsiner, and J.M. Fryxell. (2004) Extracting more out of relocation data: Building movement models as mixtures of random walks. Ecology, 85, 2436–45.

Moreau, R.E. (1972) The Palaearctic—African Bird Migration Systems, Academic Press: London, New York.

Morgan, E.R., Lundervold, M., Medley, G.F., Shaikenov, B.S., Torgerson, P., and Milner-Gulland, E.J. (2006) Assessing risks of disease transmission between wildlife and livestock: The Saiga antelope as a case study. Biological Conservation, 131, 244–54.

Morris, S. R. (1996) Mass loss and probability of stopover by migrant warblers during spring and fall migration. Journal of Field Ornithology, 67, 456–62.

Morrison, R.I.G., Davidson, N.C., and Wilson, J.R. (2007) Survival of the fattest: body stores on migration and survival in red knots *Calidris canutus islandica*. Journal of Avian Biology, 38, 4, 479–87.

Mortimer, J. A. and Carr, A. (1987) Reproduction and migrations of the Ascension-Island green turtle (Chelonia-Mydas). Copeia 1987, 103–13.

Mosser, A., and Packer, C. (2009) Group territoriality and the benefits of sociality in the African lion, Panthera leo, Animal Behaviour 78, 359–70.

Mouritsen, H. and Frost, B. J. (2002) Virtual migration in tethered flying monarch butterflies reveals their orientation mechanisms. Proceedings of the National Academy of Sciences USA, 99, 10162–66.

Mueller, T. and Fagan, W. F. (2008) Search and navigation in dynamic environments—from individual behaviors to population distributions. Oikos, 117, 654–64.

Mueller, T., Olson, K.A., Fuller, T.K., Schaller, G.B., Murray M.G., and Leimgruber, P. (2008) In search of forage: predicting dynamic habitats of Mongolian gazelles using satellite-based estimates of vegetation productivity. Journal of Applied Ecology, 45, 649–58.

Muheim, R., Phillips, J.B. and Åkesson, S. (2006) Polarized light cues underlie compass calibration in migratory songbirds. Science, 313, 837–39.

Murray, M.G. and Brown, D. (1993) Niche separation of grazing ungulates in the Serengeti—an experimental test. Journal of Animal Ecology, 62, 380–89.

Murray, M.G. and Illius, AW. (1996) Multispecies grazing in the Serengeti. In J. Hodgson and A.W. Illius (eds) The Ecology and Management of Grazing Systems, pp. 247–74, CAB International: Wallingford, UK.

Murray, M.G. and Illius, AW. (2000) Vegetation modification and resource competition in grazing ungulates. Oikos, 89, 501–508.

Murtaugh, P.A. (2009) Performance of several variable-selection methods applied to real ecological data. Ecology Letters, 12, 1061–68.

Myers, G.S. (1949) Usage of anadromous, catadromous and allied terms for migratory fishes. Copeia, 1949, 89–97.

Nakazawa, Y., Peterson, A.T., Martinez-Meyer, E., and Navarro-Siguenza, A.G. (2004) Seasonal niches of Nearctic–Neotropical migratory birds: Implications for the evolution of migration. Auk, 121, 610–18.

Nathan, R., Getz, W.M., Revilla, E., Holyoak, M., Kadmon, R., Saltz, D., and Smouse, P.E. (2008) A movement ecology paradigm for unifying organismal movement research. Proceedings of the National Academy of Sciences USA, 105, 19052–59.

Nellemann, C., Vistnes, I., Jordhøy, P., Støen, O.-G., Kaltenborn, B.P., F. Hanssen, and Helgesen, R. (2009) Effects of recreational cabins, trails and their removal for restoration of reindeer winter ranges. Restoration Ecology, 17, 1, 1–9.

Nelson, M.E. (1998) Development of migratory behavior in northern white-tailed deer. Canadian Journal of Zoology, 76, 426–32.

Nelson, M.E. and Mech, L.D. (1981) Deer Social Organization and Wolf Predation in Northeastern Minnesota, Wildlife Monographs, No.77, 1–53.

Newton, I. (2006) Advances in the study of irruptive migration. Ardea, 94, 433–60.

Newton, I. (2008) The Migration Ecology of Birds, Academic Press: Oxford.

Newton, I. and Dale, L.C. (1996a) Bird migration at different latitudes in eastern North America. Auk, 113, 626–35.

Newton, I. and Dale, L.C. (1996b) Relationship between migration and latitude among west European birds. Journal of Animal Ecology, 65, 137–46.

Ng, S.J., Dole, J.W., Sauvajot, R.M., Riley, S.P.D., and Valonec, T.J. (2004) Use of highway undercrossings by wildlife in southern California. Biological Conservation, 115, 499–507.

Niamir-Fuller, F. (1998) The resilience of pastoral herding in Sahelian Africa. In F. Berkes, C. Folke, and J. Colding (eds) Linking social and ecological systems: management practices and social mechanisms for building resilience, pp. 250–84, Cambridge University Press: Cambridge, UK.

Nilsen, T.O., Ebbesson, L.O.E., Madsen, S.S., McCormick, S.D. Andersson, E., Björnsson, B.T., Prunet, P., and Stefansson, S.O. (2007) Differential expression of gill Na+,K+-ATPase alpha- and beta-subunits, Na+,K+,2Cl(-) cotransporter and CFTR anion channel in juvenile anadromous and landlocked Atlantic salmon *Salmo salar*. Journal of Experimental Biology, 210, 2885–96.

Nilsson, A.L.K., Lindstrom, A., Jonzén, N., Nilsson, S.G., and Karlsson, L. (2006) The effect of climate change on partial migration—the blue tit paradox. Global Change Biology, 12, 10, 2014–22.

Nisbet, I.C.T. (1955) Atmospheric turbulence and bird flight. British Birds, 48, 557–59.

Nisbet, I.C.T. and Drury, W.H. (1967) Orientation of spring migrants studied by radar. Bird Banding, 38, 173–86.

Nocera, J.J., Taylor, P.D., and Ratcliffe, L.M. (2008) Inspection of mob-calls as sources of predator information: response of migrant and resident birds in the Neotropics. Behavioural Ecology and Sociobiology, 62, 1769–77.

Nolet, B.A. (2006) Speed of spring migration of Tundra Swans *Cygnus columbianus* in accordance with income or capital breeding strategy? Ardea, 94, 579–91.

Norling, B.S. (1992) Roost sites used by Sandhill Crane staging along the Platte River, Nebraska. Great Basin Naturalist, 52, 3, 253–61.

Norris, D.R. and Taylor, C.M. (2006) Predicting the consequences of carry-over effects for migratory populations. Biology Letters, 2, 148–51.

Norris, D.R., Marra, P.P., Kyser, T.K., and Ratcliffe, L.M. (2005) Tracking habitat use of a long-distance migratory bird, the American redstart *Setophaga ruticilla*, using stable-carbon isotopes in cellular blood. Journal of Avian Biology, 36, 164–70.

Norton-Griffiths, M. (2007) How many wildebeest do you need? World Economics, 8, 2, 41–64.

Norton-Griffiths, M. and Southey, C. (1995) The opportunity costs of biodiversity conservation in Kenya. Ecological Economics, 12, 125–39.

Nøttestad, L., Giske, J., Holst, J.C. and Huse, G. (1999) A length-based hypothesis for feeding migrations in pelagic fish. Canadian Journal of Fisheries and Aquatic Sciences, 56, 26–34.

Nowakowski, J.K., Remisiewicz, M., Keller, M., Busse P., and Rowiński, P. (2005) Synchronisation of the autumn mass migration of passerines: a case of Robins *Erithacus rubecula*. Acta Ornithologica, 40, 103–15.

Ohashi, K., Leslie, A., and Thomson, J.D. (2008) Trapline foraging by bumble bees: V. Effects of experience and priority on competitive performance. Behavioral Ecology, 19, 5, 936–948.

Ohashi, K. and Thomson, J.D. (2005) Efficient harvesting of renewing resources. Behavioral Ecology, 16, 592–605.

Okayasu, T., Muto, M., Jamsran, U., and Takeuchi K. (2007) Spatially heterogeneous impacts on rangeland after social system change in Mongolia. Land Degradation and Development, 18, 5, 555–66.

Økland, F., Jonsson, B., Jensen, A.J., and Hansen, L.P. (1993) Is there a threshold size regulating seaward migration of brown trout and Atlantic salmon? Journal of Fish Biology, 42, 541–50.

Okoti, M., Ng'ethe, J.C., Ekaya, W.N., and Mbuvi, D.M. (2004) Land use, ecology, and socio-economic changes in a pastoral production system. Journal of Human Ecology, 16, 2, 83–9.

Oliveira, E.G., Srygley, R.B. and Dudley, R. (1998) Do neotropical migrant butterflies navigate using a solar compass? Journal of Experimental Biology, 201, 3317–31.

Olsen, J.B., Wuttig, K., Fleming, D., Kretschmer, E.J., and Wenburg, J.K. (2006) Evidence of partial anadromy and resident-form dispersal bias on a fine scale in populations of Oncorynchus mykiss. Conservation Genetics, 7, 613–19.

Osborn, F.V. and Parker, G.E. (2003) Linking two elephant refuges with a corridor in the communal lands of Zimbabwe. African Journal of Ecology, 41, 1, 68–74.

Ouellet, J.-P., Crête, M., Maltais, J, Pelletier, C., and Huot, J. (2001) Emergency feeding of white-tailed deer: test of three feeds. Journal of Wildlife Management, 65, 129–36.

Outlaw, D.C. and Voelker, G. (2006) Phylogenetic tests of hypotheses for the evolution of avian migration: a case study using the Motacillidae. The Auk, 123, 455–66.

Owen, M. and Black, J.M. (1991) A note on migration mortality and its significance in goose populations dynamics. Ardea, 79, 195–96.

Owen, M. and Gullestad, N. (1984). Migration routes of Svalbard barnacle geese, *Branta leucopsis*, with a preliminary report on the importance of the Bjørnøya staging area. Norsk Polarinstitutt Skrifter, 81, 67–77.

Owen-Smith, N. and P. Novellie (1982) What should a clever ungulate eat? The American Naturalist, 119, 2, 151–178.

Owen-Smith, N. and Ogutu, J. (2003) Rainfall influences on ungulate population dynamics. In J. Du Toit, K.H. Rogers and H.C. Biggs (eds) The Kruger Experience, pp. 310–31, Island Press: Washington.

Packer, C., Hilborn, R., Mosser, A., Kissui, B., Borner, M., Hopcraft, G., Wilmshurst, J., Mduma, S, and Sinclair, ARE. (2005) Ecological change, group territoriality, and population dynamics in Serengeti lions. Science, 307, 390–93.

Palacín, C., Alonso, J.C., Alonso, J.A., Martín, C.A., Magaña, M., and Martin, B. (2009) Differential migration by sex in the Great Bustard: possible consequences of an extreme sexual size dimorphism. Ethology, 115, 617–26.

Papi, F., Luschi, P., Åkesson, S., Capogrossi, S., and Hays, G.C. (2000) Open-sea migration of magnetically disturbed sea turtles. Journal of Experimental Biology, 203, 3435–43.

Parmesan C. (2003) Butterflies as bio-indicators of climate change impacts. In C.L. Boggs, W.B. Watt and P.R. Ehrlich (eds) Evolution and Ecology Taking Flight: Butterflies as Model Systems, pp. 541–60, University of Chicago Press: Chicago.

Parmesan, C. and Yohe, G. (2003) A globally coherent fingerprint of climate change impacts across natural systems. Nature, 421, 37–42.

Pärt, T. (1995) The importance of local familiarity and search costs for age- and sex-biased philopatry in the collared flycatcher. Animal Behaviour, 49, 4, 1029–38.

Parvinen, K. (1999). Evolution of migration in a metapopulation. Bulletin of Mathematical Biology, 61, 531–50.

Pascual, M.A., Bentzen, P., Riva Rossi, C., Mackey, G., Kinnison, M., and Walker, R. (2001) First documented case of anadromy in a population of introduced rainbow trout in Patagonia, Argentina. Transactions of the American Fisheries Society, 130, 53–67.

Patterson, T.A., Thomas, L., Wilcox, C., Ovaskainen O., and Matthiopoulos, J. (2008) State-space models of individual animal movement. Trends in Ecology and Evolution, 23, 87–94.

Pavey, C.R. and Nano, C.E.M. (2009) Bird assemblages of arid Australia: vegetation patterns have a greater effect than disturbance and resource pulses. Journal of Arid Environments, 73, 634–42.

Pegg, M.A., Layzer, J.B., and Bettoli, P.W. (1996) Angler exploitation of anchor-tagged saugers in the lower Tennessee River. North American Journal of Fisheries Management, 16, 218–22.

Pelto, P. J. (1987) The Snowmobile Revolution. Technology and Social Change in the Arctic. Prospect Heights, Waveland Press: Illinois.

Pener M.P. and Simpson S.J. (2009) Locust phase polyphenism: An update. Advances in Insect Physiology, 36, 1–272.

Penning de Vries, F.W.T. and Djitèye, MA. (eds) (1982) La productivité des pâturages sahéliens, Centre for Agricultural Publishing and Documentation: Wageningen, The Netherlands.

Pennycuick, C. J. (1969) The mechanics of bird migration. Ibis, 111, 525–56.

Pennycuick, C. J. (1972) Animal Flight, Edward Arnold: London.

Pennycuick, C.J. (1978) Fifteen testable predictions about bird flight. Oikos, 30, 165–76.

Pennycuick, C.J. (1997) Actual and 'optimum' flight speeds: Field data reassessed. Journal of Experimental Biology, 200, 2355–61.

Pennycuick, C. J. (1998) Computer simulation of fat and muscle burn in long-distance bird migration. Journal of Theoretical Biology, 191, 47–61.

Pennycuick, C.J. (1998) Field observations of thermals and thermal streets, and the theory of cross-country soaring flight. Journal of Avian Biology, 29, 33–43.

Pennycuick, C.J., Einarsson, O., Bradbury, T.A.M., and Owen, M. (1996) Migrating Whooper Swans Cygnus cygnus: satellite tracks and flight performance calculations. Journal of Avian Biology, 27, 118–34.

Pennycuick, L. (1975) Movements of the migratory wildebeest population in the Serengeti areas between 1960 and 1973. East African Wildlife Journal, 13, 65–87.

Percival, D.P. and Walden, A.T. (2000) Wavelet methods for Time Series Analysis, Cambridge Series in Statistical and Probabilistic Mathematics, Cambridge University Press: Cambridge, UK.

Perdeck, A.C. (1967) Orientation of starlings after displacement to Spain. Ardea, 55, 194.

Peres, C.A. (2005) Why we need megareserves in Amazonia. Conservation Biology, 19, 3, 728–33.

Pess, G.R., McHenry, M.L, Beechie, T.J., and Davies, J. (2008) Biological impacts of the Elwha River dams and potential salmonid responses to dam removal. Northwest Science, 82, Special Issue, 72–90.

Peterson, C. and Messmer, T.A. (2006) Effects of winter-feeding on mule deer in Northern Utah. The Journal of Wildlife Management, 71, 5, 1440–45.

Petersons, G. (2004) Seasonal migrations of north-eastern populations of Nathusius' bat Pipistrellus nathusii (Chiroptera). Myotis, 41–42, 29–56.

Petrie, S.A. and Wilcox, K.L. (2003) Migration chronology of Eastern-population tundra swans. Canadian Journal of Zoology/Revue Canadien de Zoologie 81, 861–70.

Pettifor, R.A., Caldow, R.W.G., Rowcliffe, J.M., Goss-Custard, J.D., Black, J.M., Hodder, K.H., Houston, A.I., Lang, A. and Webb, J. (2000) Spatially explicit, individual-based, behavioural models of the annual cycle of two migratory goose populations. Journal of Applied Ecology, 37, 103–35.

Pettorelli, N., Vik, J.O., Mysterud, A., Gaillard, J.-M., Tucker, C.J. and Stenseth, N.C. (2005) Using the satellite-derived NDVI to assess ecological responses to environmental change. Trends in Ecology and Evolution, 20, 503–10.

Philips, A.R. (1975) The migration of Allen's and other hummingbirds. Condor, 77, 196–205.

Piersma, T. (1998) Phenotypic flexibility during migration: optimization of organ size contingent on the risks and rewards of fueling and flight? Journal of Avian Biology, 29, 511–20.

Piersma, T. and Gill, R.E. (1998) Guts don't fly: Small digestive organs in obese Bar-tailed Godwits. Auk, 115, 196–203.

Piersma, T., Koolhaas, A., and J. Jukema (2003) Seasonal body mass changes in Eurasian golden plovers Pluvialis apricaria staging in the Netherlands: decline in late autumn mass peak correlates with increase in raptor numbers. Ibis, 145, 565–71.

Piersma, T., Pérez-Tris, J., Mouritsen, H., Bauchinger, U., and Bairlein, F. (2005) Is there a 'migratory syndrome' common to all migrant birds? Annals of the New York Academy of Sciences, 1046, 282–93.

Pigliucci, M. (2003) Phenotypic integration: studying the ecology and evolution of complex phenotypes. Ecology Letters, 6, 265–72.

Pigliucci, M. and Murren, C.J. (2003) Genetic assimilation and a possible evolutionary paradox: can macroevolution sometimes be so fast as to pass us by? Evolution, 57, 1455–64.

Pimm, S.L. and Raven, P. (2000) Extinction by numbers. Nature, 403, 843-845.

Pinto, N. and Keitt, T.H. (2009) Beyond the least-cost path: evaluating corridor redundancy using a graph-theoretic approach. Landscape Ecology, 24, 253–66.

Poche, R.M. (1974) Notes on the roan antelope (*Hippotragus equinus* (Desmarest)) in West Africa. Journal of Applied Ecology, 11, 3, 963–68.

Poiner, I.R. and Harris, A.N.M. (1996) Incidental capture, direct mortality and delayed mortality of sea turtles in Australia's northern prawn fishery. Marine Biology, 125, 813–25.

Polis, G.A., Anderson, W.B., and Holt, R.D. (1997) Toward an integration of landscape and food web ecology: the dynamics of spatially subsidized food webs. Annual Review of Ecology and Systematics, 28, 289–316.

Polovina, J.J., Howell, E., Kobayashi, D.R., and Seki, M.P. (2001) The transition zone chlorophyll front, a dynamic global feature defining migration and forage habitat for marine resources. Progress in Oceanography, 49, 469–83.

Pomilla, C. and Rosenbaum, H.C. (2005) Against the current: an inter-oceanic whale migration event. Biology Letters, 1, 476–79.

Portugal, S.J., Green, J.A., and Butler, P.J. (2007) Annual changes in body mass and resting metabolism in captive barnacle geese (*Branta leucopsis*): the importance of wing moult. Journal of Experimental Biology, 210, 1391–97.

Post, D.M., Taylor, J.P., Kitchell, J.F., Olson, M.H., Schindler, D.E., and. Herwig, B.R. (1998) The role of migratory waterfowl as nutrient vectors in a managed wetland. Conservation Biology, 12, 910–20.

Potts, W.K. (1984) The chorus-line hypothesis of manoeuvre coordination in avian flocks. Nature, 309, 344–45.

Prins, H.H.T. (1989) East African grazing lands: overgrazed or stably degraded. In W.D. Verwey (ed.) Nature Management and Sustainable Development, IOS: Amsterdam/Tokyo, pp. 281–306.

Prins, H.H.T. (1992) The pastoral road to extinction—competition between wildlife and traditional pastoralism in East Africa. Environmental Conservation, 19, 117–23.

Pulido, F. (2007) The genetics and evolution of avian migration. Bioscience, 57, 165–74.

Pulido, F. and Berthold, P. (2003) The quantitative genetic analyses of migratory behavior. In P. Berthold and E. Gwinner (eds) Avian migration, pp. 53–77. Springer: Heidelberg.

Pulido, F., Berthold, P., and van Noordwijk, A.J. (1996) Frequency of migrants and migratory activity are genetically correlated in a bird population: evolutionary implications. Proceedings of the National Academy of Sciences, 93, 14642–47.

Quinn, T.P. (2005) The Behavior and Ecology of Pacific Salmon and Trout, American Fisheries Society, Bethesda in association with University of Washington Press: Seattle and London.

Quinn, T.P., Hodgson, S., and Peven, C. (2007) Temperature, flow, and the migration of adult sockeye salmon (*Oncorhynchus nerka*) in the Columbia River. Can. *Journal of Fisheries and Aquatic Sciences*, 54, 1349–60.

Quinn, T.P., Unwin, M.J., and Kinnison, M.T. (2000) Evolution of temporal isolation in the wild: genetic divergence in timing of migration and breeding by introduced Chinook salmon populations. Evolution, 54, 4, 1372–85.

Racey, G., Harris, A., Gerrish, L., Armstrong, E., McNicol, J., and Baker, J. (1999) Forest Management Guidelines for the Conservation of Woodland Caribou: a Landscape Approach. MS draft. Ontario Ministry of Natural Resources: Thunder Bay, Ontario.

Radakov, D.V. (1973), Schooling in the Ecology of Fish, J. Wiley & Sons: New York.

Raffalovich, L.E., Deane, G.D., Armstrong, D., and Tsao, H.S. (2008) Model selection procedures in social research: Monte-Carlo simulation results. Journal of Applied Statistics, 35, 1093–114.

Raine, A.F. (2007) The International Impact of Hunting and Trapping in the Maltese Islands, BirdLife Malta (internal report).

Ramenofsky, M. and Wingfield, J.C. (2007) Regulation of migration. Bioscience, 57, 135–44.

Ramenofsky, M., Savard, R., and Greenwood, M.R.C. (1999) Seasonal and diet transitions in physiology and behavior in the migratory dark-eyed junco. Comparative Biochemistry and Physiology A, 122, 385–97.

Rankin, MA. and Burchsted, J.C.A. (1992) The cost of migration in insects. Annual Review of Entomology, 37, 533–59.

Rannestad, O.T., Danielsen, T., Moe, S.R., and Stokke, S. (2006) Adjacent pastoral areas support higher densities of wild ungulates during the wet season than the Lake Mburo National Park in Uganda. Journal of Tropical Ecology, 22, 675–83.

Rappole, J.H. (2003) An integrative framework for understanding the origin and evolution of avian migration. Journal of Avian Biology, 34, 124–28.

Rappole, J.H. and Warner, D.W. (1976) Relationships between behavior, physiology and weather in avian transients at a migration stopover site. Oecologia, 26, 193–212.

Rappole, J.H. and Jones, P. (2002) Evolution of Old and New World migration systems. Ardea, 90, 525–37.

Ratikainen, I.L., Gill, J.A., Gunnarsson, T.G., Sutherland, W.J., and Kokko, H. (2007) When density-dependence is not instantaneous: theoretical developments and management implications. Ecology Letters, 10, 1–15.

Rautenstrauch, K.R. and Krausman, P.R. (1989) Influence of water availability and rainfall on movements of desert mule deer. Journal of Mammalogy, 70, 197–201.

Rayfield, B., James, P., Fall, A., and Fortin, M.-J. (2008) Comparing static versus dynamic protected areas in dynamic boreal ecosystems. Biological Conservation, 141, 438–49.

Reed, J. M. (2004) Recognition behavior based problems in species conservation. Annales Zoologici Fennici, 41, 859–877.

Rees, W.G., Stammler, F.M., Danks, F.S., and Vitebksy, P. (2008) Vulnerability of European reindeer husbandry to global change. Climatic Change, 87, 199–217.

Reynolds, D.R. and Riley, J.R. (1997) Flight Behaviour and Migration of Insect Pests: Radar Studies in Developing Countries, NRI Bulletin 71. Natural Resources Institute: Chatham, UK.

Reynolds, D.R. and Riley, J.R. (2002) Remote-sensing, telemetric and computer-based technologies for investigating insect movement: a survey of existing and potential techniques. Computers and Electronics in Agriculture, 35, 271–307.

Reynolds, D.R., Smith, A.D., and Chapman, J.W. (2008) A radar study of emigratory flight and layer formation by insects at dawn over southern Britain. Bulletin of Entomological Research, 98, 35–52.

Richardson, P.B., Broderick, A.C., Campbell, L.M., Godley, B.J., and Ranger, S. (2006) Marine turtle fisheries in the UK Overseas Territories of the Caribbean: domestic legislation and the requirements of multilateral agreements. Journal of International Wildlife Law and Policy, 9, 223–46.

Richardson, W.J. (1978) Timing and amount of bird migration in relation to weather: A review. Oikos, 30, 224–72.

Richardson, W.J. (1979) Southeastward shorebird migration over Nova Scotia and New Brunswick in autumn—Radar study. Canadian Journal of Zoology/Revue canadienne de zoologie, 57, 107–24.

Richardson, W.J. (1990) Timing and amount of bird migration in relation to weather: updated review. In E Gwinner (ed.) Bird Migration: Physiology and Ecophysiology, pp. 78–101, Springer: Berlin, Heidelberg.

Ricker, W. (1954) Stock and recruitment. Journal of the Fisheries Research Board of Canada, 11, 559–623.

Ricklefs, R.E. (2002) Splendid isolation: historical ecology of the South American passerine fauna. Journal of Avian Biology, 33, 207–11.

Roberge, J.M. and Angelstam, P. (2004) Usefulness of the umbrella species concept as a conservation tool. Conservation Biology, 18, 1, 76–85.

Robinson, D.W., Bowlin, M.S., Bisson, I., Shamoun-Baranes, J., Thorup, K., Diehl, R., Kunz, T., Mabey, S., and Winkler, D.W. (2010) Integrating concepts and technologies to advance the study of bird migration. Frontiers in Ecology and the Environment, doi:10.1890/080179.

Robinson, R.A., Learmouth, J.A., Hutson, A.M., MacLeod, C.D., Sparks, T.H., Leech, D.I., Pierce, G.J., Rehfische, M.M., and Crick, H.Q.P. (2005) Climate Change and Migratory Species, British Trust for Ornithology: Thetford, UK.

Robinson R.A., Crick H.Q.P., Learmonth J.A., Maclean I.M.D., Thomas C.D., Bairlein F., Forchhammer M.C., Francis C.M., Gill J.A., Godley B.J., Harwood J., Hays GC., Huntley B., Hutson A.M., Pierce G.J., Rehfisch M.M., Sims D.W., Santos B.M., Sparks T.H., Stroud D.A., and Visser M.E. (2008) Travelling through a warming world: climate change and migratory species, Endangered Species Research, 7, 87–99.

Robinson, S. and Milner-Gulland, E.J. (2003) Contraction of livestock mobility resulting from state far reorganisation. In C.K. Kerven, (ed.) Prospects for Pastoralism in Kazakhstan and Turkmenistan: From State Farms to Private Flocks, RoutledgeCurzon: London.

Robinson, S. and Milner-Gulland, E.J. (2003) Political change and factors limiting numbers of wild and domestic ungulates in Kazakhstan. Human Ecology, 31, 87–110.

Robinson, S., Milner-Gulland, E.J., and Alimaev, I. (2003) Rangeland degradation in Kazakhstan during the Soviet era: re-examining the evidence. Journal of Arid Environments, 53, 419–39.

Robinson, T. and Minton, C. (1989) The enigmatic banded stilt. Birds International, 1, 72–85.

Roed, K., Flagstad, Ø., Nieminen, M., Holand, Ø., Dwyer, M., Røv, N., and Vilà, C. (2008) Genetic analyses reveal independent domestication origins of Eurasian Reindeer. Proceedings of the Royal Society, 275, 1849–55.

Roff, D.A. (1990) The evolution of flightlessness in insects. Ecological Monographs, 60, 389–421.

Roff, D.A. and Fairbairn, D.J. (2007) The evolution and genetics of migration in insects. Bioscience, 57, 155–64.

Romanczuk, P., Couzin, I.D., and Schimansky-Geier, L. (2009) Collective motion due to individual escape and pursuit response. Physical Review Letters, 102, 4.

Ropert-Coudert, Y. and Wilson, R.P. (2005) Trends and perspectives in animal-attached remote sensing. Frontiers in Ecology and the Environment, 3, 437–44.

Rosenzweig, C., Iglesias, A., Yang, X.B., Epstein, P.R., and Chivian, E. (2001) Climate change and extreme weather events: implications for food production, plant diseases, and pests. Global Change and Human Health, 2, 2, 90–104.

Roshier, D., Asmus, M., and Klaassen, M. (2008) What drives long-distance movements in the nomadic grey teal *Anas gracilis* in Australia? Ibis, 150, 474–84.

Royle, J.A. and R.M. Dorazio. (2008) Hierarchical Modeling and Inference in Ecology: The Analysis of Data from Populations, Metapopulations and Communities, Academic Press, Elsevier: London.

Rubolini, D., Møller, A.P., Rainio, K. and Lehikoinen, E. (2007) Intraspecific consistency and geographic variability in temporal trends of spring migration phenology among European bird species. Climate Research, 35, 135–46.

Rudnick, J.A., Katzner, T.E., Bragin, E.A., and deWoody, A.J. (2008) A non-invasive genetic evaluation of population size, natal philopatry, and roosting behavior of non-breeding eastern imperial eagles (*Aquila heliaca*) in central Asia. Conservation Genetics, 9, 667–76.

Ruess, R.W. and Seagle, S.W. (1994) Landscape patterns in soil microbial processes in the Serengeti National Park, Tanzania. Ecology, 75, 892–904.

Rutz, C. and Hays, G.C. (2009) New frontiers in biologging science. Biology Letters, 5, 289–92.

Sabo, J.L. and Power, M.E. (2002) River-watershed exchange: Effects of riverine subsidies on riparian lizards and their terrestrial prey. Ecology, 83, 1860–69.

Sadovy, Y. and Domeier, M. (2005) Are aggregation-fisheries sustainable? Reef fish fisheries as a case study. Coral Reefs, 24, 2, 254–62.

Saiga News. (2009) Updates, 8, 2–7.

Saino, N. and Ambrosini, R. (2008) Climatic connectivity between Africa and Europe may serve as a basis for phenotypic adjustment of migration schedules of trans-Saharan migratory birds. Global Change Biology, 14, 250–63.

Sala, E., Ballesteros, E., and Starr, R.M. (2001) Rapid decline of Nassau grouper spawning aggregations in Belize: fishery management and conservation needs. Fisheries, 26, 23–30.

Salewski, V. and Bruderer, B. (2007) The evolution of bird migration—a synthesis. Naturwissenschaften, 94, 268–79.

Salewski, V. and Jones, P. (2006) Palearctic passerines in Afrotropical environments: a review. Journal of Ornithology, 147, 192–201.

Sanderson, F.J., Donald, P.F., Pain, D.J., Burfield, I.J., and van Bommel, F.P.J. (2006) Long-term population declines in Afro-Palearctic migrant birds. Biological Conservation, 131, 93–105.

Sapir, N. (2009) The effects of weather on Bee-eater (*Merops apiaster*) migration, PhD thesis, The Hebrew University of Jerusalem.

Sapir, N., Abramsky, Z., Shochat, E., and Izhaki, I. (2004a) Scale-dependent habitat selection in migratory frugivorous passerines. Naturwissenschaften, 91, 544–47.

Sapir, N., Tsurim, I., Gal, B., and Abramsky, Z. (2004b) The effect of water availability on fuel deposition of two staging Sylvia warblers. Journal of Avian Biology, 35, 25–32.

Sato, K., Watanuki, Y., Takahashi, A., *et al.* (2007) Stroke frequency, but not swimming speed, is related to body size in free-ranging seabirds, pinnipeds and cetaceans. Proceedings of the Royal Society B-Biological Sciences, 274, 471–77.

Satterthwaite, W.H., Beakes, M.P., Collins, E., Swank, D.R., Merz, J.E., Titus, R.G., Sogard, S.M., and Mangel, M. (2009) Steelhead life history on California's central coast: insights from a state dependent model. Transactions of the American Fisheries Society, 138, 532–48.

Satterthwaite, W.H., Beakes, M.P., Collins, E., Swank, D.R., Merz, J.E., Titus, R.G., Sogard, S.M., and Mangel, M. (2010) State-dependent life history models in a changing (and regulated) environment: steelhead in the California Central Valley. Evolutionary Applications, 3, 221–43.

Sauerbrei, W., Hollander, N., and Buchholz, A. (2008) Investigation about a screening step in model selection. Statistics and Computing, 18, 195–208.

Sawyer, H., Kauffman, M.J., Nielson, R.M., and Horne, J.S. (2009) Identifying and prioritizing ungulate migration routes for landscape-level conservation. Ecological Applications, 19, 2016–25.

SCFC (Siberian Crane Flyway Coordination) (Undated a)Reintroduction: Hang-glider Migration, [Online], Available at: http://www.sibeflyway.org/Reintroduction-Hang-glider%20Mig-web.html. [August 2009].

SCFC (Siberian Crane Flyway Coordination) (Undated b) Reintroduction. [Online], Available at: http://www.sibeflyway.org/Reintroduction-Russia-web.html#kurn. [August 2009].

Schaefer, J.A. (2003) Long-term range recession and the persistence of caribou in the taiga. Conservation Biology, 17, 1435–39.

Schaefer, J.A., Bergman, C.M., and Luttich, S.N. (2000) Site fidelity of female caribou at multiple spatial scales. Landscape Ecology, 15, 731–39.

Schaefer, J.A., Veitch, A.M., Harrington, F.H., Brown, W.K., Theberge, J.B., and Luttich, S.N. (2001) Fuzzy structure and spatial dynamics of a declining woodland caribou population. Oecologia, 126, 507–14.

Schaub, M., Jenni, L., and Bairlein, F. (2008) Fuel stores, fuel accumulation, and the decision to depart from a migration stopover site. Behavioral Ecology, 19, 657–66.

Schaub, M., Pradel, R., and Lebreton, J.-D. (2004) Is the reintroduced white stork (Ciconia ciconia) population in Switzerland self-sustainable? Biological Conservation, 119, 105–14.

Schick, R.S., S.R. Loarie, F. Colchero, B.D. Best, A. Boustany, D.A. Conde, P.N. Halpin, L.N. Joppa, C.M. McClellan, and J.S. Clark (2008) Understanding movement data and movement processes: current and emerging directions. Ecology Letters, 11, 1338–50.

Schlecht, E., P. Hiernaux, and M.D. Turner (2001) Mobilité régionale du bétail: nécessité et alternatives? In E. Tielkes, E. Schlecht, and P. Hiernaux (eds) Elevage et gestion de parcours au Sahel, implications pour le développement, Verlag Grauer: Beuren-Stuttgart.

Schlichting, C.D. and Pigliucci, M. (1998) Phenotypic evolution: a reaction norm perspective. Sinauer Associates, Inc., Sunderland.

Schmaljohann, H. and Dierschke, V. (2005) Optimal bird migration and predation risk: a field experiment with northern wheatears Oenanthe oenanthe. Journal of Animal Ecology 74, 131–38.

Schmidt, K.A. (2004) Site fidelity in temporally correlated environments enhances population persistence. Ecology Letters, 7, 176–84.

Schmidt-Nielsen, K. (1972) Locomotion: energetic cost of swimming, flying and running. Science, 177, 222–28.

Schmidt-Nielsen, K. (1975) Animal Physiology. Cambridge University Press: Cambridge.

Schmidt-Wellenburg, C.A., Engel, S., and Visser, G.H. (2008) Energy expenditure during flight in relation to body mass: effects of natural increases in mass and artificial load in Rose Coloured Starlings. Journal of Comparative Physiology B-Biochemical Systemic and Environmental Physiology, 178, 767–77.

Schmitz, J. (1986) L'état géomètre: les leydi des Peul du Fuuta Tooro (Sénégal) et du Maasina (Mali). Cahiers d'Etudes Africaines, 26, 3, 349–94.

Schoenecker, K.A., Singer, F.J., Zeigenfuss, L.C., Binkley, D., and Menezes, R.S.C. (2004) Effects of elk herbivory on vegetation and nitrogen processes. Journal of Wildlife Management, 68, 837–49.

Schuman, J.R. (1995) Environmental considerations for assessing dam removal alternatives for river restoration. Regulated Rivers: Research and Management, 11, 249–61.

Scoones, I. (1995) Exploiting heterogeneity: habitat use by cattle in dryland Zimbabwe. Journal of Arid Environments, 29, 221–37.

Seabloom, E.W., Dobson, A.P., and Stoms, D.M. (2002) Extinction rates under nonrandom patterns of habitat loss. Proceedings of the National Academy of Sciences, 99, 17, 11229–34.

Seagle, S.W. (2003) Can ungulates foraging in a multiple-use landscape alter forest nitrogen budgets? Oikos 103, 230–34.

Seagle, S.W., S.J. McNaughton, and R.W. Ruess (1992) Simulated effects of grazing on soil nitrogen and mineralization in contrasting Serengeti grasslands. Ecology, 73, 1105–23.

Secundus, C.P. (77) Historia Naturalis.

Seeley, T.D. (1985), Honeybee ecology, Princeton University Press: Princeton.

Seiler, A., Cedarlund, G, Jernelid, H, Grängstedt P., and Ringaby, E. (2003) The barrier effect of highway E4 on migratory moose (Alces alces) in the High Coast area, Sweden. Proceedings of the IENE conference on Habitat fragmentation due to transport infrastructure, Brussels, 13–14 November 2003, [Online], Available at: http://www.grimso.slu.se/research/infrastructure/Documents/HighCoast_IENE2003.pdf. [March 2009]

Senft, R.L., Coughenour, M.B., Bailey, D.W., Rittenhouse, L.R., Sala, O.E., and Swift, D.M. (1987) Large herbivore foraging and ecological hierarchies: landscape ecology can enhance traditional foraging theory. BioScience, 37, 11, 789–99.

Serneels, S. and Lambin, E. (2001) Impact of land-use change on the wildebeest migration in the northern part of the Serengeti-Mara ecosystem. Journal of Biogeography, 28, 391–407.

Shahgedanova, M. (ed.) (2003) The Physical Geography of Northern Eurasia, Oxford University Press: Oxford.

Shamoun-Baranes, J., Liechti, O., Yom-Tov, Y., and Leshem, Y. (2003a) Using a convection model to predict altitudes of white stork migration over central Israel. Boundary-Layer Meteorology, 107, 673–81.

Shamoun-Baranes, J., Leshem, Y., Yom-Tov, Y., and Liechti, O. (2003b) Differential use of thermal convection by soaring birds over central Israel. Condor, 105, 208–18.

Shannon, H., Young, G., Yates, M., Fuller, M., and Seegar, W. (2002) American white pelican soaring flight times and altitudes relative to changes in thermal depth and intensity. Condor, 104, 679–83.

Shea, R.E., Nelson, H.K., Gillette, L.N., King, J.G., and Weaver, D.K. (2002) Restoration of Trumpeter Swans in North America: A Century of Progress and Challenges. Waterbirds: The International Journal of Waterbird Biology, 25, Special Publication 1, Proceedings of the Fourth International Swan Symposium, 2001, pp. 296–300.

Shepherd, B. and Whittington, J. (2006) Response of wolves to corridor restoration and human use management. Ecology and Society, 11, 2, 1. [Online], Available: http://www.ecologyandsociety.org/vol11/iss2/art1/[November 2009].

Sherrill-Mix, S.A., James, M.C., and Myers, R.A. (2008) Migration cues and timing in leatherback sea turtles. Behavioral Ecology, 19, 2, 231–236.

Shields, E. J. and Testa, A. M. (1999) Fall migratory flight initiation of the potato leafhopper, *Empoasca fabae* (Homoptera: Cicadellidae): observations in the lower atmosphere using remote piloted vehicles. Agricultural and Forest Meteorology, 97, 317–30.

Shillinger, G.L., Palacios, D.M., Bailey, H., Bograd, S.J., Swithenbank, A.M., Gaspar, P., Wallace, B.P., Spotila, J.R., Paladino, F.V., Piedra, R., Eckert, S.A and, Block, B.A. (2008) Persistent leatherback turtle migrations present opportunities for conservation. Public Library of Science, Biology, 6, 7, 1408–16.

Shine, R. and Brown, G.P. (2008) Adapting to the unpredictable: reproductive biology of vertebrates in the Australian wet-dry tropics. Philosophical Transactions of the Royal Society of London, Series B, 363, 363–73.

Sibert, J.R. and Fournier, D.A. (2001) Possible Models for combining tracking data with conventional tagging data. Symposium on Tagging and Tracking Marine Fish with Electronic Devices, February 7–11 2000, Honolulu. In J. R. Sibert and J. L. Nielsen (eds) Electronic Tagging and Tracking in Marine Fisheries, Kluwer Academic Press.

Sih, A., Bell, A.M., Johnson, J.C., and Ziemba, R.E. (2004) Behavioral syndromes: an integrative overview. The Quarterly Review of Biology, 79, 241–77.

Sillett, T.S. and Holmes, R.T. (2002) Variation in survivorship of a migratory songbird throughout its annual cycle. Journal of Animal Ecology, 71, 296–308.

Simons, A.M. (2004) Many wrongs: the advantage of group navigation. Trends in Ecology and Evolution 19, 453–55.

Simpson, S.J. and Miller, G.A. (2007) Maternal effects on phase characteristics in the desert locust, *Schistocerca gregaria*: A review of current understanding. Journal of Insect Physiology, 53, 869–76.

Simpson, S.J. and Sword, G.A. (2008) Locusts. Current Biology, 18, 364–66.

Simpson, S.J. and Sword, G.A. (2009) Phase polyphenism in locusts: Mechanisms, population consequences, adaptive significance and evolution. In D. Whitman and T. Ananthakrishnan (eds) Phenotypic Plasticity of Insects: Mechanisms and Consequences, pp 147–90, Science Publishers Inc.: Plymouth.

Simpson, S.J., Despland, E., Hagele, B.F., and Dodgson, T. (2001) Gregarious behavior in desert locusts is evoked by touching their back legs. Proceedings of the National Academy of Sciences of the United States of America, 98, 3895–97.

Simpson, S.J., Sword, G.A., Lorch, P.D., and Couzin, I. D. (2006) Cannibal crickets on a forced march for protein and salt. Proceedings of the National Academy of Sciences of the United States of America, 103, 4152–56.

Sims, D.W., Southall, E.J., Humphries, N.E., *et al.* (2008) Scaling laws of marine predator search behaviour. Nature, 451, 1098–1102.

Sinclair, A.R.E. (1977) The African Buffalo: a Study of Resource Limitation of Populations, University of Chicago Press, Chicago.

Sinclair, A.R.E. (1978) Factors affecting food supply and breeding season of resident birds and movements of palaearctic migrants in a tropical African savanna. Ibis, 120, 480–97.

Sinclair, A.R.E. (1979) The Serengeti environment, pp. 31–45 in A.R.E. Sinclair (ed.) Serengeti—Dynamics of an Ecosystem. University of Chicago Press, Chicago.

Sinclair, A.R.E. (1983) The function of distance movement in vertebrates. In I.R. Swingland and P.J. Greenwood (eds) The Ecology of Animal Movement, pp. 248–58, Clarendon Press: Oxford.

Sinclair, A.R.E. (1985) Does interspecific competition or predation shape the African ungulate community. Journal of Animal Ecology 54, 899–918.

Sinclair, A.R.E., Dublin, H., and Borner, M. (1985) Population regulation of Serengeti wildebeest—a test of the food hypothesis. Oecologia, 65, 266–68.

Sinclair, A.R.E., Mduma, S., and Brashares, J.S. (2003) Patterns of predation in a diverse predator-prey system. Nature, 425, 288–90.

Sinclair, A.R.E., Mduma, S.A.R., Hopcraft, J.G.C., Fryxell, J.M., Hilborn, R., and Thirgood, S. (2007) Long-term ecosystem dynamics in the Serengeti: Lessons for conservation. Conservation Biology, 21, 580–90.

Singh, N., Milner-Gulland, E.J. (in press) Conserving a moving target: Planning protection for a migratory species as its distribution changes. *Journal of Applied Ecology*

Singh, N.J., Grachev, Iu.A., Bekenov, A.B., and Milner-Gulland, E.J. (2010) Tracking greenery in Central Asia: The migration of the saiga antelope. Diversity and Distributions 16, 663–75.

Skelly, D.K. and Werner, E.E. (1990) Behavioral and life-historical responses of larval American toads to an odonate predator. Ecology, 71, 2313–22.

Sladen, W.J.L. and Olsen, G.H. (2005) Teaching geese, swans and cranes pre-selected migration routes using ultralight aircraft, 1990–2004—looking into the future. In M.H. Linck and R.E. Shea (eds) Selected papers of the twentieth trumpeter swan society conference—Trumpeter Swan Restoration: Exploration and Challenges, 20–22 October 2005, Council Bluffs, Iowa, pp. 53–54.

Sladen, W.J.L., Lishman, W.A., Ellis, D.H., Shire, G.G., and Rininger, D.L. (2002) Teaching migration routes to canada geese and trumpeter swans using ultralight aircraft, 1990–2001. Waterbirds, The International Journal of Waterbird Biology, 25, Special Publication 1: Proceedings of the Fourth International Swan Symposium, 2001, pp. 132–37.

Slotte, A. and Fiksen, Ø. (2000) State-dependent spawning migration in Norwegian spring-spawning herring. Journal of Fish Biology, 56, 138–62.

Small, C.J. (2005) Regional Fisheries Management Organisations: Their Duties and Performance in Reducing Bycatch of Albatrosses and Other Species, BirdLife International: Cambridge, UK.

Smith, R.J. and Moore, F.R. (2003) Arrival fat and reproductive performance in a long-distance passerine migrant. Oecologia, 134, 325–31.

Smith, R.J. and Moore, F.R. (2005) Arrival timing and seasonal reproductive performance in a long-distance migratory landbird. Behavioral Ecology and Sociobiology, 57, 231–39.

Smith, S.B. and McWilliams, S.R. (2010) Patterns of fuel use and storage in migrating passerines in relation to fruit resources at autumn stopover sites. Auk, 127, 108–18.

Smuts, G. L. (1978) Interrelations between Predators, Prey, and Their Environment. Bioscience, 28, 316–20.

Sniegowski, P.D., Ketterson, E.D., and Nolan, V. (1988) Can experience alter the avian annual cycle—Results of migration experiments with Indigo Buntings. Ethology, 79, 333–41.

Snow, D. and Perrins, C. (1998) The Complete Birds of the Western Palearctic on CD-ROM, Oxford University Press: Oxford, England.

Sokolov, V.E. and Zhirnov, L.V. (eds) (1998) The Saiga Antelope; Phylogeny, Systematics, Ecology, Conservation and Use, Russian Academy of Sciences: Moscow.

Solbreck, C. (1978) Migration, diapause, and direct development as alternative life histories in a seed bug, Neacoryphus bicrucis. In H. Dingle (ed.) Evolution of Insect Migration and Diapause, pp.195–217, Springer: New York.

Sorokin, A., Shilina, A., Ermakov, A., and Markin, Y. (2002) Hang-glider Migration: Flight of Hope, [Online], Available at: http://www.sibeflyway.org/Reintroduction-Flight-of-Hope-Project-web.html. [August 2009].

Soulé, M.E. (1985) What is conservation biology? BioScience, 35, 727–34.

Soulé, M.E. and Kohm, K. (1989) Research Priorities in Conservation Biology, Island Press, Washington, DC.

Southwood, A. and Avens, L. (2010) Physiological, behavioural, and ecological aspects of migration in reptiles. Journal of Comparative Physiology Part B, 180, 1–23.

Southwood, A, Fritsches, K, Brill, R, and Swimmer, Y. (2008) Sound, chemical, and light detection in sea turtles and pelagic fishes: sensory-based approaches to bycatch reduction in longline fisheries. Endangered Species Research, 5, 225–38.

Southwood, T.R.E. (1962) Migration of terrestrial arthropods in relation to habitat. Biological Reviews, 37, 171–214.

Southwood, T.R.E. (1977) Habitat, the templet for ecological strategies? Journal of Animal Ecology, 46, 337–65.

Southwood, T.R.E. (1981) Ecological aspects of insect migration. In D.J. Aidley (ed.) Animal Migration, pp. 197–208, Cambridge University Press: Cambridge.

Sparks, T.H., Dennis, R.L.H., Croxton, P.J. and Cade, M., (2007) Increased migration of Lepidoptera linked to climate change. European Journal of Entomology, 104, 139–43.

Spiegelhalter, D. and Rice, K. (2009) Bayesian statistics. Scholarpedia, 4, 8, 5230.

Spotila, J., Reina, R.D., Steyermark, A.C., Plotkin, P.T., and Spotila, F.V. (2000) Pacific leatherback turtles face extinction. Nature, 405, 529–31.

Srygley, R.B. and Dudley, R. (2008) Optimal strategies for insects migrating in the flight boundary layer: mechanisms and consequences. Integrative and Comparative Biology, 48, 119–33.

Stammler, F. (2005) Reindeer Nomads Meet the Market: Culture, Property and Globalisation at the End of the Land. Vol. 6. Halle Studies in the Anthropology of Eurasia. Münster: Lit publishers.

Stammler, F.(2009) Mobile phone revolution in the tundra? Technological change among Russian reindeer nomads. In A. Ventsel (ed.) Generation P in the Tundra, Folklore, 41, 47–78, Estonian Literary Museum: Talinn.

Stammler, F. and Beach, H. (eds) (2006) People and Reindeer on the Move. Special Issue of the journal Nomadic Peoples, 10, 2, Berghahn Publishers: Oxford.

Stammler, F. and Peskov, V. (2008) Building a 'Culture of dialogue' among stakeholders in North-West Russian oil extraction. Europe-Asia Studies, 60, 5, 831–849.

Standen, E.M., Hinch, S.G., and Rand, P.S. (2004) Influence of river speed on path selection by migrating adult sockeye salmon (Oncorhynchus nerka) Canadian Journal of Fisheries and Aquatic Sciences, 61, 905–12.

Stanley, E.H. and Doyle, M.W. (2003) Trading off: the ecological effects of dam removal. Frontiers in Ecology and the Environment, 1, 15–22.

Stanley, E.H., Catalano, M.J., Mercado-Silva, N., and Orr, C.H. (2007) Short communication: effects of dam removal on brook trout in a Wisconsin stream. River Research and Applications, 23, 792–98.

Stapp, P. and Polis, G.A. (2003) Marine resources subsidize insular rodent populations in the Gulf of California, Mexico. Oecologia, 134, 496–504.

Stefansson, S.O., Björnsson, B.Th., and McCormick, S.D. (2008) Smoltification. In R. N. Finn and B. K. Kapo (eds)

Fish Larval Physiology, pp. 639–81, Science Publishers: Enfield NH, USA.

Stefansson, S.O., Nilsen, T.O., Ebbesson, L.O.E., Wargelius, A., Madsen, S.S., Björnsson, B.T., and McCormick, S.D. (2007) Molecular mechanisms of continuous light inhibition of Atlantic salmon parr-smolt transformation. Aquaculture, 273, 235–45.

Stenning, D.J. (1957) Transhumance, migratory drift, migration: patterns of pastoral Fulani nomadism. Journal of the Royal Anthropological Institute, 87, 57–73.

Stephens, D.M. and Krebs, J.R. (1986) Foraging Theory, Princeton University Press, Princeton, New Jersey.

Stephens, D.W., Brown, J.S., and Ydenberg, R.C. (2007) Foraging Behavior and Ecology. University of Chicago Press: Chicago.

Stige, L.C., Chan, K.-S., Zhang, Z., Frank, D., and Stenseth, N.C. (2007) Thousand-year-long Chinese time series reveals climatic forcing of decadal locust dynamics. Proceedings of the National Academy of Sciences USA, 104, 16188–93.

Stoddard, P.K., Marsden, J.E., and Williams, T.C. (1983) Computer simulation of autumnal bird migration over the western North Atlantic. Animal Behaviour, 31, 173–80.

Strandberg, R. (2008) Migration strategies of raptors: Spatio-temporal adaptations and constraints in travelling and foraging. Ph.D. thesis, Lund University: Lund.

Strandberg, R. and Alerstam, T. (2007) The strategy of fly-and-forage migration, illustrated for the osprey (Pandion haliaetus). Behavioral Ecology and Sociobiology, 61, 1865–75.

Strandberg, R., Klaassen, R.H.G., Hake, M., and Alerstam, T. (2009) How hazardous is the Sahara Desert crossing for migratory birds? Indications from satellite tracking of raptors. Biology Letters, doi:10.1098/rsbl.2009.0785.

Strandberg, R., Klaassen, R.H.G., Hake, M., Olofsson, P., and Alerstam, T. (2009) Converging migration routes of Eurasian hobbies Falco subbuteo crossing the African equatorial rain forest. Proceedings of the Royal Society B-Biological Sciences, 276, 727–33.

Strindberg, S. and Buckland, S.T. (2004) Zigzag Survey Designs in Line Transect Sampling. Journal of Agricultural, Biological, and Environmental Statistics, 9, 443–61.

Studds, C.E. and Marra, P.P. (2005) Nonbreeding habitat occupancy and population processes: an upgrade experiment with a migratory bird. Ecology, 86, 2380–85.

Studds, C.E. and Marra, P.P. (2007) Linking fluctuations in rainfall to nonbreeding season performance in a long-distance migratory bird, Septophaga ruticilla. Climate Research, 35, 115–22.

Stull, R.B. (1988) An Introduction to Boundary Layer Meteorology, Kluwer Academic Publishers: Boston.

Stutchbury, B.J.M., Tarof, S.A., Done, T., Gow, E., Kramer, P.M., Tautin, J., Fox, J.W., and Afanasuev, V. (2009) Tracking long-distance songbird migration by using geolocators. Science, 323, 896.

Sunnucks, P. (2000) Efficient genetic markers for population biology. Trends in Ecology and Evolution, 15, 199–203.

Sutherland, W.J. (1983) Aggregation and the 'Ideal Free' distribution. Journal of Animal Ecology, 52, 821–28.

Sutherland, W.J. (1996) From Individual Behaviour to Population Ecology, Oxford University Press: Oxford.

Sutherland, W.J. (1998) Evidence for flexibility and constraint in migration systems. Journal of Avian Biology, 29, 441–46.

Sutherland, W.J., Clout, M., Cote, I.M., Daszak, P., Depledge, M.H., Fellman, L., Fleishman, E., Garthwaite, R., Gibbons, D.W., De Lurio, J., Impey, A.J., Lickorish, F., Lindenmayer, D., Madgwick, J., Margerison, C., Maynard, T., Peck, L.S., Pretty, J., Prior, S., Redford, K.H., Scharlemann, J.P.W., Spalding, M., and Watkinson, A.R. (2010) A horizon scan of global conservation issues for 2010. Trends in Ecology and Evolution, 25, 1–7.

Svedang, H. and Wickstrom, H. (1997) Low fat contents in female silver eels: indications of insufficient energetic stores for migration and gonadal development. Journal of Fish Biology, 50, 475–86.

Switzer, P.V. (1997) Past reproductive success affects future habitat selection. Behavioral Ecology and Sociobiology, 40, 307–12.

Sword, G.A. (2003) To be or not to be a locust? A comparative analysis of behavioral phase change in nymphs of Schistocerca americana and S. gregaria. Journal of Insect Physiology, 47, 709–17.

Sword, G.A., Lorch, P.D., and Gwynne, D.T. (2005) Migratory bands give crickets protection. Nature, 433, 703.

Takakura, H. (2002) An institutionalized human-animal relationship and the aftermath: the reproductive process of horse-bands and husbandry in Northern Yakutia, Siberia. Human Ecology, 30, 1–19.

Takakura, H. (2004) Gathering and Releasing Animals: Reindeer herd control activities of the indigenous peoples of the Verkhoyansky Region, Siberia. Bulletin of the National Museum of Ethnology, 29, 1, 43–70.

Takimoto, G., Iwata, T., and Murakami, M. (2002) Seasonal subsidy stabilizes food web dynamics: Balance in a heterogeneous landscape. Ecological Research, 17, 433–39.

Talbot, L.M. and Talbot M.H., (1963) The wildebeest in western Masailand, East Africa. Wildlife Monographs, 12, 1–88.

Taylor, C.M. and Norris, D.R. (2007) Predicting conditions for migration: effects of density dependence and habitat quality. Biology Letters, 3, 280–83.

Taylor, L. R. (1974) Insect migration, flight periodicity and boundary-layer. Journal of Animal Ecology, 43, 225ff.

Taylor, L.R. (1986) Synoptic ecology, migration of the second kind, and the Rothamssted Insect Survey. Presidential address. Journal of Animal Ecology, 55, 1–38.

Teo, S.L.H., Boustany, A., Dewar, H., *et al.* (2007) Annual migrations, diving behavior, and thermal biology of Atlantic bluefin tuna, *Thunnus thynnus*, on their Gulf of Mexico breeding grounds. Marine Biology, 151, 1–18.

Terborgh, J. and van Schaik, C. (2002) Why the world needs parks. In J. Terborgh, C. van Schaik, L. Davenport, and M. Rao (eds) Making Parks Work: Strategies for Preserving Tropical Nature, pp. 3–14, Island Press: Washington, DC.

Tevis, L., Jr. and Newell, I.M. (1962) Studies on the biology and seasonal cycle of the Giant Red Velvet Mite, *Dinothrombium pandorae* (Acari, Trombidiidae). Ecology, 43, 497–505.

Thirgood, S., Mosser, A., Tham, S., Hopcraft, G., Mwangomo, E., Mlengeya, T., Kilewo, M., Fryxell, J., Sinclair, A.R.E., and Borner, M. (2004) Can parks protect migratory ungulates? The case of the Serengeti wildebeest. Animal Conservation, 7, 113–20.

Thomas, C.D., Cameron, A., Green, R.E., Bakkenes, M., Beaumont, L.J., Collingham, Y.C., Erasmus, B. F. N., Ferreira de Siqueira, M., Grainger, A., Hannah, L., Hughes, L., Huntley, B., van Jaarsveld, A.S., Midgley, G.F., Miles, L., Ortega-Huerta, M.A., Peterson, A. T., Phillips, O.L., and Williams, S.E. (2004) Extinction risk from climate change. Nature, 427, 8, 145–48.

Thomson, A.L. (1926) Problems of Bird-Migration, Witherby: London.

Thorpe, J. E. (1977) Bimodal distribution of length of juvenile Atlantic salmon (*Salmo salar* L) under artificial rearing conditions. Journal of Fish Biology, 11, 175–84.

Thorpe, J.E. (1988) Salmon migration. Scientific Progress, 72, 345–70.

Thorup, K., Alerstam, T., Hake, M., and Kjellen, N. (2003) Bird orientation: compensation for wind drift in migrating raptors is age dependent. Proceedings of the Royal Society of London Series B-Biological Sciences, 270, S8–S11.

Thorup, K., Bisson, I. A., Bowlin, M. S., Holland, R. A., Wingfield, J. C., Ramenofsky, M., and Wikelski, M. (2007) Evidence for a navigational map stretching across the continental US in a migratory songbird. Proceedings of the National Academy of Sciences USA, 104, 18115–19.

Thrower, F.P., Hard, J.J., and Joyce, J.E. (2004) Genetic architecture of growth and early life-history transitions in anadromous and derived freshwater populations of steelhead. Journal of Fish Biology, 65 (Supplement A), 286–310.

Thrush, S.F., Pridmore, R.D., Hewitt J.E., and Cummings, V.J. (1994) The Importance of Predators on a Sandflat—Interplay between Seasonal-Changes in Prey Densities and Predator Effects, Marine Ecology-Progress Series, 107, pp. 211–22.

Tilman, D. (1994) Competition and Biodiversity in Spatially Structured Habitats. Ecology, 75, 2–16.

Torney, C.N.Z., Neufeld, Z., and Couzin, I.D. (2009) Context-dependent interaction leads to emergent search behavior in social aggregates. Proceedings of the National Academy of Sciences USA, 106, 22055–60.

Torre-Bueno, J.R. (1976) Temperature regulation and heat dissipation during flight in birds. Journal of Experimental Biology, 65, 471–82.

Trakimas, G. and Sidaravičius, J. (2008) Road mortality threatens small northern populations of the European pond turtle, *Emys orbicularis*. Acta Herpetologica, 3, 2, 161–66.

Treherne, J.E. and Foster, W.A. (1981) Group transmission of predator avoidance-behaviour in a marine insect—the Trafalgar effect. Animal Behaviour, 29, 911–17.

Tremblay, Y., Robinson, P.W., and Costa, D.P. (2009) A parsimonious approach to modeling animal movement data. PLoS ONE, 4, e4711.

Tremblay, Y., S.A. Shaffer, S.L. Fowler, C.E. Kuhn, B.I. McDonald, M.J. Weise, C.-A. Bost, H. Weimerskirch, D. E. Crocker, M.E. Goebel, and D.P. Costa. (2006) Interpolation of animal tracking data in a fluid environment. Journal of Experimental Biology, 209, 128–40.

Tsurim, I., Sapir, N., Belmaker, J., Shanni, I., Izhaki, I., Wojciechowski, M.S., Karasov, W.H., and Pinshow, B. (2008) Drinking water boosts food intake rate, body mass increase and fat accumulation in migratory blackcaps (*Sylvia atricapilla*). Oecologia, 156, 21–30.

Tucker, G. and Goriup, P. (2005) Assessment of the merits of an instrument under the Convention on Migratory Species covering migratory raptors in the African–Eurasian region: status report, DEFRA: Bristol, UK.

Tuisku, T. (2002), Reindeer herding. In Ildikó Lehtinen (ed.) Siberia: Life on the Taiga and Tundra, pp. 100–107, National Board of Antiquities: Helsinki.

Turchin, P. (1998) Quantitative Analysis of Movement, Sinauer Associates, Inc. Publishers: Sunderland, Massachusetts.

Turner, M. D. (1999a) The role of social networks, indefinite boundaries and political bargaining in maintaining the ecological and economic resilience of the transhumance systems of Sudano-Sahelian West Africa. In M. Niamir-Fuller (ed.) Managing Mobility in African Rangelands: the Legitimization of Transhumance, pp. 97–123, Intermediate Technology Publications: London.

Turner, M.D. (1999b) No space for participation: Pastoralist narratives and the etiology of park-herder conflict in southwestern Niger. Land Degradation and Development, 10, 4, 343–61.

Turner, M.D. (1999c) Spatial and temporal scaling of grazing impact on the species composition and productivity of Sahelian annual grasslands. Journal of Arid Environments, 41, 3, 277–97.

Turner, M.D. (2009) Capital on the move: The changing relation between livestock and labor in Mali, West Africa. Geoforum 40, 746–55.

Turner, M.D. and Hiernaux, P. (2008) Changing access to labor, pastures, and knowledge: The extensification of grazing management in Sudano-Sahelian West Africa. Human Ecology, 26, 1, 59–80.

Turner, M. D., Hiernaux, P., and Schlecht, E. (2005) The distribution of grazing pressure in relation to vegetation resources in semi-arid West Africa: the role of herding. Ecosystems, 8, 1432–9840.

USGS Alaska science center bar-tailed godwit updates website (http://alaska.usgs.gov/science/biology/shore-birds/barg_updates.html), accessed 1 February 2009.

Valenza, J. (1981) Surveillance continue des pâturages naturels sahéliens. Revue de l'élevage et médecine vétérinaire des pays tropicaux, 34, 1, 83–100.

van Asch, M. and Visser, M.E. (2007) Phenology of forest caterpillars and their host trees: The importance of synchrony. Annual Review of Entomology, 52, 37–55.

van Bael, S.A., Philpott, S.M., Greenberg, R., Bichier, P., Barber, N.A,. Mooney, K.A., and Gruner, D.S. (2008) Birds as predators in tropical agroforestry systems. Ecology, 89, 928–34.

van der Jeugd, H.P., Eichhorn, G., Litvins, K.E., Stahl, J., Larsson, K., van der Graaf, A.J., and Drent, R.H. (2009) Keeping up with early springs: rapid range expansion in an avian herbivore incurs a mismatch between reproductive timing and food supply. Global Change Biology, 15, 1057–71.

van Dyne, G.M., Brockington, N.R., Szocs, Z., Duek, J., and Ribic, C.A. (1980) Large herbivore subsystem. In Grasslands, Systems, Analysis and Man. International Biological Programme 19, pp. 269–537, Cambridge University Press: Cambridge.

van Gils, J. A., Beekman, J. H., Coehoorn, P., Corporaal, E., Dekkers, T., Klaassen, M., van Kraaij, R., de Leeuw, R., and de Vries, P.P. (2008) Longer guts and higher food quality increase energy intake in migratory swans. Journal of Animal Ecology, 77, 1234–41.

van Gils, J.A., Piersma, T., Dekinga, A., and Battley, P.F. (2006) Modelling phenotypic flexibility: an optimality analysis of gizzard size in red knots Calidris canutus. Ardea, 94, 409–20.

van Ginneken, V.J.T. and van den Thillart, G. (2000) Physiology—Eel fat stores are enough to reach the Sargasso. Nature, 403, 156–57.

van Ginneken, V., Antonissen, E., Muller, U.K., Booms, R., Eding, E., Verreth, J., and van den Thillart, G. (2005) Eel migration to the Sargasso: remarkably high swimming efficiency and low energy costs. Journal of Experimental Biology, 208, 1329–35.

van Moorter, B., Visscher, D., Benhamou, S., Börger, L., Boyce, M.S., and Gaillard, J.-M. (2009) Memory keeps you at home: a mechanistic model for home range emergence. Oikos, 118, 641–52.

van Noordwijk, A.J., Pulido, F., Helm, B., Coppack, T., Delingat, J., Dingle, H., Hedenström, A., van der Jeugd, H., Marchetti, C., Nilsson, A., and Pérez-Tris, J. (2006) A framework for the study of genetic variation in migratory behavior. Journal of Ornithology, 147, 221–33.

van Soest, P.J. (1982) Nutritional Ecology of the Ruminant, O and B Books: Corvallis, OR.

Vanni, M.A., DeAngelis, D.L., Schindler, D.E., and Huxel, G.R. (2004) Overview: cross-habitat flux of nutrients and detritus. In G. E. Polis, M. E. Power, and G. R. Huxel (eds) Food Webs at the Landscape Level, pp. 3–11, University of Chicago Press: Chicago.

Varpe, Ø., Fiksen, Ø., and Slotte, A. (2005) Meta-ecosystems and biological energy transport from ocean to coast: the ecological importance of herring migration. Oecologia, 146, 443–51.

Veeranagoudar, D.K., Shanbhag, B.A., and Saidapur, S.K. (2004) Foraging behaviour in tadpoles of the bronze frog Rana temporalis: experimental evidence for the Ideal Free Distribution. Journal of Biosciences, 29, 2, 201–207.

Ventsel, A. (2006) Hunting–herding continuum in Sakha. In F. Stammler and H. Beach (eds) People and Reindeer on the Move. Special Issue of the journal Nomadic Peoples, 10, 2, 68–86, Berghahn Publishers: Oxford.

Videler, J. and Groenewold, A. (1991) Field measurements of hanging flight aerodynamics in the kestrel Falco tinnurculus. Journal of Experimental Biology, 155, 519–30.

Visser M.E., Holleman, L.J.M., and Gienapp, P. (2006) Shifts in caterpillar biomass phenology due to climate change and its impact on the breeding biology of an insectivorous bird. Oecologia, 147, 164–72.

Vitebsky, P. (2005) Reindeer People. Living with Animals and Spirits in Siberia, Harper Collins: London.

Vogel, S. (1994) Life in Moving Fluids: The Physical Biology of Flow, Princeton University Press: Princeton.

von Essen, L. (1991) A note on the Lesser White-fronted goose. Ardea, 79, 305–306.

von Haartman, L. (1968) The evolution of resident versus migratory habit in birds. Some considerations. Ornis Fennica, 45, 1–6.

Vors, L.S., Schaefer, J.A., Pond, B.A., Rodgers, A.R., and Patterson, B.R. (2007) Woodland Caribou Extirpation and Anthropogenic Landscape Disturbance in Ontario. Journal of Wildlife Management, 71, 4, 1249–56.

Vrieze, L.A. and Sorensen, P.W. (2001) Laboratory assessment of the role of a larval pheromone and natural stream odor in spawning stream localization by migratory sea lamprey (*Petromyzon murinus*). Canadian Journal of Fisheries and Aquatic Sciences, 58, 2374–85.

Wacher, T.J, Rawlings, P., and W.F. Snow. (1993) Cattle migration and stocking densities in relation to tsetse-trypanosomiasis challenge in the Gambia. Annals of Tropical Medicine and Parasitology, 87, 517–24.

Walker, M.M., Diebel, C.E., Haugh, C.V., Pankhurst, P.M., Montgomery, J.C., and Green, C.R. (1997) Structure and function of the vertebrate magnetic sense. Nature, 390, 371–76.

Walker, N.A., Henry, H.A.L., Wilson, D.J., and Jefferies, R.L. (2003) The dynamics of nitrogen movement in an Arctic salt marsh in response to goose herbivory: a parameterized model with alternate stable states. Journal of Ecology, 91, 637–50.

Wall, J., Douglas-Hamilton, I., and Vollrath, F. (2006) Elephants avoid costly mountaineering. Current Biology, 16, R527–29.

Wallraff, H.G. (2004) Avian olfactory navigation: Its empirical foundation and conceptual state. Animal Behaviour, 67, 189–204.

Wallraff, H.G. and Andreae, M.O. (2000) Spatial gradients in ratios of atmospheric trace gases: a study stimulated by experiments on bird navigation. Tellus Series B-Chemical And Physical Meteorology, 52, 1138–57.

Walther, G.-R., Post, E., Convey, P., Menzel, A., Parmesan, C., Beebee, T.J.C., Fromentin, J.-M.,Hoegh-Guldberg, O., and Bairlein, F. (2002) Ecological responses to recent climate change. Nature, 416, 389–395.

Wang, W.L. (2003) China case study I: Studies on traditional transhumance and a system where herders return to settled winter bases in Burjin County, Altai Prefecture, Xinjiang, China. In J.M. Suttie and S.G. Reynolds (eds) Transhumant grazing systems in temperate Asia, pp. 115–42, Food and Agriculture Organization of the United Nations (FAO): Rome.

Ward, E.J. (2008) A review and comparison of four commonly used Bayesian and maximum likelihood model selection tools. Ecological Modelling, 211, 1–10.

Ward, J.F., Austin R.M., and Macdonald, D.W. (2000) A simulation model of foraging behaviour and the effect of predation risk. Journal of Animal Ecology, 69, 16–30.

Ward, S., Bishop, C.M., Woakes, A.J., and Butler, P.J. (2002) Heart rate and the rate of oxygen consumption of flying and walking barnacle geese (*Branta leucopsis*) and bar-headed geese (*Anser indicus*). Journal of Experimental Biology, 205, 3347–56.

Watkinson, A. and Sutherland, W.J. (1995) Sources, sinks and pseudo-sinks. Journal of Animal Ecology, 64, 126–30.

Watson, J.W., Epperly, S.P., Shah, A.K., and Foster, D.G. (2005) Fishing methods to reduce sea turtle mortality associated with pelagic longlines. Canadian Journal of Fisheries and Aquatic Sciences, 62, 965–81.

Watson, M., Wilson, J.M., Koshkin, M., Sherbakov, B., Karpov, F., Gavrilov, A., Schielzeth, H., Brombacher, M., Collar, N.J., and Cresswell, W. (2006) Nest survival and productivity of the critically endangered sociable lapwing, *Vanellus gregarius*. Ibis, 148, 489–502.

Weber, J.M. (2009) The physiology of long-distance migration: extending the limits of endurance metabolism. Journal of Experimental Biology, 212, 593–97.

Weber, R.E., Jessen, T.H., Malte, H., and Tame, J. (1993) Mutant hemoglobins (α119-ala and β55-ser) functions related to high-altitude respiration in geese. Journal of Applied Physiology, 75, 2646–55.

Weber, T.P. and Hedenström, A. (2001) Long-distance migrants as a model system of structural and physiological plasticity. Evolutionary Ecology Research, 3, 255–71.

Weber, T.P. and Houston, A. I. (1997) Flight costs, flight range and stopover ecology of migrating birds. Journal of Animal Ecology, 66, 297–306.

Weber, T.P., Fransson, T., and Houston, A.I. (1999) Should I stay or should I go? Testing optimality models of stopover decisions in migrating birds. Behavioral Ecology and Sociobiology, 46, 280–86.

Webster, M.S., Marra, P.P., Haig, S.M., Bensch, S., and Holmes, R.T. (2002) Links between worlds: unraveling migratory connectivity. Trends in Ecology and Evolution, 17, 76–83.

Wedin, W.F., Lingvall, P., Thorsell, B, and Jonsson, N. (1984) Quality of first-growth forage during maturation at diverse latitudes. In H. Riley and A.O Skjelvag (eds) The Impact of Climate on Grass Production and Quality, Proceedings of the 10th General Meeting of the European Grassland Federation, As, Norway, pp. 46–50.

Weimerskirch, H. (2010) Editorial. Argos Forum, 68, 3. http://argos-system.clsamerica.com/documents/publications/newsletter/anl_68.pdf.

Weimerskirch, H. and Wilson, R.P. (2000) Oceanic respite for wandering albatrosses. Nature, 406, 955–56.

Weiner, D.R. (2000) Models of Nature: Ecology, Conservation, and Cultural Revolution in Soviet Russia, University of Pittsburgh Press: Pittsburgh, 324 pp.

Weng, K.C., Boustany, A.M., Pyle, P. Anderson, S.D., Brown, A., and Block, B.A. (2007) Migration and habitat of white sharks (*Carcharodon carcharias*) in the eastern Pacific Ocean. Marine Biology, 152, 877–94.

Western, D. (1975) Seasonality in water availability and its influence on the structure, dynamics and efficiency of a savannah large mammal community. East African Wildlife Journal, 13, 265–86.

White, P.J., Davis, T.L., Barnowe-Meyer, K.K., Crabtree, R.L., and Garrott, R.A. (2007) Partial migration and philopatry of Yellowstone pronghorn. Biological Conservation, 135, 502–10.

White, R.G. (1983) Foraging patterns and their multiplier effects on productivity of northern ungulates. Oikos, 40, 377–84.

Whyte, I.J. and Joubert, S.C.J. (1988) Blue wildebeest population trends in the Kruger National-Park and the effects of fencing. South African Journal of Wildlife Research, 18, 78–87.

Widmer, M. (1999) Altitudinal variation of migratory traits in the garden warbler *Sylvia borin*. PhD thesis, University of Zürich.

Wiens, J.A. (1989) The Ecology of Bird Communities. Volume 1. Foundations and Patterns. Cambridge Studies in Ecology, Cambridge University Press: Cambridge.

Wiens, J.A. (1991) Ecological similarity of shrub-desert avifaunas of Australia and North America. Ecology, 72, 479–95.

Wikelski, M. and Cooke S.J., (2006) Conservation physiology. Trends in Ecology and Evolution, 21, 38–46.

Wikelski, M., Kays, R.W., Kasdin, N.J., Thorup, K., Smith, J.A., and Swenson, G.W., Jr. (2007) Going wild: what a global small-animal tracking system could do for experimental biologists. Journal of Experimental Biology, 210, 181–86.

Wikelski, M., Moskowith, D., Adelman, J.S., Cochran, J., Wilcove, D.S. and May, M.L. (2006) Simple rules guide dragonfly migration. Biology Letters, 2, 325–29.

Wikelski, M., Tarlow, E., Raim, A., Diehl, R.H., Larkin, R.P., and Visser, G.H. (2003) Costs of migration in free-flying songbirds. Nature, 423, 704.

Wilcove, D. (2007) No Way Home: The Decline of the World's Great Migrations. Island Press: Covello, CA, 256 pp.

Wilcove, D. S. (2008) Animal migration: An endangered phenomenon? Issues in Science and Technology, 24, 71–78.

Wilcove, D.S. and Wikelski, M. (2008) Going, going gone: Is animal migration disappearing. PLoS Biology, 6, e188. doi:10.1371/journal.pbio.0060188.

Wilcove, D.S., Rothstein, D., Dubow, J., Phillips, A., and Losos, E. (1998) Quantifying Threats to Imperiled Species in the United States. Bioscience, 48, 8, 607–15.

Williams, R., Lusseau, D., and Hammond, P.S. (2009) The role of social aggregations and protected areas in killer whale conservation: the mixed blessing of critical habitat. Biological Conservation, 142, 4, 709–19.

Williams, T.C., Ireland, L.C., and Williams, J.M. (1973) High-altitude flights of free-tailed bat, Tadarida brasiliensis, observed with radar. Journal of Mammalogy, 54, 807–21.

Williams, T.C. and Williams, J. M. (1978) The orientation of transatlantic migrants. In K. Schmidt-Koenig and W. Keeton (eds) Animal migration, orientation and homing, pp. 239–51, Springer-Verlag: Berlin.

Williamson, D., Williamson, J., and Ngwamotsoko, K.T. (1988) Wildebeest migration in the Kalahari. African Journal of Ecology, 26, 269–80.

Wilmshurst, J.F., Fryxell, J.M., and Hudson, R.J. (1995) Forage quality and patch choice by wapiti (*Cervus elaphus*) Behavioral Ecology, 6, 2, 209–17.

Wilmshurst, J.F., Fryxell, J.M., and Bergman, C.M. (2000) The allometry of patch selection in ruminanats. Proc. Roy. Sco. Lond. (B), 267, 345–49.

Wilmshurst, J.F., Fryxell, J.M., Farm, B.P., Sinclair, A.R.E., and C.P. Henschel (1999) Spatial distribution of Serengeti wildebeest in relation to resources. Canadian Journal of Zoology, 77, 1223–32.

Wilps, H. and Diop, B. (1997) Field investigations on *Schistocerca gregaria* (Forskal) adults, hoppers and hopper bands. In S. Krall, R. Peveling, and D. Ba Diallo (eds) New Strategies in Locust Control., pp. 117–28, Birkhäuser-Verlag: Basel.

Wilson, C. and Tisdell, C. (2003) Conservation and economic benefits of wildlife-based marine tourism: sea turtles and whales as case studies. Human Dimensions of Wildlife, 8, 1, 49–58.

Wilson, R. P., White, C.R., Quintana, F., Halsey, L.G., Liebsch, N., Martin, G.R., and Butler, P.J. (2006) Moving towards acceleration for estimates of activity-specific metabolic rate in free-living animals: the case of the cormorant. Journal of Animal Ecology, 75, 1081–90.

Wilson, S.G., Taylor, J.G., and Pearce, A.F. (2001) The seasonal aggregation of whale sharks at Ningaloo Reef, Western Australia: Currents, migrations and the El Nino/Southern Oscillation. Environmental Biology of Fishes, 61, 1–11.

Wiltschko, R. and Wiltschko, W. (2003) Mechanisms of orientation and navigation in migratory birds. In P. Berthold, E. Gwinner and E. Sonnenschein (eds) Avian Migration, pp. 433–56, Springer Verlag: Berlin.

Wiltschko, W. and Wiltschko, R. (1972) Magnetic compass of European robins. Science, 176, 62–64.

Wingfield, J.C. (2008) Organization of vertebrate annual cycles: implications for control mechanisms. Philosophical Transactions of the Royal Society of London, Series B, 363, 425–41.

Winkler, D.W. (2005) How do migration and dispersal interact? In R. Greenberg and P.P. Marra (eds) Birds of

Two Worlds: the Ecology and Evolution of Migration, pp. 401–13, Johns Hopkins University Press: Baltimore.

Winkler, D.W. and Allen, P.E. (1996) The seasonal decline in tree swallow clutch size: Physiological constraint or strategic adjustment? Ecology, 77, 922–32.

Winkler, K. (2000) Migration and speciation. Nature, 404, 36.

Wirestam, R., Fagerlund, T., Rosén, M., and Hedenström, A. (2008) Magnetic resonance imaging for non-invasive analysis of fat storage in migratory. Auk, 125, 965–71.

Witey, J.C., Bloxton, T.D., and Marzluff, J.M. (2001) Effects of tagging and location error in wildlife radiotelemetry studies. In J. J. Millspaugh and J. M. Marzluff (eds) Radiotracking and animal populations, pp. 45–75, Academic Press: San Diego.

Woinarsky, J.C.Z. (2006) Predictors of nomadism in Australian birds: a reanalysis of Allen and Saunders (2002) Ecosystems. 9, 689–93.

Wolanski, E. and Gereta, E. (2001) Water quality and quantity as the factors driving the Serengeti ecosystem. Tanzania. Hydrobiologia. 458, 169–80.

Wolf, C.M., Griffith, B., Reed, C., and Temple, S.A. (1996) Avian and mammalian translocations: update and reanalysis of 1987 survey data. Conservation Biology, 10, 1142–54.

Wood, C.R., Chapman, J.W., Reynolds, D.R., Barlow, J.F., Smith, A.D., and Woiwod, I.P. (2006) The influence of the atmospheric boundary layer on nocturnal layers of noctuids and other moths migrating over southern Britain. International Journal of Biometeorology, 50, 193–204.

Wood, P. and Wolfe, M.L. (1988) Intercept feeding as a means of reducing deer-vehicle collisions. Wildlife Society Bulletin, 16, 376–80.

Woodbury, A.M. (1941) Animal migration: periodic-response theory. The Auk, 58, 463–505.

Woodrey, M.S. and Chandler, C.R. (1997) Age-related timing of migration: Geographic and interspecific patterns. Wilson Bulletin, 109, 52–67.

Woodrey, M.S. and Moore, F.R. (1997) Age-related differences in the stopover of fall landbird migrants on the coast of Alabama. Auk, 114, 695–707.

Woods, J. (1991) Ecology of a partially migratory elk population Ph.D. thesis, University of British Columbia: Vancouver, BC.

Worm, B., Barbier, E.B., Beaumont, N., Duffy, J.E., Folke, C., Halpern, B.S., Jackson, J.B.C., Lotze, H.K., Micheli, F., Palumbi, S.R., Sala, E., Selkoe, K.A., Stachowicz, J.J., and Watson, R. (2006) Impacts of Biodiversity Loss on Ocean Ecosystem Services. Science, 314, 3, 787–90.

Wright, S. (1931) Evolution in Mendelian populations. Genetics, 16, 97–159.

Xia, L., Yang, Q., Li, Z., Wu, Y., and Feng, Z. (2007) The effect of the Qinghai-Tibet railway on the migration of Tibetan antelope (Pantholops hodgsonii) in Hoh-xil National Nature Reserve, China. Oryx, 41, 3, 352–57.

Yan Zhaoli, Ning, Wu, Dorji, Yeshi, and Jia, Ru (2005) A review of rangeland privatisation and its implications in the Tibetan Plateau, China. Nomadic Peoples, 9 (1 & 2), 31–52.

Yates, C.A., R. Erban, C. Escudero, I.D. Couzin, J. Buhl, I. G. Kevrekidis, P.K. Maini, and D.J.T. Sumpter (2009) Inherent noise can facilitate coherence in collective swarm motion. Proceedings of the National Academy of Sciences, 106, 5464–69.

Ydenberg, R.C., Butler, R.W., Lank, D.B., Guglielmo, C.G., Lemon, M., and Wolf, N. (2002) Trade-offs, condition dependence and stopover site selection by migrating sandpipers. Journal of Avian Biology, 33, 47–55.

Yeh, E.T. (2005) Green governmentality and pastoralism in Western China: 'Converting pastures to grasslands'. Nomadic Peoples, 9, 9–30.

Yong, W., Finch, D.M., Moore, F.R., and Kelly, J.F. (1998) Stopover ecology and habitat use of migratory Wilson's warblers. Auk, 115, 829–42.

Yoshida, A. (1997) Kul'tura Pitania Gydanskikh Nentsev (Interpretatsia i sotsial'naia adaptatsiia) Novye Issledovania po etnologii i antropologii, Russian Academy of Sciences: Moscow.

Yoshihara, Y., Ito, T.Y., Lhagvasuren, B., and Takatsuki, S. (2008) A comparison of food resources used by Mongolian gazelles and sympatric livestock in three areas in Mongolia. Journal of Arid Environments, 72, 1, 48–55, DOI: 10.1016/j.jaridenv.2007.05.001.

Young, K.A. (1999) Managing the decline of Pacific salmon: Metapopulation theory and artificial recolonization as ecological mitigation. Canadian Journal of Fisheries and Aquatic Sciences, 56, 9, 1700–706.

Yuzhakov, A. and Mukhachev, A. (2001) Etnicheskoe Olenevodstvo Zapadnoi Sibiri: Nenetskii Tip, Agricultural Science Publishers: Novosibirsk.

Zabrodin, V.A., Borozdin, E.K., Vostriakov, P.N., D'iachenko, N.O., Kriuchkov, V.V., and Andreev, V.N. (1979) Severnoe Olenevodstvo, Kolos: Moscow.

Zink, R.M. (2002) Towards a framework for understanding the evolution of avian migration. Journal of Avian Biology, 33, 436.

Zwarts, L. and Dirksen, S. (1990) Digestive bottleneck limits the increase in food-intake of whimbrels preparing for spring migration from the Banc-Darguin, Mauritania. Ardea, 78, 257–78.

Index